ADENOSINE RECEPTORS IN NEURODEGENERATIVE DISEASES

ADENOSINE RECEPTORS IN NEURODEGENERATIVE DISEASES

Edited by

DAVID BLUM
Inserm UMR-1172, University of Lille, Lille, France

LUÍSA V. LOPES
Institute of Molecular Medicine, Faculty of Medicine, University of Lisbon, Portugal

ELSEVIER

ACADEMIC PRESS

An imprint of Elsevier
elsevier.com

Academic Press is an imprint of Elsevier
125 London Wall, London EC2Y 5AS, United Kingdom
525 B Street, Suite 1800, San Diego, CA 92101-4495, United States
50 Hampshire Street, 5th Floor, Cambridge, MA 02139, United States
The Boulevard, Langford Lane, Kidlington, Oxford OX5 1GB, United Kingdom

Notices

Knowledge and best practice in this field are constantly changing. As new research and experience broaden our understanding, changes in research methods, professional practices, or medical treatment may become necessary.

Practitioners and researchers must always rely on their own experience and knowledge in evaluating and using any information, methods, compounds, or experiments described herein. In using such information or methods they should be mindful of their own safety and the safety of others, including parties for whom they have a professional responsibility.

To the fullest extent of the law, neither the Publisher nor the authors, contributors, or editors, assume any liability for any injury and/or damage to persons or property as a matter of products liability, negligence or otherwise, or from any use or operation of any methods, products, instructions, or ideas contained in the material herein.

British Library Cataloguing-in-Publication Data
A catalogue record for this book is available from the British Library

Library of Congress Cataloging-in-Publication Data
A catalog record for this book is available from the Library of Congress

ISBN: 978-0-12-803724-9

Publisher: Mara Conner
Acquisition Editor: Melanie Tucker
Editorial Project Manager: Kristi Anderson
Production Project Manager: Karen East and Kirsty Halterman
Designer: Maria Inês Cruz

Typeset by MPS Limited, Chennai, India

Dedication

We hope the present book illustrates the complexity and importance of adenosine universe in the nervous system and will convince you to integrate its role in your future projects. You will feel like Captain Kirk, embarking boldly where no man has gone before!

Contents

7. Neuronal vs Glial Cell Contribution to Adenosine A_{2A} Receptor-Induced Neurodegeneration

ANTONELLA FERRANTE, MARIA T. TEBANO, ALBERTO MARTIRE, MARIA R. DOMENICI AND PATRIZIA POPOLI

8. Adenosine and Oxygen/Glucose Deprivation in the Brain

FELICITA PEDATA, ILARIA DETTORI, IRENE FUSCO, ELISABETTA COPPI, ANNA M. PUGLIESE AND ALESSIA MELANI

9. Adenosine Receptors and Memory Disorders

ALEXANDRE DE MENDONCA, DAVID BLUM AND JONATHAN D. GEIGER

10. Control of Motor Function by Adenosine A_{2A} Receptors in Parkinson's and Huntington's Disease

ANNALISA PINNA, JADWIGA WARDAS, MARIA R. DOMENICI, PATRIZIA POPOLI, GIOVANNI COSSU AND MICAELA MORELLI

11. Adenosine Receptors Oligomers in Parkinson's Disease

VÍCTOR FERNÁNDEZ-DUEÑAS AND FRANCISCO CIRUELA

12. Adenosine Control of Striatal Function—Implications for the Treatment of Apathy in Basal Ganglia Disorders

SERGI FERRÉ

List of Contributors

Marta Barrachina Bellvitge Biomedical Research Institute, Barcelona, Spain

Christophe Bernard Institute of Neurosciences of Systems, Marseille, France

Theresa E. Bjorness University of Texas Southwestern Medical Center, Dallas, TX, United States; Veterans Administration Medical Center, Dallas, TX, United States

David Blum University of Lille, Lille, France

Detlev Boison Legacy Research Institute, Portland, OR, United States

Yijuang Chern Academia Sinica, Taipei, Taiwan

Francisco Ciruela University of Barcelona, Barcelona, Spain

Elisabetta Coppi University of Florence, Florence, Italy

Giovanni Cossu AOB S. Michele General Hospital, Cagliari, Italy

Sofia Cristóvão-Ferreira University of Lisbon, Lisbon, Portugal

Ilaria Dettori University of Florence, Florence, Italy

Raquel B. Dias University of Lisbon, Lisbon, Portugal

Maria R. Domenici Istituto Superiore di Sanità, Rome, Italy; Italian National Institute of Health, Rome, Italy

Víctor Fernández-Dueñas University of Barcelona, Barcelona, Spain

Antonella Ferrante Istituto Superiore di Sanità, Rome, Italy

Sergi Ferré National Institutes of Health, Baltimore, MD, United States

Michael Freissmuth University of Vienna, Vienna, Austria

Irene Fusco University of Florence, Florence, Italy

Jonathan D. Geiger University of North Dakota, Grand Forks, ND, United States

Robert W. Greene University of Texas Southwestern Medical Center, Dallas, TX, United States; Veterans Administration Medical Center, Dallas, TX, United States

Karl-Norbert Klotz University of Würzburg, Würzburg, Germany

Chien-Yu Lin Academia Sinica, Taipei, Taiwan

Luísa V. Lopes University of Lisbon, Lisbon, Portugal

Alberto Martire Istituto Superiore di Sanità, Rome, Italy

Alessia Melani University of Florence, Florence, Italy

Alexandre de Mendonca University of Lisbon, Lisbon, Portugal

Micaela Morelli Neuroscience Institute, Cagliari, Italy; University of Cagliari, Cagliari, Italy

Felicita Pedata University of Florence, Florence, Italy

Annalisa Pinna Neuroscience Institute, Cagliari, Italy

Traci Plumb Legacy Research Institute, Portland, OR, United States

Patrizia Popoli Istituto Superiore di Sanità, Rome, Italy; Italian National Institute of Health, Rome, Italy

Paula Pousinha Institute of Molecular and Cellular Pharmacology, CNRS UMR 7275, France

Anna M. Pugliese University of Florence, Florence, Italy

Joaquim A. Ribeiro University of Lisbon, Lisbon, Portugal

Ana M. Sebastião University of Lisbon, Lisbon, Portugal

Maria T. Tebano Istituto Superiore di Sanità, Rome, Italy

Jadwiga Wardas Polish Academy of Sciences, Krakow, Poland

Stefania Zappettini Institute of Neurosciences of Systems, Marseille, France

Foreword and History of Adenosine in the Central Nervous System

Adenosine was shown early on to have inhibitory actions in the central nervous system (CNS)[1–3] and to stimulate cyclic AMP production,[4] well before ATP was recognized as playing major roles in the CNS (see Ref. 5). It functions via four G protein–coupled receptor subtypes, A_1, A_{2A}, A_{2B}, and $_{A3}$. (see Refs. 6 and 7). The A_1 subtype of P1 receptors was localized in rat brains using [^{3}H] cyclohexyladenosine in most regions.[8] P1 receptors were then shown to be located on the axon terminals of excitatory neurons.[9] A_1 receptors have a similar distribution in the human brain.[10]

The A_2 receptor mRNA was identified in the human brain, particularly in the basal ganglia.[11] Immunohistochemical detection of both A_1[12,13] and A_{2A}[14,15] receptors was later reported. The distribution of adenosine deaminase[16] and 5′-nucleotidase[17] and the sites of adenosine uptake[18] were described. There is evidence that some neurons and astrocytes in the brain release adenosine directly as well as being produced extracellularly after breakdown of released ATP.[19–20]

Interactions between adenosine and benzodiazepines (inhibiting uptake of adenosine) and methylxanthines (adenosine receptor antagonists) in the CNS have been reported[21] and adenosine has been shown to play a prejunctional neuromodulatory role, inhibiting excitatory transmitter release.[22,23] Adenosine is also a regulator of cerebral blood flow.[24] Adenosine stimulates glutamate release from astrocytes via A_{2a} adenosine receptors.[25] The ontogenesis of adenosine receptors in the rat CNS has been described.[26,27] Adenosine acts as an antiinflammatory agent and plays an important role in neuroprotection.[28] Recent research has shown that A_{2A} receptors regulate human blood–brain barrier permeability.[29]

Clinical aspects of adenosinergic actions in the CNS have been reviewed.[30–41] Adenosine agonists gave sedative, anticonvulsant and anxiolytic actions, while adenosine antagonists, such as caffeine, have stimulant and anxiogenic effects. Adenosine–dopamine interactions in the brain were recognized[42] and this has been prominent for the treatment of Parkinson's disease.[43,44]

Adenosine not only acts as a short-term signaling molecule in the CNS, but also has long-term (trophic) roles mediating both long-term potentiation and long-term depression in the hippocampus.[45] A_1 receptors are expressed by oligodendrocyte progenitor cells and mediate oligodendrocyte maturation and inhibit their proliferation.[46]

Reviews concerned with adenosine receptors in the mammalian CNS are available.[47–57]

This book provides a timely update of the roles of adenosine in the CNS in health and disease by leading figures in the field.

Geoffrey Burnstock[1,2]

[1]*University College Medical School, London, United Kingdom*

[2]*The University of Melbourne, Melbourne, Australia*

References

1. Phillis JW, Kostopoulos GK, Limacher JJ. Depression of corticospinal cells by various purines and pyrimidines. *Can J Physiol Pharmacol.* 1974;52:1226–1229.
2. Kuroda Y, Saito M, Kobayashi K. Concomitant changes in cyclic AMP level and postsynaptic potentials of olfactory cortex slices induced by adenosine derivatives. *Brain Res.* 1976;109:196–201.
3. Schubert P, Mitzdorf U. Analysis and quantitative evaluation of the depressive effect of adenosine on evoked potentials in hippocampal slices. *Brain Res.* 1979;172:186–190.
4. Sattin A, Rall TW. The effect of adenosine and adenine nucleotides on the cyclic adenosine 3′,5′-phosphate content of guinea-pig cerebral cortex slices. *Mol Pharmacol.* 1970;6:13–23.
5. Burnstock G, Verkhratsky A. *Purinergic Signalling and the Nervous System.* Heidelberg/Berlin: Springer; 2012:1–715.
6. Tucker AL, Linden J. Cloned receptors and cardiovascular responses to adenosine. *Cardiovasc Res.* 1993;27:62–67.
7. Ralevic V, Burnstock G. Receptors for purines and pyrimidines. *Pharmacol Rev.* 1998;50:413–492.
8. Goodman RR, Synder SH. Autoradiographic localization of adenosine receptors in rat brain using [^{3}H]cyclohexyladenosine. *J Neurosci.* 1982;2:1230–1241.
9. Goodman RR, Kuhar MJ, Hester L, Snyder SH. Adenosine receptors: autoradiographic evidence for their location on axon terminals of excitatory neurons. *Science.* 1983;220:967–969.
10. Fastbom J, Pazos A, Probst A, Palacios JM. Adenosine A_1-receptors in human brain: characterization and autoradiographic visualization. *Neurosci Lett.* 1986;65:127–132.
11. Schiffmann SN, Libert F, Vassart G, Vanderhaeghen JJ. Distribution of adenosine A_2 receptor mRNA in the human brain. *Neurosci Lett.* 1991;130:177–181.
12. Rivkees SA, Price SL, Zhou FC. Immunohistochemical detection of A_1 adenosine receptors in rat brain with emphasis on localization in the hippocampal formation, cerebral cortex, cerebellum, and basal ganglia. *Brain Res.* 1995;677:193–203.
13. Schindler M, Harris CA, Hayes B, Papotti M, Humphrey PP. Immunohistochemical localization of adenosine A_1 receptors in human brain regions. *Neurosci Lett.* 2001;297:211–215.
14. Svenningsson P, Le Moine C, Kull B, Sunahara R, Bloch B, Fredholm BB. Cellular expression of adenosine A_{2A} receptor messenger RNA in the rat central nervous

system with special reference to dopamine innervated areas. *Neuroscience.* 1997;80:1171–1185.
15. Rosin DL, Robeva A, Woodard RL, Guyenet PG, Linden J. Immunohistochemical localization of adenosine A_{2A} receptors in the rat central nervous system. *J Comp Neurol.* 1998;401:163–186.
16. Nagy JI, La Bella LA, Daddona PE. Immunohistochemistry of adenosine deaminase: implications for adenosine neurotransmission. *Science.* 1984;224:166.
17. Fastbom J, Pazos A, Palacios JM. The distribution of adenosine A1 receptors and 5′-nucleotidase in the brain of some commonly used experimental animals. *Neuroscience.* 1987;22:813–826.
18. Jarvis SM, Ng AS. Identification of the adenosine uptake sites in guinea pig brain. *J Neurochem.* 1985;44:183–188.
19. Wall MJ, Dale N. Neuronal transporter and astrocytic ATP exocytosis underlie activity-dependent adenosine release in the hippocampus. *J Physiol.* 2013;591:3853–3871.
20. Sims RE, Dale N. Activity-dependent adenosine release may be linked to activation of Na^{+}-K^{+} ATPase: an in vitro rat study. *PLoS One.* 2014;9:e87481.
21. Phillis JW, Wu PH. Interactions between the benzodiazepines, methylxanthines, and adenosine. *Can J Neurol Sci.* 1980;7:247–249.
22. Phillis JW, Wu PH. The role of adenosine and its nucleotides in central synaptic transmission. *Prog Neurobiol.* 1981;16:187–239.
23. Ribeiro JA. Purinergic inhibition of neurotransmitter release in the central nervous system. *Pharmacol Toxicol.* 1995;77:299–305.
24. Dunwiddie TV, Fredholm BB. Adenosine neuro-modulation. In: Jacobson KA, Jarvis MF, eds. *Purinergic Approaches in Experimental Therapeutics.* New York: Wiley-Liss; 1997:359–382.
25. Nishizaki T. ATP- and adenosine-mediated signaling in the central nervous system: adenosine stimulates glutamate release from astrocytes via A_{2A} adenosine receptors. *J Pharmacol Sci.* 2004;94:100–102.
26. Geiger JD, LaBella FS, Nagy JI. Ontogenesis of adenosine receptors in the central nervous system of the rat. *Brain Res Dev Brain Res.* 1984;13:97–104.
27. Morgan PF, Marangos PJ. Ontogenetic appearance of the adenosine receptor precedes *N*-protein coupling in rat forebrain. *Brain Res.* 1987;432:269–274.
28. Cieslak M, Czarnecka J, Roszek K. The roles of purinergic signaling in psychiatric disorders. *Acta Biochim Pol.* 2016;63:1004.
29. Kim DG, Bynoe MS. A_{2A} adenosine receptor regulates the human blood-brain barrier permeability. *Mol Neurobiol.* 2015;52:664–678.
30. Marangos PJ, Boulenger JP. Basic and clinical aspects of adenosinergic neuromodulation. *Neurosci Biobehav Rev.* 1985;9:421–430.
31. Chin JH. Adenosine receptors in brain: neuromodulation and role in epilepsy. *Ann Neurol.* 1989;26:695–698.
32. Haas HL, Selbach O. Functions of neuronal adenosine receptors. *Naunyn Schmiedebergs Arch Pharmacol.* 2000;362:375–381.
33. Ribeiro JA, Sebastião AM, de Mendonça A. Adenosine receptors in the nervous system: pathophysiological implications. *Prog Neurobiol.* 2003;68:377–392.
34. Fredholm BB, Chen JF, Masino SA, Vaugeois JM. Actions of adenosine at its receptors in the CNS: insights from knockouts and drugs. *Annu Rev Pharmacol Toxicol.* 2005;45:385–412.
35. Ferré S, Diamond I, Goldberg SR, et al. Adenosine A_{2A} receptors in ventral striatum, hypothalamus and nociceptive circuitry implications for drug addiction, sleep and pain. *Prog Neurobiol.* 2007;83:332–347.

36. Benarroch EE. Adenosine and its receptors: multiple modulatory functions and potential therapeutic targets for neurologic disease. *Neurology*. 2008;70:231–236.
37. Boison D. Adenosine as a neuromodulator in neurological diseases. *Curr Opin Pharmacol*. 2008;8:2–7.
38. Paul S, Elsinga PH, Ishiwata K, Dierckx RA, van Waarde A. Adenosine A_1 receptors in the central nervous system: their functions in health and disease, and possible elucidation by PET imaging. *Curr Med Chem*. 2011;18:4820–4835.
39. Kovács Z, Dobolyi Á, Kékesi KA, Juhász G. 5′-nucleotidases, nucleosides and their distribution in the brain: pathological and therapeutic implications. *Curr Med Chem*. 2013;20:4217–4240.
40. Chiu GS, Freund GG. Modulation of neuroimmunity by adenosine and its receptors: metabolism to mental illness. *Metabolism*. 2014;63:1491–1498.
41. Yamada K, Kobayashi M, Kanda T. Involvement of adenosine A2A receptors in depression and anxiety. *Int Rev Neurobiol*. 2014;119:373–393.
42. Ferré S, Fuxe K, von Euler G, Johansson B, Fredholm BB. Adenosine-dopamine interactions in the brain. *Neuroscience*. 1992;51:501–512.
43. Fuxe K, Marcellino D, Borroto-Escuela DO, et al. Adenosine-dopamine interactions in the pathophysiology and treatment of CNS disorders. *CNS Neurosci Ther*. 2010;16: e18–e42.
44. Jörg M, Scammells PJ, Capuano B. The dopamine D_2 and adenosine A_{2A} receptors: past, present and future trends for the treatment of Parkinson's disease. *Curr Med Chem*. 2014;21:3188–3210.
45. de Mendonça A, Ribeiro JA. Adenosine and synaptic plasticity. *Drug Dev Res*. 2001;52:283–290.
46. Coppi E, Cellai L, Maraula G, et al. Role of adenosine in oligodendrocyte precursor maturation. *Front Cell Neurosci*. 2015;9:155.
47. Williams M. Adenosine receptors in the mammalian central nervous system. *Prog Neuro-Psychopharmacol Biol Psychiatry*. 1983;7:443–450.
48. Daly JW. Adenosine receptors. *Adv Cyclic Nucleotide Protein Phosphorylation Res*. 1985;19:29–46.
49. Dunwiddie TV. The physiological role of adenosine in the central nervous system. *Int Rev Neurobiol*. 1985;27:63–139.
50. Greene RW, Haas HL. The electrophysiology of adenosine in the mammalian central nervous system. *Prog Neurobiol*. 1991;36:329–341.
51. Fredholm BB. Adenosine receptors in the central nervous system. *News Physiol Sci*. 1995;10:122–128.
52. Latini S, Pazzagli M, Pepeu G, Pedata F. A_2 adenosine receptors: their presence and neuromodulatory role in the central nervous system. *Gen Pharmacol*. 1996;27:925–933.
53. Sebastião AM, Ribeiro JA. Adenosine A_2 receptor-mediated excitatory actions on the nervous system. *Prog Neurobiol*. 1996;48:167–189.
54. Sebastião AM, Ribeiro JA. Tuning and fine-tuning of synapses with adenosine. *Curr Neuropharmacol*. 2009;7:180–194.
55. Cunha RA. Adenosine as a neuromodulator and as a homeostatic regulator in the nervous system: different roles, different sources and different receptors. *Neurochem Int*. 2001;38:107–125.
56. Kovács Z, Dobolyi Á, Juhász G, Kékesi KA. Nucleoside map of the human central nervous system. *Neurochem Res*. 2010;35:452–464.
57. Chen JF, Lee CF, Chern Y. Adenosine receptor neurobiology: overview. *Int Rev Neurobiol*. 2014;119:1–49.

Acknowledgments

Luísa V. Lopes and David Blum are grateful to all the colleagues who kindly accepted to contribute to this book. Our aim was to be inclusive and thorough on the selection of themes and contributors. Some fundamental scientists in the field originally invited could not contribute at this stage, due to prior commitments. Our words of friendship and admiration to them that we, nevertheless, tried to honor along these pages.

We also want to express our gratitude to our lab members, students, and colleagues that over the years have challenged our views of science and pushed us forward.

SECTION I

PHYSIOLOGY

CHAPTER

1

Adenosine Receptor Biology in the Central Nervous System

Michael Freissmuth[1] *and Karl-Norbert Klotz*[2]

[1]University of Vienna, Vienna, Austria [2]University of Würzburg, Würzburg, Germany

OUTLINE

Adenosine Receptors in Neurodegenerative Diseases.
DOI: http://dx.doi.org/10.1016/B978-0-12-803724-9.00001-6

SOURCES OF ADENOSINE IN THE CNS

Adenosine is ubiquitously present in body fluids; however, it is turned over very rapidly. Accordingly, under physiological conditions, adenosine levels are low both within cells and in the extracellular compartment. Intracellular adenosine can arise from the breakdown of AMP by cytosolic 5′ nucleotidase-IA, which is muscle specific, and by cytosolic 5′ nucleotidase-II, which is ubiquitously expressed but less specific.[1] Alternatively, adenosine is formed by *S*-adenosylhomocysteine hydrolase, which performs its eponymous action, i.e., the hydrolytic cleavage of *S*-adenosylhomocysteine. The principal source of extracellular adenosine is the breakdown of adenine nucleotides; ATP is taken up into synaptic vesicles by a vesicular nucleotide transporter (vNuT, SLC17A9[2]) and released as cotransmitter. In addition, ATP can be released through pannexin-1 hemichannels and/or P2X7 receptors[3] and CALMH1 (calcium homeostasis modulator 1).[4] Finally, cell death results in the release of ATP. Thus, in the brain, both neurons and glial cells act as the cellular sources of extracellular ATP. Extracellular ATP is rapidly broken down to adenosine by the sequential action of ectonucleotidases, i.e., the ectonucleoside triphosphate diphosphohydrolases (E-NTPDases), ecto-5′-nucleotidase (eN), ectonucleotide pyrophosphatase/phosphodiesterases (E-NPPs), and alkaline and acid phosphatases. It is not clear if the latter two enzyme families play a (physiological) role in generating adenosine. In contrast, ectonucleoside triphosphate diphosphohydrolases (E-NTPDases) and ecto-5′-nucleotidase (eN) are known to be important in supporting purinergic signaling[5]; of the eight known E-NTPDases, three are expressed in brain, i.e., E-NTPDase-1 (CD39, apyrase), E-NTPDase-2 (CD39L1, ecto-ATPase), and E-NTPDase-3 (CD39L3, HB6). They cleave ATP to AMP (E-NTPDase-1 and -2) or ADP (E-NTPDase-2) with K_M values in the range of 50–100 μM for ATP. Adenosine is formed from AMP by the action of ecto-5′-nucleotidase (eN, CD73), a glycolipid-anchored protein. CD73 operates with a substrate affinity in the low μM range and with a high turnover number (about 180 s^{-1}).[6] Hence, CD73 can rapidly convert extracellular AMP to adenosine. In fact, the current evidence indicates that CD73 is the only relevant enzyme which generates extracellular adenosine in the brain.[7] Purinergic receptors (i.e., receptors for ATP/UTP, ADP/UDP and adenosine) and ectonucleotidases may be organized in larger complexes, which are at least in part supported by the ability of the proteins to form heterooligomers.[8]

Because ATP is continuously released as a neurotransmitter in the brain, adenosine is continuously produced within the extracellular space. Its physiological levels are therefore thought to oscillate with neuronal activity. Adenosine is removed by degradation via adenosine deaminase-1 and by cellular uptake. Adenosine deaminase-1 is the predominant one of the two isoforms, which is ubiquitously expressed and hence also present in the brain. Deficiency in adenosine deaminase-1 gives rise to severe combined immunodeficiency (because adenosine is converted to a toxic metabolite, which kills T-lympocytes). However, deficiency in adenosine deaminase rarely gives rise to neurological deficiencies.[9] This suggests that cellular uptake of adenosine is more important for terminating the action of extracellular adenosine. Adenosine can be translocated over cell membranes by concentrative (CNT1-3 = SLC28A1-A3) or equilibrative transporters (ENT1−4 = SLC29A1−4).[10] Concentrative transporters operate at the blood brain barrier but are otherwise only found at low levels in the brain. ENT3 is found primarily in intracellular compartments, ENT4 operates at low pH and also acts as a plasmalemmal transporter for monoamines. Accordingly, cellular uptake of adenosine is thought to be predominantly accomplished by the equilibrative nucleoside tranporters-1 and -2 (ENT1 & ENT2). Under physiological conditions, ENT1 & ENT2 operate in a relay with adenosine kinase, i.e., the rapid phosphorylation of adenosine by intracellular adenosine kinase maintains the gradient, which supports clearance of extracellular adenosine. The importance of adenosine kinase in the central nervous system (CNS) can be gauged from the fact that its deficiency leads to global mental retardation.[11] This phenotypic consequence is presumably unrelated to excessive signaling via adenosine receptors because the elimination of ENT1 or of ENT2 does not produce any gross abnormality in genetically deficient ("knock-out") mice.[12,13]

Adenosine accumulates during hypoxia. This is a central tenet of the retaliatory metabolite concept: hypoxia-induced adenosine formation leads to extracellular accumulation of adenosine, which engages its cognate receptor to restore homeostasis by orchestrating responses, which proceed on different time scales—from reducing oxygen demand by dampening cellular activity via A_1-receptors to increasing oxygen delivery via A_{2A}-dependent vasodilation, to modulation of the inflammatory response and neo-angiogenesis, which is the result of a concerted activation of A_{2A}-, A_{2B}- and—to a lesser extent—A_3-receptors.[14] Originally, hypoxia-induced accumulation of adenosine in heart muscle was shown to result from intracellular breakdown of AMP (as a result of ATP consumption) via cytosolic 5′-nucleotidase and/or *S*-adenosylhomocysteine hydrolase resulting in outflow of adenosine.[15] It has, however, been questioned whether this also holds true for the brain, because both neurons and glial cells contribute to extracellular ATP via different mechanisms and because the brain does not contain high levels of cytosolic AMP-specific 5′-nucleotidase-Ia (see above). Surprisingly, there is circumstantial evidence that mice deficient in ENT1 have reduced extracellular levels of adenosine in the brain.[16] Conversely, transgenic neuronal overexpression of ENT1 in neurons renders the mice less sensitive to the behavioral effects of the adenosine receptor antagonist caffeine, indicating that endogenous extracellular levels of adenosine are higher in the presence of elevated ENT1. This indicates that under physiological conditions, at least some of the extracellular adenosine originates from a pool, which is generated within neurons and is subject to ENT1-mediated efflux. This interpretation is substantiated by the observation that neuronal release of adenosine rather than (glial or neuronal) ATP is crucial for A_1 receptor-mediated, activity-dependent depression during high-frequency repetitive firing.[7] Even more surprisingly, in hypoxia, ENT1 apparently operates predominantly in the influx mode: if hippocampal slices from ENT1 overexpressing mice are exposed to hypoxia, excitatory postsynaptic responses are less inhibited than in slices derived from wild type control animals.[17] This suggests that there is a net uptake of adenosine by neurons in hypoxia. By inference, it appears that, in the hypoxic brain, adenosine is predominantly formed from extracellular sources. At the very least, these observations indicate that the retaliatory metabolite concept is somewhat simplistic. Regardless of the sources of adenosine, it is clear that during hypoxia intracellular and extracellular adenosine tend to equilibrate, because adenosine kinase is inhibited.[18,19] Intracellular adenosine can also act on its cognate G protein-coupled receptors (GPCRs) by promoting their maturation: hypoxia increases the expression of A_1 receptors, because intracellular adenosine promotes their folding in the endoplasmic reticulum (ER).[20] Thus, adenosine acts in a manner analogous to a pharmacochaperone.[21] A similar mechanism may operate in the A_{2A}-receptor for the following reasons: (1) similar to the A_1 adenosine receptor,[22] the A_{2A} receptor incurs a folding problem in the ER,[23,24] and (2) oxygen deprivation also promotes the surface expression by mobilizing an intracellular pool of the A_{2A} receptor.[25]

RECEPTOR SUBTYPES AND THEIR CNS DISTRIBUTION

Adenosine receptors are members of the large superfamily of seven transmembrane-spanning GPCRs. The four known subtypes are denoted as A_1, A_{2A}, A_{2B}, and A_3.[26,27] Early hints at the existence of adenosine receptors in the brain resulted from the work by Sattin and Rall[28] who observed an adenosine-mediated increase of cAMP production in guinea pig brain slices, which was antagonized by methylxanthines like caffeine. Later on, it was discovered that adenosine might trigger also an inhibitory response suggesting that different receptor subtypes exist. These inhibitory and stimulatory subtypes were named A_1 and A_2, respectively.[29] Both adenosine receptor subtypes showed a distinct tissue distribution and proved to be responsible for many specific actions in almost all organs. Both subtypes were also found in the CNS where they are expressed in a well-defined pattern in different brain areas. A_1 adenosine receptors are primarily located in the cortex, hippocampus, and cerebellum.[26] One of the most important localizations of the A_2 receptor was found to be the striatum.[26]

The situation became more complicated when additional adenosine receptor subtypes were discovered. It was long suspected that two varieties of the stimulatory A_2 receptor might exist.[30] With the cloning of A_{2A} and A_{2B} receptors their existence was finally established.[31–33] At about the same time, the unexpected A_3 subtype was discovered and cloned.[34,35]

A_1 Adenosine Receptors

The inhibitory A_1 adenosine receptor is found in many organs and plays a critical role in the regulation of the cellular energy balance. The most prominent expression sites outside the CNS are the heart and kidneys where it helps to avoid excessive energy consumption. It does so in the heart by reducing the heart rate and the force of contraction and, thus, opposing the effect of input from the sympathetic nervous system. In kidneys the glomerular filtration rate is reduced, thereby leading to a diminished need for energy-dependent reabsorption of solutes from the filtrate. Blockade of A_1 adenosine receptors explains the well-known effects of the naturally occurring antagonist caffeine on the heart and kidneys.[36]

The significance of regulatory functions of adenosine in the CNS were long appreciated as caffeine also shows striking CNS effects as a result of antagonism at A_1 and A_{2A} receptors. Stimulation of A_1 adenosine receptors elicits numerous effects on brain function which are typically inhibitory in nature.[36] Activation of A_1 adenosine receptors results in postsynaptic inhibition of many different neurons and in inhibition of neurotransmitter release, primarily by acting on presynaptic receptors in excitatory neurons.[37]

Administration of adenosine into the brain results in sleep induction. Regulation of the sleep-wake cycle seems to be related to adenosine production caused by increased neuronal and metabolic activity and enhanced ATP catabolism. Other important CNS functions that are regulated via A_1 receptor stimulation are locomotor activity, learning and memory, pain, and food intake.[38]

In addition to such physiological activities, stimulation of A_1 adenosine receptors plays an equally important role in pathophysiological conditions. In general, adenosine has been shown to act as a neuroprotective factor when ATP and adenosine are released as a consequence of injury or hypoxic conditions. The damage caused by ischemic conditions can be limited by stimulation of A_1 receptors whereas A_1 antagonists worsen the outcome.[39,40]

Pathophysiological conditions possibly involving A_1 adenosine receptors include neurodegenerative and neuropsychiatric disorders. In patients who died from Alzheimer's disease a loss of hippocampal A_1 receptors was observed.[41] There are hints that stimulation of A_1 receptors might help to process β-amyloid peptides into soluble forms,[42] making this receptor subtype an interesting target for drug therapy in Alzheimer's disease. Similarly, A_1 adenosine receptors were shown to play a role in mood disorders. An important contribution to this role is made by nonneuronal cells, most importantly by astrocytes.[43]

For a comprehensive overview of adenosine receptor signaling in the CNS.[40,44]

A_{2A} Adenosine Receptors

In general, the A_{2A} adenosine receptor is typically found in blood vessels and therefore usually involved in an adenosine-induced vasodilation mediated by activation of G_s and the resulting increase in cAMP. As an important coronary vasodilator adenosine complements the A_1-mediated effect of limiting energy consumption (see above) by improving the oxygen supply of the heart. Blockage of the A_{2A} adenosine receptor in cerebral vessels is thought to contribute to pain relief by caffeine in migraine headaches.[45] In addition, stimulation of A_{2A} receptors promotes endothelial cell proliferation[46] via stimulation of MAP kinase.[47] This does not require stimulation of G_s but relies on an alternative signaling pathway comprising RAS[48] and ARNO.[49,50]

A prominent CNS effect transmitted by stimulation of the A_{2A} receptor is reduced locomotion. This response is accounted for by prominent expression of A_{2A} receptors in the striatum (see below), where it activates the $G\alpha_s$ paralogue $G\alpha_{olf}$ (because it is more abundant than $G\alpha_s$).

Inhibition of neurotransmitter release is an important mechanism by which stimulation of A_1 adenosine receptors protects the brain against injury from diverse causes.[40] On the contrary, A_{2A} adenosine receptors may contribute to neuronal injury by facilitating neurotransmitter release. Consequently, protection from excitotoxicity has been shown in the presence of A_{2A} receptor antagonists.[40,51] Experiments utilizing A_{2A} knockout mice confirmed the role of A_{2A} adenosine receptors in the development of excitotoxicity.[52]

A_{2B} Adenosine Receptors

The initially defined brain distribution of A_1 and A_2 receptors still represents the relevant neuronal distribution of these subtypes where A_2 embodies for the most part the A_{2A} receptor. A_{2B} receptors are primarily found outside the brain, e.g., in colon, mast cells, and blood vessels, but are detectable in pretty much every cell and tissue.[53,54] Their role in inflammation and immune responses is well documented by the effects of adenosine on mast cells and other cells with important functions in immune modulation.[54] After A_2 receptors were subdivided into A_{2A} and A_{2B} it was realized that the A_{2B} subtype contributes importantly to the regulation of vascular tone. It needs to be pointed out though that the affinity of adenosine for the A_{2B} is low such that physiological concentrations of adenosine do not cause A_{2B} activation. This suggests that the A_{2B} receptor is primarily an "emergency" receptor that is activated in pathophysiological conditions associated with massive adenosine release.[54,55] However, consistent evidence for a significant role in the CNS is not available.[44]

A_3 Adenosine Receptors

In functional terms the A_3 adenosine receptor resembles the A_1 subtype as it also represents an inhibitory receptor mediating a decrease of cAMP levels upon stimulation. In addition, it may also trigger a Ca^{2+} response[56–58] or MAP kinase signaling.[59,60] The A_3 adenosine receptor shares the least similarity with the other subtypes, and is the least conserved between species. In addition, the A_3 receptor is typically expressed at low density and its expression pattern is also very much species dependent. Although a role in an ischemic insult of the brain was suggested,[61] clear evidence of the presence of A_3 receptors is lacking. A recent study using tissue autoradiography in various rat tissues and human postmortem brain confirmed the presence of A_3 receptors in all investigated brain areas at low level.[62]

CROSSTALK WITH OTHER RECEPTORS

In many receptor systems it has been shown that signaling is not just a simple on/off mechanism but rather a process that is tightly modulated in many ways. One such mechanism is crosstalk between receptors for different ligands. In the CNS it became apparent already several decades ago that adenosine and dopamine signaling were mutually influencing each other. One of the first observations indicating such an interaction showed that some phosphodiesterase (PDE) inhibitors, now known to also be adenosine receptor antagonists, had effects on dopamine signaling that could not be solely explained by PDE inhibition.[63]

Over the years it became clear that A_{2A} adenosine receptors and D_2 dopamine receptors are both expressed on striato-pallidal neurons where they functionally interact.[64] Adenosine produces an inhibitory signal on dopaminergic neurons, which may occur on different levels. One obvious explanation would be the opposing effects on cAMP production due to G_s coupling of the A_{2A} and G_i coupling of the D_2 receptor. The reverse holds true for the interaction between A_1 and D_1 receptors. In a variety of studies an antagonistic interaction between adenosine and dopamine receptors was found, including the modulation of agonist affinity of one receptor by stimulation of the other one.[65] This antagonistic interplay was also observed in early behavioral investigations when it was shown that adenosine receptor antagonists enhanced the motor-activating effects of dopamine.[66]

More recent biochemical analyses allowed for a direct investigation of the physical interaction of such receptor pairs. In coimmunoprecipitation studies with transfected A_{2A} and D_2 receptors it was shown that they interacted in the absence of the agonist, suggesting a constitutive interaction between these receptors.[67] Substantial advances have been made in microscopy-based approaches, which allow for the investigation of protein–protein interactions by employing fluorescently labeled receptors. Accordingly, numerous studies confirmed the formation of heteromers between A_{2A} and D_2 receptors.[67,68] Novel fluorescence-based methods also enabled detection of oligomer formation in intact cells.[69,70] These heteromers are believed to be the basis for the therapeutic efficacy of A_{2A} receptor antagonists in the treatment of Parkinson's disease.

Interaction of A_{2A} and D_2 receptors with additional receptors leading to higher order oligomers was expected from mutual effects of activation of a given receptor on signaling pathways of other receptors. Such functional interaction was observed with $mGlu_5$ and CB_1 cannabinoid receptors.[71] Further evidence for the existence of higher order heteromers of A_{2A} and D_2 receptors with glutamate and cannabinoid receptors was provided by studies of protein–protein interaction utilizing various fluorescence and bioluminescence resonance energy transfer techniques (for review see Ref. 68). The modulation of signals originating from one neurotransmitter by the activation state of other receptors may play an important role not only for the development and treatment of Parkinson's disease but also for other neurodegenerative disorders (see Chapter 9: Adenosine Receptors and Memory Disorders; Chapter 10: Control of Motor Function by Adenosine A2A Receptors in Parkinson's and Huntington's Disease; Chapter 11: Adenosine Receptors Oligomers in Parkinson's Disease; Chapter 12: Adenosine Control of Striatal Function—Implications for the Treatment of Apathy in Basal Ganglia Disorders) or conditions like schizophrenia or drug addiction.

ADENOSINE AND CNS DISEASES

The phenotype of adenosine receptor-deficient mice provides some insights that allow for inferring the role of adenosine in brain disorders (and that of other organs). Mice rendered deficient in the A_1 adenosine receptor are more prone to prolonged epileptic seizures,[72,73] they are anxious (and hence aggressive), they have a reduced life span (presumably due to reduced renal and cardiac adaptive capacity)[74] and male mice sire less offspring, because their sperm are less likely to be capacitated.[75] Expression of the A_1 receptor in brain and spermatogonia is driven from an alternative promoter, which supports high-level expression, i.e., up to >1 pmol/mg.[76] In contrast, the A_1 receptor is only present at <10% of these levels in the other tissues (e.g., heart and kidney).

Deletion of the A_{2A} receptor results in hypertension, enhanced platelet aggregation, reduced exploratory behavior, increased anxiety, and male aggressiveness.[77] The phenotype is less pronounced in mouse strains other than CD1.[52] Surprisingly, the absence of A_{2A} receptors protects neurons against ischemia.[52] Another surprise is the finding that the absence of A_{2A} receptors renders mice more susceptible to the stimulant action of amphetamine and cocain.[78] This finding is difficult to reconcile with the mutually antagonistic interaction of D_2 and A_{2A} receptors (see above), but the absence of the A_{2A} receptor results in some traits that are consistent with reduced dopaminergic tone.[79]

Given that the A_{2B} receptor requires high levels of adenosine for its activation, its absence was not predicted to have any physiological repercussions. However, mice deficient the A_{2B} receptor are susceptible to inflammation, because expression of proinflammatory cytokines (TNF-α, IL-6) is enhanced and because the endothelium supports leucocyte transmigration by expressing adhesion molecules (e.g., E-selectin, P-selectin, ICAM-1).[80] This suggests that the antiinflammatory actions of adenosine are not only brought about by A_{2A} receptor activation but that A_{2B} receptors play an important role in mitigating macrophage activation. The A_{2B}-receptor is not expressed in the brain. However, the absence or the blockage of A_{2B} receptors protects against experimental autoimmune encephalitis by precluding the expansion of Th17 cells.[81] Taken together, these observations suggest that A_{2B}-receptor activation is a double-edged sword, which can elicit both, pro- and antiinflammatory responses.

The A_3 adenosine receptor is predominantly found in granulocytes and (rodent) mast cells; accordingly, abrogation of the receptor eliminates the agonist-induced potentiation of mast cell degranulation.[82] The absence of the A_3 receptor affects the circadian activity pattern but the underlying mechanisms are poorly understood.[83] In the brain, A_3 receptors are expressed in serotoninergic raphe neurons, where they are thought to regulate the surface levels of the serotonin transporter; stimulation of the A_3 receptor increases serotonin uptake by promoting insertion of the serotonin into the plasma membrane.[84] The effect is more pronounced if the serotonin transporter is coexpressed with the variant A_3 receptor-L90V. This variant is overrepresented in patients with autism.[85]

Because of its action on neuronal A_1 receptors (see above), adenosine has been proposed as an endogenous antiepileptic.[86] Similarly, the adenosinergic tone is thought to exert a disease modifying effect in many common disorders of the brain including Alzheimer's disease, sleep dysfunction, depression, Parkinson's disease, etc.[87] While this may be the case, it is more gratifying to pursue those avenues where there is pharmacoepidemiological and/or genetic/genomic evidence for an association between a disease and a given receptor. There are two instances where the evidence is strong: (1) Parkinson's disease, and (2) Huntington's disease. Both, retrospective[88,89] and prospective cohort studies[90,91] show an inverse association between coffee consumption and Parkinson's disease. This is consistent with the conjecture that caffeine intake reduces the risk of people developing Parkinson's disease. In addition, blockage of A_{2A} receptors protects against α-synuclein-induced cell death.[92] These observations support the concept that A_{2A} adenosine receptor antagonists ought to ameliorate the symptoms of Parkinson's disease (see below).

Huntington's disease is caused by the accumulation of huntingtin aggregates, which results in the loss of the medium size spiny neurons of the striatum. Huntingtin contains a polyglutamine tract in its amino terminus (encoded by CAG repeats). Disease onset is defined by the number of CAG repeats—the more the earlier. Modifying genes have an additional effect. A single nucleotide polymorphism in the A_{2A} receptor (rs5751876; 1976 C/T) shortens disease latency. When corrected for repeat numbers, disease onset in patients with the T/T genotype is on average 3.8 years earlier than in patients with the C/C genotype.[93] It is not clear how this effect is brought about because the polymorphism is synonymous—both genotypes encode Y^{361} in the C-terminus of the A_{2A} receptor. The link is, however, supported by an independent cohort study on a different continent[94] and two additional lines of evidence: (1) in a retrospective analysis, caffeine intake >190 mg/day was also associated with earlier onset of disease after correction for repeat length and smoking,[95] and (2) stimulation of the A_{2A} receptor by an agonist ameliorates the symptoms of Huntington's disease in a transgenic mouse model (R6/2) and reduces DNA damage in GABAergic neuron-like cells derived from induced pluripotent stem cells (iPSCs), which were derived from patients harboring mutated huntingtin.[96] It is somehow surprising though, that in a different, transgenic mouse line (R6/1), the

A_{2A} (and D_1) receptor-dependent activation of protein kinase A was found to be augmented[97] and that the disease course was found to be mitigated if two lines of R6/2 mice were crossed with A_{2A} receptor-deficient mice. Thus, it is currently not clear whether activation or inhibition of A_{2A}-receptors is the more promising modality. It is also worth pointing out that the focus on cAMP-dependent signaling may be too narrow: signaling by the A_{2A} receptor is remarkable for several reasons. The A_{2A}-receptor has a very long C-terminus (122 amino acids), which supports the recruitment of many signaling molecules other than its cognate G proteins G_s/G_{olf}.[98] The A_{2A}-receptor undergoes restricted collision coupling,[50,99] because it is confined in microdomains.[99,100] The C-terminus of the A_{2A} receptor lacks the cysteine residue at the end of the C-terminal helix 8, which runs parallel to the membrane. This cysteine is the site of palmitoylation. Confinement can be relieved by engineering a palmitoylation site into the receptor.[99] The C-terminus of the A_{2A}-receptor also defines the basal activity of the receptor and its signaling bias.[101] Finally, the A_{2A} adenosine receptor can transactivate the neurotrophin receptor TrkB and thus promote neuronal survival[102]; obviously, this action is beneficial in a neurodegenerative disease. It is therefore conceivable that the unliganded A_{2A}-receptor exerts an action, which is detrimental in the presence of huntingtin aggregates. Agonist-binding may redistribute the receptor into a different membrane compartment and thus mitigate its detrimental action, while antagonist binding may block the activity of the receptor. This hypothetical model addresses the observed discrepancy and makes testable predictions: (1) changing receptor mobility ought to affect outcome. This can for instance be examined by reintroducing wild type and palmitoylated receptor into A_{2A} receptor-deficient striatal neurons harboring mutated huntingtin differently, and (2) the effect of a biased agonist ought to differ from that of agonist or antagonists. Inosine has been proposed to act as a low affinity biased agonist at the A_{2A} receptor.[103]

RECENT CLINICAL TRIALS

Although over the years a plethora of adenosine receptor ligands were developed only a few made it into clinical practice. Recent clinical trials are mostly in the fields of cardiovascular conditions and CNS diseases like Parkinson's disease. Adenosine (or its precursor ATP) has been the first-line agent in the acute treatment of supraventricular tachycardia for some three decades. It is also employed as a coronary vasodilator in myocardial perfusion imaging. Caffeine (and to a lesser extent theophylline) has been used to treat neonatal apnoea for some 40 years.

Its beneficial actions were documented in adequately powered long-term trials.[104–106] The swift action of adenosine in supraventricular tachycardia arises from A_1 receptor stimulation; the transient flushing and hypotension reflect the nonselective nature of adenosine. Similarly, the action of caffeine in sleep apnea is thought to reflect blockage of A_{2A}-receptors, but caffeine poorly discriminates between A_{2A} and A_1 receptors. Currently, several clinical trials are being conducted with agonists and antagonists, which are designed to target each of the four adenosine receptor subtypes in a selective manner.[107,108]

A_1 Adenosine Receptors

Due to the diuretic effect caused by blockade of renal A_1 adenosine receptors it was long thought that A_1 antagonist might be beneficial in the treatment of acute heart failure. One such antagonist that made it into clinical testing was rolofylline, an A_1 selective xanthine derivative. The development of the drug was abandoned after phase III of the PROTECT study as it showed no benefit for renal function as compared to placebo treatment.[109]

The A_1 antagonist PBF-680 with a nonxanthine structure is under phase II clinical investigation (www.clinicaltrials.gov identifier NCT02635945) for the treatment of asthma and COPD after a safety (NCT01845181) and a proof-of-concept study (NCT01939587) were completed.

Treatment with A_1 agonists was envisioned to be beneficial in a number of conditions including cardiac arrhythmias, cardioprotection, pain, and diabetes type 2.[110,111] Attempts to establish such therapies frequently suffered from tachyphylaxis, which is a known problem in treatments with agonists. Also, cardiovascular and noncardiovascular side effects limited the success of targeting A_1 receptors with agonists.

A_{2A} Adenosine Receptors

As outlined above, there is mutual antagonism between adenosine and dopamine signaling in the striatum, which is consistent with the stimulatory effect of caffeine on locomotor activity. In addition, in schizophrenic patients, the antipsychotic activity of the D_2 receptor blocker haloperidol is augmented by concomitant administration of the ENT1-inhibitor dipyridamole.[112] These observations further substantiate the proposed mutual antagonism between D_2 and A_{2A} receptors. Accordingly, a number of A_{2A} antagonists are under development for this indication.[113,114] Istradefylline is the first compound that received approval in Japan in 2013. Approval in the United States and European Union is pending.

The nonxanthine A_{2A} antagonist PBF-509, which is also in clinical trial to treat Parkinson's disease, is investigated in a phase I clinical trial as a treatment for Non-Small Cell Lung Cancer (NSCLC) (NCT02403193). This idea was pursued based on the observation that adenosine helps cancer cells to evade T cell-mediated rejection and promotes angiogenesis. In addition, it was shown that NSCLC tissue expresses A_{2A} adenosine receptors and that antagonizing these receptors inhibited tumor cell growth.[115]

The A_{2A} selective agonist regadenoson, a 2-substituted adenosine derivative, was approved as a coronary vasodilator in myocardial perfusion imaging in recent years.[116,117] Regadenoson is also under investigation in a phase II clinical trial (NCT01788631) for the treatment of sickle cell disease. Another study with regadenoson is planned for the treatment of pulmonary hypertension (NCT02220634).

A_{2B} Adenosine Receptors

Targeting the A_{2B} adenosine receptor as a drug target has been considered for the treatment of asthma and gastrointestinal disorders. In humans the A_{2B} receptor is thought to be involved in adenosine-induced airway constriction. Therefore, blockage of A_{2B} receptors is predicted to be beneficial in the treatment of asthma. Stimulation of intestinal A_{2B} adenosine receptors may lead to diarrhea and/or intestinal inflammation which, again, suggests A_{2B} antagonists as a useful treatment.[107] Earlier phase I trials with the xanthine CVT-6883 were not pursued, no recent or current trials are listed at *ClinicalTrials.gov*.

A_3 Adenosine Receptors

The most diverse indications are currently explored for the A_3 adenosine receptor as a target. The agonist CF101 (IB-MECA) is projected for a phase III clinical trial for the treatment of rheumatoid arthritis (NCT02647762). Another ongoing phase II trial is studying the use of CF101 for ocular hypertension and glaucoma (NCT01033422). The same compound was considered for the treatment of psoriasis (phase II, NCT00428974). The closely related compound CF102 (Cl-IB-MECA) is currently scrutinized in a phase II trial for the treatment of hepatocellular carcinoma (NCT02128958).

Interestingly, treatment of glaucoma and inflammatory bowel disease is also planned to be studied in a phase I trial with the A_3 antagonist PDF-677 (Palobiofarma, Barcelona, Spain, received approval from the Spanish Regulatory Agency to start a First-in-Man clinical trial in healthy volunteers in December 2015). It will be interesting to compare the outcomes for agonist versus antagonist treatment of glaucoma and inflammatory diseases.

References

1. Bianchi V, Spychala J. Mammalian 5′-nucleotidases. *J Biol Chem*. 2003;278(47): 46195–46198.
2. Sawada K, Echigo N, Juge N, Miyaji T, Otsuka M, Omote H, et al. Identification of a vesicular nucleotide transporter. *Proc Natl Acad Sci U S A*. 2008;105(15):5683–5686.
3. Cisneros-Mejorado A, Pérez-Samartín A, Gottlieb M, Matute C. ATP Signaling in brain: release, excitotoxicity and potential therapeutic targets. *Cell Mol Neurobiol*. 2015;35(1): 1–6.
4. Ma Z, Tanis JE, Taruno A, Foskett JK. Calcium homeostasis modulator (CALHM) ion channels. *Pflugers Arch*. 2016;468(3):395–403.
5. Zimmermann H, Zebisch M, Sträter N. Cellular function and molecular structure of ecto-nucleotidases. *Purinergic Signal*. 2012;8(3):437–502.
6. Heuts DP, Weissenborn MJ, Olkhov RV, Shaw AM, Gummadova J, Levy C, et al. Crystal structure of a soluble form of human CD73 with ecto-5′-nucleotidase activity. *ChemBioChem*. 2012;13(16):2384–2391.
7. Lovatt D, Xu Q, Liu W, Takano T, Smith NA, Schnermann J, et al. Neuronal adenosine release, and not astrocytic ATP release, mediates feedback inhibition of excitatory activity. *Proc Natl Acad Sci U S A*. 2012;109(16):6265–6270.
8. Schicker K, Hussl S, Chandaka GK, Kosenburger K, Yang JW, Waldhoer M, et al. A membrane network of receptors and enzymes for adenine nucleotides and nucleosides. *Biochim Biophys Acta*. 2009;1793(2):325–334.
9. Nofech-Mozes Y, Blaser SI, Kobayashi J, Grunebaum E, Roifman CM. Neurologic abnormalities in patients with adenosine deaminase deficiency. *Pediatr Neurol*. 2007;37 (3):218–221.
10. Young JD, Yao SY, Baldwin JM, Cass CE, Baldwin SA. The human concentrative and equilibrative nucleoside transporter families, SLC28 and SLC29. *Mol Aspects Med*. 2013;34(2–3):529–547.
11. Staufner C, Lindner M, Dionisi-Vici C, Freisinger P, Dobbelaere D, Douillard C, et al. Adenosine kinase deficiency: expanding the clinical spectrum and evaluating therapeutic options. *J Inherit Metab Dis*. 2016;39(2):273–283.
12. Chen J, Rinaldo L, Lim SJ, Young H, Messing RO, Choi DS. The type 1 equilibrative nucleoside transporter regulates anxiety-like behavior in mice. *Genes Brain Behav*. 2007;6(8):776–783.
13. Morote-Garcia JC, Köhler D, Roth JM, Mirakaj V, Eldh T, Eltzschig HK, et al. Repression of the equilibrative nucleoside transporters dampens inflammatory lung injury. *Am J Respir Cell Mol Biol*. 2013;49(2):296–305.
14. Fredholm BB. Adenosine, an endogenous distress signal, modulates tissue damage and repair. *Cell Death Differ*. 2007;14(7):1315–1323.
15. Schütz W, Schrader J, Gerlach E. Different sites of adenosine formation in the heart. *Am J Physiol*. 1981;240(6):H963–H970.
16. Choi DS, Cascini MG, Mailliard W, Young H, Paredes P, McMahon T, et al. The type 1 equilibrative nucleoside transporter regulates ethanol intoxication and preference. *Nat Neurosci*. 2004;7(8):855–861.
17. Zhang D, Xiong W, Albensi BC, Parkinson FE. Expression of human equilibrative nucleoside transporter 1 in mouse neurons regulates adenosine levels in physiological and hypoxic-ischemic conditions. *J Neurochem*. 2011;118(1):4–11.
18. Decking UK, Schlieper G, Kroll K, Schrader J. Hypoxia-induced inhibition of adenosine kinase potentiates cardiac adenosine release. *Circ Res*. 1997;81(2):154–164.
19. Lynch 3rd JJ, Alexander KM, Jarvis MF, Kowaluk EA. Inhibition of adenosine kinase during oxygen-glucose deprivation in rat cortical neuronal cultures. *Neurosci Lett*. 1998;252(3):207–210.

20. Kusek J, Yang Q, Witek M, Gruber CW, Nanoff C, Freissmuth M. Chaperoning of the A_1-adenosine receptor by endogenous adenosine-an extension of the retaliatory metabolite concept. *Mol Pharmacol.* 2015;87(1):39–51.
21. Málaga-Diéguez L, Yang Q, Bauer J, Pankevych H, Freissmuth M, Nanoff C. Pharmacochaperoning of the A1 adenosine receptor is contingent on the endoplasmic reticulum. *Mol Pharmacol.* 2010;77(6):940–952.
22. Pankevych H, Korkhov V, Freissmuth M, Nanoff C. Truncation of the A1 adenosine receptor reveals distinct roles of the membrane-proximal carboxyl terminus in receptor folding and G protein coupling. *J Biol Chem.* 2003;278(32):30283–30293.
23. Milojevic T, Reiterer V, Stefan E, Korkhov VM, Dorostkar MM, Ducza E, et al. The ubiquitin-specific protease Usp4 regulates the cell surface level of the A2A receptor. *Mol Pharmacol.* 2006;69(4):1083–1094.
24. Bergmayr C, Thurner P, Keuerleber S, Kudlacek O, Nanoff C, Freissmuth M, et al. Recruitment of a cytoplasmic chaperone relay by the A2A-adenosine receptor. *J Biol Chem.* 2013;288(40):28831–28844.
25. Arslan G, Kull B, Fredholm BB. Anoxia redistributes adenosine A2A receptors in PC12 cells and increases receptor-mediated formation of cAMP. *Naunyn Schmiedebergs Arch Pharmacol.* 2002;365(2):150–157.
26. Fredholm BB, IJzerman AP, Jacobson KA, Klotz K-N, Linden J. International Union of Pharmacology. XXV. Nomenclature and classification of adenosine receptors. *Pharmacol Rev.* 2001;53(4):527–552.
27. Fredholm BB, IJzerman AP, Jacobson KA, Linden J, Müller CE. International Union of Pharmacology. LXXXI. Nomenclature and classification of adenosine receptors – an update. *Pharmacol Rev.* 2011;63(1):1–34.
28. Sattin A, Rall TW. The effect of adenosine and adenine nucleotides on the cyclic adenosine 3′, 5′-phosphate content of guinea pig cerebral cortex slices. *Mol Pharmacol.* 1970;6(1):13–23.
29. van Calker D, Müller M, Hamprecht B. Adenosine regulates via two different types of receptors, the accumulation of cyclic AMP in cultured brain cells. *J Neurochem.* 1979;33(5):999–1005.
30. Bruns RF. Adenosine receptor activation in human fibroblasts: nucleoside agonists and antagonists. *Can J Physiol Pharmacol.* 1980;58(6):673–691.
31. Libert F, Parmentier M, Lefort A, Dinsart C, van Sande J, Maenhaut C, et al. Selective amplification and cloning of four new members of the G protein-coupled receptor family. *Science.* 1989;244(4904):569–572.
32. Maenhaut C, van Sande J, Libert F, Abramowicz M, Parmentier M, Vanderhaegen J-J, et al. RDC8 codes for an adenosine A2 receptor with physiological constitutive activity. *Biochem Biophys Res Commun.* 1990;173(3):1169–1178.
33. Pierce KD, Furlong TJ, Selbie L-A, Shine J. Molecular cloning and expression of an adenosine A2b receptor from human brain. *Biochem Biophys Res Commun.* 1992;187(1):86–93.
34. Zhou Q-Y, Li C, Olah ME, Johnson RA, Stiles GL, Civelli O. Molecular cloning and characterization of an adenosine receptor: the A3 adenosine receptor. *Proc Natl Acad Sci U S A.* 1992;89(16):7432–7436.
35. Salvatore CA, Jacobson MA, Taylor HE, Linden J, Johnson RG. Molecular cloning and characterization of the human A3 adenosine receptor. *Proc Natl Acad Sci U S A.* 1993;90(21):10365–10369.
36. Fredholm BB, Bättig K, Holmén J, Nehlig A, Zvartau EE. Actions of caffeine in the brain with specific reference to factors that contribute to its widespread use. *Pharmacol Rev.* 1999;51(1):83–133.
37. Haas BB, Seelbach O. Functions of neuronal adenosine receptors. *Naunyn-Schmiedebergs Arch Pharmacol.* 2000;362(4–5):375–381.

38. Dunwiddie TV, Masino SA. The role and regulation of adenosine in the central nervous system. *Annu Rev Neurosci.* March 2001;24:31–55.
39. von Lubitz DK, Lin RC, Melman N, Ji XD, Carter MF, Jacobson KA. Chronic administration of selective adenosine A_1 receptor agonist or antagonist in cerebral ischemia. *Eur J Pharmacol.* 1994;256(2):161–167.
40. Fredholm BB, Chen J-F, Cunha RA, Svenningsson P, Vaugeois J-M. Adenosine and brain function. *Int Rev Neurobiol.* March 2005;63:191–270.
41. Ulas J, Brunner LC, Ngyen L, Cotman CW. Reduced density of adenosine A_1 receptors and preserved coupling of adenosine A_1 receptors to G proteins in Alzheimer hippocampus: a quantitative autoradiographic study. *Neuroscience.* 1993;52(4):843–854.
42. Angulo E, Casado V, Mallol J, Canela EI, Vinals F, Ferrer I, et al. A_1 adenosine receptors accumulate in neurodegenerative structures in Alzheimer disease and mediate both amyloid precursor processing and tau phosphorylation and translocation. *Brain Pathol.* 2003;13(4):440–451.
43. van Calker D, Biber K. The role of glial adenosine receptors in neural resilience and the neurobiology of mood disorders. *Neurochem Res.* 2005;30(10):1205–1217.
44. Burnstock G, Krügel U, Abbracchio MP, Illes P. Purinergic signalling: from normal behaviour to pathological brain function. *Prog Neurobiol.* Oct. 2011;95:229–274.
45. Diener HC, Pfaffenrath V, Pageler L, Peil H, Aicher B. The fixed combination of acetylsalicylic acid, paracetamol and caffeine is more effective than single substances and dual combination for the treatment of headache: a multicentre, randomized, double-blind, single-dose, placebo-controlled parallel group study. *Cephalalgia.* 2005;25 (10):776–787.
46. Sexl V, Mancusi G, Parzer S, Schütz W, Freissmuth M. Stimulation of human umbilical venous endothelial cell proliferation by A_{2A}-adenosine and β_2-adrenergic receptors. *Brit J Pharmacol.* 1995;114(8):1577–1586.
47. Sexl V, Mancusi G, Höller C, Gloria-Maercker E, Schütz W, Freissmuth M. Stimulation of the mitogen-activated-protein kinase (MAP kinase) via the A_{2A}-adenosine receptor in primary human endothelial cells. *J Biol Chem.* 1997;272(9):5792–5799.
48. Seidel MG, Klinger M, Freissmuth M, Höller C. Activation of MAP kinase by the A_{2A}-adenosine receptor via a rap1-dependent and via a $p21^{ras}$-dependent pathway. *J Biol Chem.* 1999;274(36):25833–35841.
49. Gsandtner I, Charalambous C, Stefan E, Ogris E, Freissmuth M, Zezula J. Heterotrimeric G protein independent signaling of a G protein-coupled receptor: direct binding of ARNO/cytohesin-2 to the carboxyl terminus of the A_{2A}-adenosine receptor is necessary for sustained activation of the ERK/MAP-kinase pathway. *J Biol Chem.* 2005;280(36):31898–31905.
50. Charalambous C, Gsandtner I, Keuerleber S, Milan-Lobo L, Kudlacek O, Freissmuth M, et al. Restricted collision coupling of the A_{2A}-receptor revisited: evidence for physical separation of two signaling cascades. *J Biol Chem.* 2008;283(14):9276–9288.
51. Ongini E, Adami M, Ferri C, Bertorelli R. Adenosine A_{2A} receptors and neuroprotection. *Ann NY Acad Sci.* Oct. 1997;825:31–48.
52. Chen JF, Huang Z, Ma J, Zhu J, Moratalla T, Standaert D, et al. A_{2A} adenosine receptor deficiency attenuates brain injury induced by transient focal ischemia in mice. *J Neurosci.* 1999;19(21):9192–9200.
53. Feoktistov I, Biaggioni I. Adenosine A_{2B} receptors. *Pharmacol Rev.* 1997;49(4):381–402.
54. Haskó G, Csóka B, Németh ZH, Vizi ES, Pacher P. A_{2B} adenosine receptors in immunity and inflammation. *Trends Immunol.* 2009;30(6):263–270.
55. Borea PA, Gessi S, Merighi S, Varani K. Adenosine as a multi-signalling guardian angel in human diseases: when, where and how does it exert its protective effects? *Trends Pharmacol Sci.* 2016;37(6):419–434.

56. Kohno Y, Ji X, Mawhorter SD, Koshiba M, Jacobson KA. Activation of A3 adenosine receptors on human eosinophils elevates intracellular calcium. *Blood*. 1996;88(9):3569–3574.
57. Englert M, Quitterer U, Klotz K-N. Effector coupling of stably transfected human A_3 adenosine receptors in CHO cells. *Biochem Pharmacol*. 2002;64(1):69–73.
58. Fossetta J, Jackson J, Deno G, Fan X, Du XK, Bober L, et al. Pharmacological analysis of calcium responses mediated by the human A_3 adenosine receptor in monocyte-derived dendritic cells and recombinant cells. *Mol Pharmacol*. 2003;63(2):342–350.
59. Schulte G, Fredholm BB. Human adenosine A_1, A_{2A}, A_{2B}, and A_3 receptors expressed in Chinese hamster ovary cells all mediate the phosphorylation of extracellular-regulated kinase 1/2. *Mol Pharmacol*. 2000;58(3):477–482.
60. Graham S, Combes P, Crumiere M, Klotz K-N, Dickenson JM. Regulation of p42/p44 mitogen-activated protein kinase by the human adenosine A_3 receptor in transfected CHO cells. *Eur J Pharmacol*. 2001;420(1):19–26.
61. von Lubitz DK, Lin RC, Popik P, Carter MF, Jacobson KA. Adenosine A_3 receptor stimulation and cerebral ischemia. *Eur J Pharmacol*. 1994;263(1–2):59–67.
62. Haeusler D, Grassinger L, Fuchshuber F, Hörleinsberger WJ, Höftberger R, Leisser I, et al. Hide and seek: a comparative autoradiographic in vitro investigation of the adenosine A_3 receptor. *Eur J Nucl Med Mol Imaging*. 2015;42(6):928–939.
63. Fredholm BB, Fuxe K, Agnati L. Effect of some phosphodiesterase inhibitors on central dopamine mechanisms. *Eur J Pharmacol*. 1976;38(1):31–38.
64. Ongini E, Fredholm BB. Pharmacology of adenosine A_{2A} receptors. *Trends Pharmacol*. 1996;17(10):364–372.
65. Ferré S, von Euler G, Johansson B, Fredholm BB, Fuxe K. Stimulation of high affinity adenosine A_2 receptors decreases the affinity of dopamine D_2 receptors in rat striatal membranes. *Proc Natl Acad Sci U S A*. 1991;88(16):7238–7241.
66. Ferré S, Fuxe K, von Euler G, Johansson B, Fredholm BB. Adenosine-dopamine interactions in the brain. *Neuroscience*. 1992;51(3):501–512.
67. Fuxe K, Ferré S, Canals M, Torvinen M, Terasmaa A, Marcellino D, et al. Adenosine A_{2A} and dopamine D_2 heteromeric receptor complexes and their function. *J Mol Neurosci*. 2005;26(2–3):209–220.
68. Fuxe K, Marcellino D, Borroto-Escuela DO, Guescini M, Fernández-Dueñas V, Tanganelli S, et al. Adenosine-dopamine interactions in the pathophysiology and treatment of CNS disorders. *CNS Neurosci Ther*. 2010;16(3):e18–e42.
69. Hillon J, Canals M, Torvinen M, Casadó V, Scott R, Terasmaa A, et al. Coaggregation, cointernalization, and codesensitization of adenosine A_{2A} receptors and dopamine D_2 receptors. *J Biol Chem*. 2002;277(20):18091–18097.
70. Kamiya T, Saitoh O, Yoshioka K, Nakata H. Oligomerization of adenosine A_{2A} and dopamine D_2 receptors in living cells. *Biochem Biophys Res Comm*. 2003;306(2):544–549.
71. Schiffmann SN, Fisone G, Moresco R, Cunha RA, Ferré S. Adenosine A_{2A} receptors and basal ganglia physiology. *Prog Neurobiol*. 2007;83(5):277–292.
72. Fedele DE, Li T, Lan JQ, Fredholm BB, Boison D. Adenosine A_1 receptors are crucial in keeping an epileptic focus localized. *Exp Neurol*. 2006;200(1):184–190.
73. Kochanek PM, Vagni VA, Janesko KL, Washington CB, Crumrine PK, Garman RH, et al. Adenosine A_1 receptor knockout mice develop lethal status epilepticus after experimental traumatic brain injury. *J Cereb Blood Flow Metab*. 2006;26(4):565–575.
74. Gimenez-Llort L, Fernandez-Teruel A, Escorihuela RM, Fredholm BB, Tobena A, Pekny M, et al. Mice lacking the adenosine A_1 receptor are anxious and aggressive, but are normal learners with reduced muscle strength and survival rate. *Eur J Neurosci*. 2002;16(3):547–550.
75. Minelli A, Liguori L, Bellazza I, Mannucci R, Johansson B, Fredholm BB. Involvement of A_1 adenosine receptors in the acquisition of fertilizing capacity. *J Androl*. 2004;25 (2):286–292.

76. Ren H, Stiles GL. Separate promoters in the human A_1 adenosine receptor gene direct the synthesis of distinct messenger RNAs that regulate receptor abundance. *Mol Pharmacol*. 1995;48(6):975–980.
77. Ledent C, Vaugeois JM, Schiffmann SN, Pedrazzini T, El Yacoubi M, Vanderhaeghen JJ, et al. Aggressiveness, hypoalgesia and high blood pressure in mice lacking the adenosine A_{2a} receptor. *Nature*. 1997;388(6643):674–678.
78. Chen JF, Beilstein M, Xu YH, Turner TJ, Moratalla R, Standaert DG, et al. Selective attenuation of psychostimulant-induced behavioral responses in mice lacking A_{2A} adenosine receptors. *Neuroscience*. 2000;97(1):95–104.
79. Dassesse D, Massie A, Ferrari R, Ledent C, Parmentier M, Arckens L, et al. Functional striatal hypodopaminergic activity in mice lacking adenosine A_{2A} receptors. *J Neurochem*. 2001;78(1):183–198.
80. Yang D, Zhang Y, Nguyen HG, Koupenova M, Chauhan AK, Makitalo M, et al. The A_{2B} adenosine receptor protects against inflammation and excessive vascular adhesion. *J Clin Invest*. 2006;116(7):1913–1923.
81. Wei W, Du C, Lv J, Zhao G, Li Z, Wu Z, et al. Blocking A_{2B} adenosine receptor alleviates pathogenesis of experimental autoimmune encephalomyelitis via inhibition of IL-6 production and Th17 differentiation. *J Immunol*. 2013;190(1):138–146.
82. Salvatore CA, Tilley SL, Latour AM, Fletcher DS, Koller BH, Jacobson MA. Disruption of the A_3 adenosine receptor gene in mice and its effect on stimulated inflammatory cells. *J Biol Chem*. 2000;275(6):4429–4434.
83. Yang JN, Wang Y, Garcia-Roves PM, Björnholm M, Fredholm BB. Adenosine A3 receptors regulate heart rate, motor activity and body temperature. *Acta Physiol (Oxf)*. 2010;199(2):221–230.
84. Zhu CB, Lindler KM, Campbell NG, Sutcliffe JS, Hewlett WA, Blakely RD. Colocalization and regulated physical association of presynaptic serotonin transporters with A_3 adenosine receptors. *Mol Pharmacol*. 2011;80(3):458–465.
85. Campbell NG, Zhu CB, Lindler KM, Yaspan BL, Kistner-Griffin E, et al. Rare coding variants of the adenosine A_3 receptor are increased in autism: on the trail of the serotonin transporter regulome. *Mol Autism*. 2013;4(1):28.
86. Boison D. Adenosinergic signaling in epilepsy. *Neuropharmacology*. May 2016;104: 131–139.
87. Boison D, Aronica E. Comorbidities in neurology: is adenosine the common link? *Neuropharmacology*. Oct 2015;97:18–34.
88. Benedetti MD, Bower JH, Maraganore DM, McDonnell SK, Peterson BJ, Ahlskog JE, et al. Smoking, alcohol, and coffee consumption preceding Parkinson's disease: a case-control study. *Neurology*. 2000;55(9):1350–1358.
89. Ragonese P, Salemi G, Morgante L, Aridon P, Epifanio A, Buffa D, et al. A case-control study on cigarette, alcohol, and coffee consumption preceding Parkinson's disease. *Neuroepidemiology*. 2003;22(5):297–304.
90. Ross GW, Abbott RD, Petrovitch H, Morens DM, Grandinetti A, Tung KH, et al. Association of coffee and caffeine intake with the risk of Parkinson disease. *JAMA*. 2000;283(20):2674–2679.
91. Ascherio A, Zhang SM, Hernan MA, Kawachi I, Colditz GA, Speizer FE, et al. Prospective study of caffeine consumption and risk of Parkinson's disease in men and women. *Ann Neurol*. 2001;50(1):56–63.
92. Ferreira DG, Batalha VL, Vicente Miranda H, Coelho JE, Gomes R, Gonçalves FQ, et al. Adenosine A2A receptors modulate α-synuclein aggregation and toxicity. *Cereb Cortex* 2015;pii:bhv268.
93. Dhaenens CM, Burnouf S, Simonin C, Van Brussel E, Duhamel A, Defebvre L, et al. Huntington French Speaking Network. A genetic variation in the ADORA2A gene modifies age at onset in Huntington's disease. *Neurobiol Dis*. 2009;35(3):474–476.

94. Vital M, Bidegain E, Raggio V, Esperon P. Molecular characterization of genes modifying the age at onset in Huntington's Disease in Uruguayan patients. *Int J Neurosci.* May 2015;22:1–17.
95. Simonin C, Duru C, Salleron J, Hincker P, Charles P, Delval A, et al. REGISTRY Study of the European Huntington's Disease Network, Bachoud-Levi AC, Richard F, Blum D, Krystkowiak P; Huntington French Speaking Network. Association between caffeine intake and age at onset in Huntington's disease. *Neurobiol Dis.* Oct. 2013;58: 179–182.
96. Chiu FL, Lin JT, Chuang CY, Chien T, Chen CM, Chen KH, et al. Elucidating the role of the A_{2A} adenosine receptor in neurodegeneration using neurons derived from Huntington's disease iPSCs. *Hum Mol Genet.* 2015;24(21):6066–6079.
97. Tyebji S, Saavedra A, Canas PM, Pliassova A, Delgado-García JM, Alberch J, et al. Hyperactivation of D1 and A2A receptors contributes to cognitive dysfunction in Huntington's disease. *Neurobiol Dis.* Feb. 2015;74:41–57.
98. Keuerleber S, Gsandtner I, Freissmuth M. From cradle to twilight: the carboxyl terminus directs the fate of the A2A-adenosine receptor. *Biochim Biophys Acta.* 2011;1808 (5):1350–1357.
99. Keuerleber S, Thurner P, Gruber CW, Zezula J, Freissmuth M. Reengineering the collision coupling and diffusion mode of the A_{2A}-adenosine receptor: palmitoylation in helix 8 relieves confinement. *J Biol Chem.* 2012;287(50):42104–42118.
100. Thurner P, Gsandtner I, Kudlacek O, Choquet D, Nanoff C, Freissmuth M, et al. A two-state model for the diffusion of the A_{2A} adenosine receptor in hippocampal neurons: agonist-induced switch to slow mobility is modified by synapse-associated protein 102 (SAP102). *J Biol Chem.* 2014;289(13):9263–9274.
101. Klinger M, Kuhn M, Just H, Stefan E, Palmer T, Freissmuth M, et al. Removal of the carboxy terminus of the A2A-adenosine receptor blunts constitutive activity: differential effect on cAMP accumulation and MAP kinase stimulation. *Naunyn Schmiedebergs Arch Pharmacol.* 2002;366(4):287–298.
102. Wiese S, Jablonka S, Holtmann B, Orel N, Rajagopal R, Chao MV, et al. Adenosine receptor A_{2A}-R contributes to motoneuron survival by transactivating the tyrosine kinase receptor TrkB. *Proc Natl Acad Sci U S A.* 2007;104(43):17210–17215.
103. Welihinda AA, Kaur M, Greene K, Zhai Y, Amento EP. The adenosine metabolite inosine is a functional agonist of the adenosine A_{2A} receptor with a unique signaling bias. *Cell Signal.* 2016;28(6):552–560.
104. Schmidt B, Roberts RS, Davis P, Doyle LW, Barrington KJ, Ohlsson A, et al. Caffeine for Apnea of Prematurity Trial Group. Long-term effects of caffeine therapy for apnea of prematurity. *N Engl J Med.* 2007;357(19):1893–1902.
105. Marcus CL, Meltzer LJ, Roberts RS, Traylor J, Dix J, D'ilario J, et al. Caffeine for Apnea of Prematurity–Sleep Study. Long-term effects of caffeine therapy for apnea of prematurity on sleep at school age. *Am J Respir Crit Care Med.* 2014;190(7):791–799.
106. Lodha A, Seshia M, McMillan DD, Barrington K, Yang J, Lee SK, et al. Canadian Neonatal Network. Association of early caffeine administration and neonatal outcomes in very preterm neonates. *JAMA Pediatr.* 2015;169(1):33–38.
107. Gessi S, Merighi S, Varani K, Borea PA. Adenosine receptors in health and disease. *Adv Pharmacol.* 2011;61:41–75.
108. Müller CE, Jacobson KA. Recent developments in adenosine receptor ligands and their potential as novel drugs. *Biochim Biophys Acta.* 2011;1808(5):1290–1308.
109. Massie BM, O'Connor CM, Metra M, Ponikowski P, Teerlink JR, Cotter G, et al. for the PROTECT Investigators and Committees. Rolofylline, an adenosine A_1-receptor antagonist, in acute heart failure. *N Engl J Med.* 2010;363(15):1419–1428.
110. Elzein E, Zablocki J. A_1 adenosine receptor agonists and their potential therapeutic applications. *Expert Opin Investig Drugs.* 2008;17(12):1901–1910.

111. Gao Z-G, Jacobson KA. Emerging adenosine receptor agonists – an update. *Expert Opin Emerg Drugs*. 2011;16(4):597–602.
112. Akhondzadeh S, Shasavand E, Jamilian H, Shabestari O, Kamalipour A. Dipyridamole in the treatment of schizophrenia: adenosine-dopamine receptor interactions. *J Clin Pharm Ther*. 2000;25(2):131–137.
113. Jenner P. An overview of adenosine A_{2A} receptor antagonists in Parkinson's disease. *Int Rev Neurobiol*. 2014;119:71–86.
114. Pinna A. Adenosine A_{2A} receptor antagonists in Parkinson's disease: progress in clinical trials from newly approved istradefylline to drugs in early development and those already discontinued. *CNS Drugs*. 2014;28(5):455–474.
115. Mediavilla-Varela M, Luddy K, Noyes D, Khalil FK, Neuger AM, Soliman H, et al. Antagonism of adenosine A_{2A} receptor expressed by lung adenocarcinoma tumor cells and cancer associated fibroblasts inhibits their growth. *Cancer Biol Ther*. 2013;14(9):860–868.
116. Farzaneh-Far A, Shaw LK, Dunning A, Oldan JD, O'Connor CM, Borges-Neto S. Comparison of the prognostic value of regadenoson and adenosine myocardial perfusion imaging. *J Nucl Cardiol*. 2015;22(4):600–607.
117. Reyes E. Regadenoson stress for myocardial perfusion imaging. *Future Cardiol*. 2016;12(1):59–67.

CHAPTER

2

Adenosine Signaling Throughout Development

Stefania Zappettini and Christophe Bernard
Institute of Neurosciences of Systems, Marseille, France

OUTLINE

Adenosine Receptors in Neurodegenerative Diseases.
DOI: http://dx.doi.org/10.1016/B978-0-12-803724-9.00002-8

INTRODUCTION

Adenosine, a degradation product of adenosine-5′-triphosphate (ATP), is a prototypical paracrine signal.[1] As constituent of the energy metabolite ATP and of RNA, adenosine has therefore been described as a "retaliatory metabolite" that protects cells against the effects of excessive energy deficits, maintaining energy-homeostasis in most organ systems including the brain.[2,3] Within cells, adenosine is an intermediate for the synthesis of nucleic acids and ATP. It is generated from 5′-adenosine monophosphate (AMP) by the 5′-nucleotidase and can be converted back to AMP by adenosine kinase. Adenosine can also be derived from *S*-adenosylhomocysteine (SAH) due to the activity of SAH hydrolase. Intracellular adenosine is metabolized into inosine by adenosine deaminase (ADA) and into AMP by adenosine kinase. Inosine formed by deamination can exit the cell intact or can be degraded to hypoxanthine, xanthine, and ultimately uric acid. A low level of cellular adenosine can be quickly released in the extracellular space via equilibrative nucleoside transporters (ENTs). This release increases when intracellular adenosine concentration is augmented (e.g., during ischemia, hypoxia, and seizures). Beyond its role in energy-homeostasis, adenosine also serves a neuromodulatory role in the central nervous system (CNS); changing neuronal excitability,[4,5] controlling the release of various neurotransmitters including glutamate,[6,7] γ-aminobutyric acid (GABA),[8] acetylcholine,[9] and dopamine,[10] and modulating synaptic plasticity.[11] Unlike neurotransmitters, however, adenosine does not appear to be stored and released from vesicles but rather is generated by the highly regulated intracellular metabolism of AMP and transported out of the cell through bi-directional facilitated diffusion transporters,[12,13] or by the rapid (1 ms) extracellular conversion of locally released ATP and/or cAMP through a series of ectonucleotidases such as CD73 and CD39.[14–16] Extracellular adenosine, the levels of which are estimated to fall within the 25–250 nanomolar range under basal conditions,[14,17] exerts a tonic inhibitory effect on synaptic transmission. This means that under basal conditions, adenosine is sufficient to tonically activate relevant ARs subtypes.[18] This equilibrium might be compromised under pathologic conditions when extracellular adenosine levels can rise as much as 100-fold, activating the lower-affinity ARs subtypes, which then alter cellular function.[19–21] Adenosine interacts with four receptors (A_1, A_{2A}, A_{2B}, and A_3 receptors) which are seven-membrane-spanning proteins that couple to heterotrimeric G proteins to several intracellular signaling pathways.[22,23] Adenosine mainly signals through high affinity A_1R and $A_{2A}R$ receptors[24] in the adult brain.[1] They are predominately located in presynaptic terminals where they inhibit (A_1R)[25] or facilitate ($A_{2A}R$)[19] glutamate release,[26] through presynaptic and postsynaptic mechanisms by coupling with G_i[25] or G_s[19] to inhibit or potentiate the

AC-cAMP-protein kinase A (PKA) signaling pathway. In a pathological context, where excessive glutamate release is deleterious, bolstering A_1R activation or blocking $A_{2A}R$ affords neuroprotection.[27,28] Today the role of adenosine as endogenous anticonvulsant and neuroprotectant is well recognized.[1,29–31] Adenosine provides potent seizure control in all experimental model systems studied, including a model of pharmacoresistant epilepsy.[32] The anticonvulsant effects of adenosine are largely mediated by activation of pre- and postsynaptic $G_{i/o}$ protein coupled (A_1Rs), which induce presynaptic inhibition by reducing the inward flows of calcium, and which reduce excitability of the postsynaptic membrane by enhancing postsynaptic outflow of potassium through G-protein coupled inwardly rectifying potassium channels.[29] In addition to the global inhibitory tone provided by A_1R activation, adenosine further fine-tunes neurotransmission via all four types of adenosine receptors (A_1, A_{2A}, A_{2B}, A_3), which interact, at least in part, by heterodimerization with other G-protein coupled receptors, and thereby affect all major neurotransmitter and neurotrophin systems.[33] Therefore, endogenous adenosine is uniquely able to control neuronal excitability on multiple levels, and—consequently—any pathological disruption of adenosine homeostasis is likely to affect network excitability.[3] For example, in epilepsy, adenosine deficiency is also likely to affect psychomotor control by insufficient activation of A_{2A} receptors,[34] sleep regulation,[35] and might contribute to depression.[36] If adenosine deficiency is a pathological hallmark of epilepsy or other neurological disorders,[34,37] then therapeutic adenosine augmentation should be an effective tool to control epileptic seizures. Unfortunately, systemic augmentation of adenosine signaling is not a therapeutic option due to widespread, mainly cardiovascular, side effects, and due to liver toxicity of ADK disruption.[1,38] As a primordial biochemical regulator, adenosine could thus have direct influence on epigenetic functions, which are thought to be involved in epileptogenesis, and which form the basis of the "methylation hypothesis," suggesting that seizures by themselves can induce epigenetic chromatin modifications, thereby aggravating the epileptogenic condition.[39] Adenosine is available early in development, and this nucleoside may influence development of specific neuronal systems by interacting with transiently expressed populations of adenosine receptors.[40]

EFFECTS OF ADENOSINE SIGNALING AND ITS DYSFUNCTION IN CNS

Within the CNS, adenosine is released by both neurons and glial cells. It regulates multiple physiological functions such as sleep, arousal, neuroprotection, learning and memory, and cerebral blood circulation,[41] as well as pathological activities such as seizures. These effects involve adenosine modulation of neuronal excitability, vasodilatation, release of neurotransmitters, synaptic plasticity/function and local inflammatory processes.[14,31,33] In situ hybridization using 35*S*-labeled RNA probes generated from the rat A_{2A} receptor cDNA revealed receptor gene expression in the striatum on gestational day (GD) 14. Developmental alterations in the pattern of receptor gene expression within the striatum suggest that this receptor mRNA is expressed by striatal neurons soon after they complete neurogenesis. Transient expression of the A_{2A} receptor mRNA has been observed in the cerebral cortex, subiculum, parafascicularis nucleus of the thalamus, facial nucleus, trigeminal nucleus, locus coeruleus, area postrema, anterior pituitary gland and in the fetal cerebral vasculature. The ganglia of cranial nerves V, VII, VIII, IX, and X express A_{2A} receptor mRNA at a fetal stage; but expression levels in adults have not been examined. A_{2A} receptor mRNA is expressed in the carotid body and intermediate lobe of the pituitary during development and also in adult rats. Northern blot analysis revealed that the A_{2A} receptor transcript is consistent in size (c. 2.5 kb) across the developmental period examined (GD 14 through adult).[40] If the A_{2A} receptor mRNA is translated into receptor protein shortly after the mRNA is expressed, adenosine could influence neuronal differentiation, migration, synaptogenesis, and angiogenesis.

Adenosine receptor antagonists, including caffeine and theophylline, may influence brain development through interaction with A_{2A} receptors.[40] Few studies have reported the effects of substances naturally containing caffeine (e.g., coffee and tea), an antagonist at adenosine receptors, on fetal neural development. Mild to moderate caffeine consumption (<200 mg/day) by mothers during pregnancy and lactation is not considered harmful to the fetus and neonate from the perspective of miscarriage or risk of prematurity. This is reflected in the current guidelines for caffeine consumption during pregnancy by the American College of Obstetricians and Gynecologists.[42–44] However, numerous studies report that neural development is strongly influenced by external (e.g., environmental) factors including certain psychoactive drugs,[45] and caffeine is the most frequently psychoactive substance consumed both among the general population and pregnant women.[46,47] These psychoactive substances are known to modify cell migration[48,49] and synapse formation,[50,51] resulting in improper wiring of neuronal circuits, ultimately leading to behavioral modifications in offspring later in life.[52] Both A_1Rs and A_{2A}Rs are present at early stages of development[53–55]: A_1Rs control axon development,[56] while A_{2A}Rs, as we have shown, are able to control the migration speed of some GABA neurons,[57] and their blockade via exposure to caffeine during gestation, may have deleterious consequences on offspring.[52,57] Several studies show that dysfunction of adenosine homeostasis during development might be implicated in neurological disorders.[3,37] Adenosine provides a critical homeostatic control over the brain immune system, and its dysfunction would offset the balance between pro and antiinflammatory cytokines necessary for normal brain development.[58]

In order to test how interfering with this signaling pathway during early life may affect brain circuits, we mimicked coffee consumption during pregnancy in humans, providing pregnant female mice with 0.3 g/L of caffeine dissolved in tap water (the equivalent of 3 espresso cups/day in humans). We unraveled unsuspected roles of adenosine signaling, thus opening new and original research avenues in the field of neurobiology of development. We showed that caffeine treatment of dams delayed GABA neuron migration in their offspring,[57] via its antagonistic action on $A_{2A}Rs$. Such mechanism allowed demonstration of the deleterious consequences of nontoxic doses of caffeine consumption during pregnancy on the developing brain. Human studies also suggest that maternal caffeine intake during pregnancy has detrimental effects on subsequent brain development in offspring, since there is an association between caffeine intake during pregnancy and impaired cognitive development in offspring.[59] The phenotypic traits observed after caffeine exposure show some similarities to those found in models of neuronal migration disorders,[60] and models of psychiatric disorders linked to GABA neuron dysfunction.[61] This suggests that interfering with GABA neurons during development may have adverse consequences. The altered balance between excitation and inhibition found in adult caffeine-exposed offspring may contribute to the impaired memory observed on several memory tests.[57] Our results indicate that interfering with the adenosine pathway during pregnancy and lactation in mice delays the insertion of GABA/somatostatin (SST) interneurons into the offspring's hippocampus,[57] and visual cortex (unpublished), which results in an increased hyperexcitability in vivo and in vitro, and in increased susceptibility to febrile seizures, the most common type of seizures in human infants.[62] Caffeine exposure during early life thus leaves a permanent trace in neuronal networks: There is a significant loss of GABA neurons, a decreased GABAergic drive received by hippocampal neurons (associated with a disturbed excitation/inhibition balance in vitro and a significantly altered neuronal network activity in vivo), and significant cognitive deficits, with alterations of spatial memory in adult offspring[57] which may sensitize offspring to neurological disorders.[63,64] This effect is likely due to a delayed migration caused by the caffeine mediated blockade of $A_{2A}Rs$,[57] although we cannot rule out contributions of other targets of caffeine as, e.g., adenosine type A_1 receptors.[65,66] Not only did early life exposure to caffeine delay GABA neurons migration and maturation into the neocortex, it also induced morphological alterations such as varicosities reflecting dendritic swelling of interneurons.[57]

EFFECTS OF ADENOSINE IN OLIGODENDROCYTES DEVELOPMENT

Purines, in addition to their neurotransmitter and neuromodulator functions, can also act as growth and trophic factors, thus influencing development of neuronal[67,68] and glial cells.[69] Studies performed on neural crest (NC) cultures (a temporary group of cells unique to vertebrates that arise from the embryonic ectoderm cell layer, and in turn give rise to a diverse cell lineage—including melanocytes, craniofacial cartilage and bone, smooth muscle, peripheral and enteric neurons and glia[70]), which express A_{2A} and A_{2B} receptors, show that adenosine is a physiological signal in neuronal differentiation of the CNS-derived catecholaminergic CAD cell line and suggest that adenosine signaling is involved in NC cell development in vivo.[71] Oligodendrocyte progenitor cells (OPCs) express (1) functional adenosine receptors, which are activated in response to action potential firing,[69,72,73] (2) the equilibrative nucleoside transporters ENT1 and ENT2, as well as (3) the adenosine degrading enzymes ADA and adenosine kinase.[74] Adenosine acts as a potent neuron-glial transmitter to inhibit OPC proliferation, stimulate differentiation, and promote the formation of myelin. Via its activation of A_1 receptors, adenosine promotes oligodendrocyte maturation and inhibits their proliferation.[69] Conversely, when activating A_{2A} receptors, adenosine inhibits oligodendrocyte maturation.[73] As evaluated directly on cultured OPCs, adenosine acts as a dual modulator of OPC development since premature differentiation and reduced proliferation of OPCs might compromise the process of myelination as in preterm infants. These results identify adenosine as a potent axonal signal promoting OPC differentiation into myelinating oligodendrocytes. Only a brief exposure of OPCs to adenosine was sufficient to induce significant morphological changes in OPCs, which subsequently accelerated OPC-axon interactions and myelination at later stages. The diverse array of adenosine receptors on OPCs, which act through multiple intracellular signaling pathways, opens new areas of research on signals regulating oligodendrocyte development and interactions with neurons. Possibly, the ability of adenosine to promote OPC differentiation and increase myelination through activation of adenosine A_1 receptors may offer new approaches to the treatment of demyelinating diseases in the CNS, such as multiple sclerosis.[69]

EFFECTS OF ADENOSINE ON CARDIOVASCULAR SYSTEM DURING EMBRYOGENESIS

In different species, adenosine has been shown to exert potent effects on the developing cardiovascular system during embryogenesis. For example, intravenous infusion of adenosine analogues into fetal lambs produced dose-dependent bradycardia and hypotension.[75] Although adenosine causes cardiovascular changes in pregnant ewes, the effects are well tolerated and do not significantly affect the normal arterial blood gases and pH values of the fetus.[76] Ligand binding shows that A_{1A}Rs are expressed at days 5 and 6 in the heart chicken embryos and adversely influence embryogenesis, which differs from that seen in mammals[78] where it exerts dramatic protective effects.[79–81] In the rat heart, at GDs 8–11, A_1R mRNA expression was detected in the atrium, but not in other fetal structures.[82] Adenosine plays a key role in protecting the embryo against intrauterine stress, and adenosine exerts protective effects through A_{1A}Rs expressed in the heart.[80] When pregnant mice lacking A_1 receptors (A_1 −/−) were exposed to a hypoxic atmosphere, embryos (at embryonic day 7.5 or 8.5) had much more severe heart growth retardation than embryos from A_1 +/+ or A_1 +/− mice.[78] In addition to A_1 receptors, adenosine influences cardiac prenatal development via other adenosine receptor subtypes both in mammalian and nonmammalian species. A study of ventricular cells cultured from chick embryos 14 days in ovo showed that a functional A_{2A} receptor is expressed and mediates augmentation of myocyte contractility.[82] The A_{2A} receptor, which coexists with A_{2B} receptor, plays an important modulatory role in the presence of low levels of adenosine, while the low affinity A_{2B} receptor becomes functionally important when adenosine levels are high. Because caffeine is widely consumed, potential effects of caffeine on the developing fetus have been examined in animals and humans.[83] Following administration to pregnant rodents, embryo and fetal caffeine levels are 90% of maternal levels.[83,84] In rats, teratogenic effects of caffeine on the fetal heart are observed at doses in excess of 50 mg/kg.[85] The most common cardiovascular malformations are ventricular defects.[86] Cardiac morphogenesis has been found to be impaired in embryos from mothers treated with both ethanol and caffeine,[87] showing that caffeine can amplify effects of other toxic substances.

POSTNATAL EFFECTS OF ADENOSINE ON CARDIOVASCULAR SYSTEM

Adenosine plays a fundamental role in the regulation of heart function also during postnatal development. In preinnervated immature rat myocytes (from 1- to 3-day old pups) A_1Rs are functionally coupled to their effector sites and are present in greater numbers as compared to myocytes obtained from weaning and adult animals, possibly serving as a cardiac autocrine protective mechanism to offset the effects of the surge in catecholamine at birth.[88] In accordance, A_1R density, as measured by standard receptor binding assay, increased in whole rat hearts during late gestation (14 and 19 days of gestation), peaked during the newborn period and decreased in the adult age. A_1R mRNA levels changed in parallel with receptor protein expression, suggesting a possible role for A_1Rs in the transition to the extrauterine circulation.[89] Within the heart, adenosine has regulatory functions that include control of cardiac contractility/adrenergic responsiveness, impulse generation and conduction, coronary vascular tone, and cardiac substrate utilization. In cardiomyocytes isolated from newborn rabbits, adenosine inhibits β-adrenergic-stimulated L-type Ca^{++} currents as well as cAMP levels equipotently to carbachol, but is almost ineffective in cells obtained from adult animals.[90] In rat myocytes adenosine-induced K^+ current increased from gestation day 12 reaching a maximum at postnatal day 10 and decreasing in adult age.[91] Other regulatory functions may be linked to adaptation (e.g., regulation of fibroblast activity) or protection, and are addressed in subsequent sections. Adenosine also plays a key role in regulating inflammatory responses in a wide range of cells.[92] Adenosine primarily modifies cardiac contractility in an indirect fashion, through the modulation of adrenergic responses.[93] The antiadrenergic effects of adenosine in animal models are mediated via $A_{1A}Rs$, involve $G\alpha_i$ inhibition of PKA activation by ß-adrenoceptors,[93,94] and modulation of ß1-adrenoceptor stimulated G_s cycling.[95] In addition to effects on cardiomyocyte adrenoceptor responses, adenosine and $A_{1A}Rs$ inhibit release of noradrenaline from cardiac nerves,[96] reducing its concentration during ischemia and reperfusion.[97,98] These effects are protective,[97] and contribute to inhibition of cardiac activation during periods of enhanced adenosine release. Differing A_1/A_{2A} dependent actions of adenosine may act to limit cardiac overstimulation, matching metabolic capacity of the tissue with the level of adrenergic drive.[99] One of the earliest observed actions of adenosine was a slowing of heart rate and impulse conduction.[100] Adenosine inhibits impulse generation in supraventricular tissue (SA node, AV node) and the HisPurkinje system.[99] Adenosine and its receptors modify vascular tone, and may play a role in vasculogenesis/angiogenesis and vascular remodeling.[99] The A_{2A} and $A_{2B}Rs$ are expressed in both smooth muscle and endothelium, and contribute to what appears to be primarily endothelial dependent, but also direct smooth muscle, relaxation. The A_{2A} and $A_{2B}Rs$ affect coronary vasodilatation in a species-specific manner—$A_{2A}Rs$ play a role in human,[101] pig,[102] guinea pig,[103,104] and mouse,[105] whereas $A_{2B}Rs$ are also active in humans[106] and possibly in rats.[107] Exogenous adenosine modifies vascular tone in peripheral vessels, and studies of gene deletion support AR control of peripheral tone and blood pressure: $A_{2A}R$ deletion significantly elevates systemic pressure,[108] although whether this phenotype arises from local vascular or central/renal changes is uncertain, as the $A_{2A}R$ may modify nervous control of pressure.[109]

POSTNATAL EFFECT OF ADENOSINE ON RESPIRATORY SYSTEM

Most early studies on the role of nucleotides in development have been discussed in terms of their intracellular signaling and metabolic roles. However, since it is now widely accepted that purines and pyrimidines have potent extracellular actions mediated by specific membrane receptors, many of these previous studies have been reinterpreted.[110] ATP and adenosine play an important role in the regulation of respiration, both at a central and peripheral level during development.[111,112] Studies of transgenic mice indicate that adenosine exerts suppressive effects on respiration via A_{1A}Rs.[113] Stimulatory effects of caffeine on respiration are not observed in A_{1A}R knock-out mice, identifying A_{1A}Rs as mediating caffeine-stimulated respiratory drive.[113] R-PIA (a selective adenosine A_1R antagonist)[114] caused a considerably more pronounced effect in 1- to 3-day old animals than in 8-day old animals and was shown to bind with higher affinity in brains from newborn animals as compared to older animals. Adenosine actions on hypoxic ventilatory depression have been shown to undergo postnatal regulation also in piglets. At postnatal day 3, adenosine released from the CNS is a major contributor to the secondary fall in ventilation, whereas at 3 weeks adenosine makes little contribution to the ventilator responses.[115] In developing lambs, A_{2A} receptors seem to play a critical role in hypoxic ventilatory decline, while A_1 receptors are involved in cardiorespiratory control during normoxia.[116] An important beneficial effect of caffeine observed in the CAP study was a reduction in rates of bronchopulmonary dysplasia.[117,118] At present, the biological basis of this observation is not known. Yet, it likely that this effect reflects blockade of A_{2A}Rs and/or A_{2B}Rs, which influence capillary permeability, inflammation, and lung remodeling, as these receptor subtypes promote capillary leak.[119,120] In rats, at postnatal days 4 and 12, systemic administration of the adenosine antagonist, caffeine, has no effect on the early phase of hypoxic ventilatory response or metabolism, but enhances ventilation during the depression phase at both ages.[112] Accordingly, theophylline abolishes the hypoxic ventilator decline phase in 3-day old piglets,[115] but not in 3-week-old piglets, whereas aminophylline reverses the respiratory depression induced by intravenous adenosine injection in newborn rabbits.[121] Although many reports agree on the ability of A_1 and A_2Rs located in the brainstem to control postnatal ventilatory responses to hypoxia,[122] the role of adenosine on peripheral chemoreceptors located in the carotid body of developing subjects is still debated.[111,112] In the rat carotid body from postnatal days 0–22, A_{2A}Rs mRNA was abundantly expressed, while A_1R mRNA could not be found at any postnatal age investigated, although it was essentially expressed in all cells of the petrosal ganglion, which contains the cell bodies for chemoafferents.[123]

USE OF METHYLXANTHINES IN PREMATURE BABIES

During early development, adenosine contributes to the occurrence of respiratory depression and recurrent apneas.[124,125] Recent physiological studies indicate that GABAergic mechanisms may be involved in this inhibitory action of adenosine, via their A_{2A} receptors.[126] Recurrent apnea is common in preterm infants, particularly at very early gestational ages.[127] These episodes of loss of effective breathing can lead to hypoxemia and bradycardia, which may be severe enough to require resuscitation including use of positive pressure ventilation. Methylxanthines may help reduce the incidence of events in a neonate with central apnea of prematurity[128] as a useful adjunct in weaning from a mechanical ventilator.[129] Two forms of methylxanthine (caffeine and theophylline) have been used to stimulate breathing in order to prevent apnea and its consequences. Several studies show that after theophylline administration, the patients had higher heart rates, faster respiratory rates, and a decrease in transcutaneous CO_2.[130] These results confirm that theophylline produces an increase in minute ventilation, a decrease in $PaCO_2$,[131,132] and an increase in most indices of neural respiratory drive.[130] Lung mechanics measurements showed a significant improvement in respiratory compliance after treatment with theophylline. The underlying mechanism may be related to enhanced excursion of the diaphragm[133] and inspiratory eVort.[134] However, caffeine has been proposed as an adjunct treatment for successful extubation from the ventilator during first week of life of a very low birth weight premature neonate.[127,135] The advantages of caffeine are not only that it is the most acceptable prophylactic agent to facilitate successful extubation in preterm infants,[136] but it may also reduce the rate of bronchopulmonary dysplasia in very low birth weight infants.[117] In addition, caffeine has a therapeutic margin wider than that of other methylxanthines, such as theophylline, and an overdose is less likely to occur than with other drugs in its class.[135] Note that such acute treatment may not be sufficient to produce deleterious consequences on brain development, as compared to the chronic exposure to caffeine through the drinking water via the dam.

POSTNATAL EFFECTS OF ADENOSINE ON THE GENITOURINARY SYSTEM

ATP and acetylcholine are cotransmitters in parasympathetic nerves targeting the bladder.[137] The rate pattern of breakdown of ATP and adenosine by ectoenzymes in the rat urinary bladder was shown to be identical in neonates and adults, indicating that the marked differences in potency to ATP and adenosine during development is likely to be due to changes in receptor number and/or agonist affinity or efficacy.[138] Responses to adenosine (inhibitory) were more potent in the neonate than in the adult. ATP and adenosine play key signaling roles from the very beginning of life, representing a fundamental sperm-to-egg signal already in the process of fertilization.[139] Examination of the ontogeny of P1 purinergic receptors showed that inhibitory prejunctional A_1 and postjunctional A_2-like receptors were present from days 10 and 15, the earliest age studied, while excitatory postjunctional A_1 receptors did not appear until after day 20.[140]

POSTNATAL EFFECT OF ADENOSINE ON THE ENTERIC SYSTEM

A_1 and $A_{2A}R$ proteins are abundantly expressed in petrosal ganglion neurons, with a different pattern of distribution: A_1 receptors showing a cytoplasmic localization while A_2 receptors are prevalently localized in the nucleus of the same neurons.[123] Adenosine and its breakdown enzymes play an important role in the regulation of growth and development of sympathetic neurons.[141] In rat duodenal segments, ATP and ADP produced contractile responses on postnatal day 1. This response increased with age, peaked on day 7, was followed by a gradual decrease, and was nonexistent by day 21.[142] Adenosine or AMP did not elicit responses before day 14, which were then small relaxations that increased with age[143] and it was reported that if the tissues were precontracted with carbachol, low concentrations of ATP could produce relaxations from day 2 increasing with age, while higher concentrations of ATP (3 μmol/L and above) were excitatory, but only until day 15 and they postulated that P2Y receptors mediated both relaxations and contractions. Responses to adenosine, ATP and α,β-meATP have been examined in the rat colon longitudinal muscle and muscularis mucosa during postnatal development.[144] The longitudinal muscle relaxes via A_{2B} and P2Y receptors, while the muscularis mucosa contracts through A_1 and probably P2Y receptors, with UTP also causing contraction. The contractile responses of the muscularis mucosa to all three agonists were observed from the day after birth, but much lower than in the adult.[145] The distribution and localization of the breakdown enzyme for adenosine, ADA, were studied during postnatal development of the mouse alimentary tract.[146] ADA was predominantly localized to the keratinized squamous epithelium that lines the mucosal layer of the esophagus, fore-stomach and surface of the tongue and the simple columnar epithelium of proximal small intestine. ADA in these tissues were low at birth, but achieved very high levels within the first 2 weeks of postnatal life and, in these tissues, must be subjected to strong cell-specific developmental regulation during functional differentiation of certain foregut derivatives in mice.[147]

POSTNATAL EFFECT OF ADENOSINE ON OTHER SYSTEMS

Adenosine is also important in regulating the action of insulin on the rat fat cell metabolism during postnatal development and aging.[148] ATP and ADP and, to a much lesser extent, AMP and adenosine, increase insulin secretion from the isolated, perfused newborn dog pancreas.[149] In the piglet at postnatal day 3, adenosine released within the kidney is a major contributor to renal vasoconstriction, whereas at 3 weeks adenosine makes little contribution on the contractile response.[115] A_{2A} and A_3 adenosine receptor mRNAs undergo marked changes in the rat kidney between postnatal days 2 and 60, which may be related to functional maturation and morphological development, while A_1 and A_{2B} adenosine receptor mRNA levels remain substantially unchanged.[150] In situ hybridization and RT-PCR studies have shown that A_1 and A_{2A} receptor mRNAs are present in rat petrosal ganglion (which contains the bodies of general somatic sensory neurons (GSA fibers) that innervate the pharynx, tonsils, tongue, middle ear, auditory tube and the ear canal) neurons from postnatal days 0–22; A_{2A} mRNA receptor levels increase with age while A_1 mRNA levels, although higher than those of A_{2A} receptors, are not influenced by increasing age, possibly explaining the progressive shift of adenosine effect on the ventilatory function during development: from inhibitory (mediated by A_1 receptors) to excitatory (mediated by A_2 receptors).[123] It is also known that deletion of A_{2A}Rs strongly impacts naive T cell development and survival in C57BL/6 mice.[151] A_{2A}Rs couple to the heterotrimeric G protein, primarily G_s and partially to G_{olf} in the CNS.[152] These G proteins activate adenylyl cyclase and increase cAMP and PKA activity, which in turn inhibits AKT,[153] which stimulates T cell proliferation and reduces IL-7Rα expression downstream of TCR stimulation,[154] required for naive T cell survival. Interestingly, adenosine concentrations are higher in the thymus than other organs.[155] A new role for adenosine in the regulation of T cell homeostasis through A_{2A}R signaling has been proposed.[151] In the immune system, adenosine dampens excessive inflammation after tissue injury and activates tissue remodeling responses such as angiogenesis and fibrosis, thereby helping to establish long-term homeostasis after local or systemic disturbances. Ongoing studies indicate that $A_{2A}R$ signaling helps to maintain T cells in adenosine-rich hypoxic solid tumor microenvironments (unpublished data) parallel to our observation that adenosine may regulate CD127 expression even in effector/memory phenotype T cells.[151]

Over the last decade, the presence and function of adenosine receptors on the plasma membrane of mesenchymal stem cells (MSCs) has been investigated in several studies. Upon stimulation, MSCs have been reported to actively secrete nucleotides such as ATP and NAD+ in order to modulate MSC functions, e.g., proliferation, migration, and immunosuppression of activated T lymphocytes.[156] It has been demonstrated that adenosine receptor signaling via cyclic AMP contributes to MSC differentiation toward chondrocytes and osteoblasts.[157] Furthermore, it has been reported in several in vitro or in vivo studies that adenosine acts as an autocrine/paracrine signaling molecule that induces osteogenic differentiation of murine as well as human BM-MSCs via adenosine A_{2B} receptor stimulation.[158]

Expression of adenosine receptors has been investigated during MSC adipogenesis. A_1 receptors are mainly involved in the lipogenic activity of adipocytes, whereas A_{2A} receptors enhanced adipocytic differentiation and lipid accumulation.[159]

In addition to its prominent role in regulating MSC osteogenesis versus adipogenesis, adenosine has also been shown to induce the expression of hepatocyte-specific genes in mouse and human BM-MSCs in vitro.[160] Moreover, topical application of an exogenous adenosine A_{2A} receptor agonist has been reported to promote wound healing via enhancement of local vessel sprouting and vasculogenesis in the early stages of tissue regeneration through recruitment of bone marrow-derived endothelial cells.[161]

CONCLUSION

An expanding body of data shows that adenosine plays an important role during pre- and postnatal development of central and peripheral nervous system. Endogenous adenosine is a widely distributed upstream regulator of a broad spectrum of neurotransmitters, receptors, and signaling pathways that converge to contribute to the expression of an array of important and complex functions, ranging from neuronal plasticity, motor, motivation, sleep–wake cycle, cognition, and emotion in the normal brain to neuroprotection, neuroinflammation, and maladaptive behavioral and neuropsychiatric disorders in the pathologic brain. This nucleoside is available early in development, and it influences the development of specific neuronal/nonneuronal systems by interacting with populations of adenosine receptors already expressed during gestation and the fist postnatal weeks. To conclude, adenosine plays important modulatory roles in mammalian development, conferring protective or deleterious effects depending on the timing of exposure and site of action. Methylxanthines, including caffeine and theophylline, which are widely consumed during pregnancy and lactation, may have deleterious consequences for the embryo or the newborn, with possible negative outcomes later in life, including favoring the appearance of pathological states.

References

1. Fredholm BB, Chen JF, Cunha RA, Svenningsson P, Vaugeois JM. Adenosine and brain function. *Int Rev Neurobiol.* 2005;63:191–270.
2. Newby AC, Worku Y, Holmquist CA. Adenosine formation. Evidence for a direct biochemical link with energy metabolism. *Adv Myocardiol.* 1985;6:273–284.
3. Boison D. Adenosine dysfunction in epilepsy. *Glia.* 2012;60:1234–1243.
4. Dunwiddie T. Endogenously released adenosine regulates excitability in the in vitro hippocampus. *Epilepsia.* 1980;21:541–548.
5. Fredholm BB, Hedqvist P. Modulation of neurotransmission by purine nucleotides and nucleosides. *Biochem Pharmacol.* 1980;29:1635–1643.
6. Obrietan K, Belousov AB, Heller HC, van den Pol AN. Adenosine pre- and postsynaptic modulation of glutamate-dependent calcium activity in hypothalamic neurons. *J Neurophysiol.* 1995;74:2150–2162.
7. Popoli P, Frank C, Tebano MT, Potenza RL, Pintor A, Domenici MR, et al. Modulation of glutamate release and excitotoxicity by adenosine A2A receptors. *Neurology.* 2003;61:69–71.
8. Ochi M, Shiozaki S, Kase H. Adenosine A(2A) receptor-mediated modulation of GABA and glutamate release in the output regions of the basal ganglia in a rodent model of Parkinson's disease. *Neuroscience.* 2004;127:223–231.
9. Rodrigues RJ, Canas PM, Lopes LV, Oliveira CR, Cunha RA. Modification of adenosine modulation of acetylcholine release in the hippocampus of aged rats. *Neurobiol Aging.* 2008;29:1597–1601.
10. Quarta D, Borycz J, Solinas M, Patkar K, Hockemeyer J, Ciruela F, et al. Adenosine receptor-mediated modulation of dopamine release in the nucleus accumbens depends on glutamate neurotransmission and N-methyl-D-aspartate receptor stimulation. *J Neurochem.* 2004;91:873–880.
11. Sebastiao AM, Ribeiro JA. Adenosine A2 receptor-mediated excitatory actions on the nervous system. *Prog Neurobiol.* 1996;48:167–189.
12. Sala-Newby GB, Skladanowski AC, Newby AC. The mechanism of adenosine formation in cells. Cloning of cytosolic 5'-nucleotidase-I. *J Biol Chem.* 1999;274:17789–17793.
13. Meghji P, Rubio R, Berne RM. Intracellular adenosine formation and its carrier-mediated release in cultured embryonic chick heart cells. *Life Sci.* 1988;43:1851–1859.
14. Dunwiddie TV, Masino SA. The role and regulation of adenosine in the central nervous system. *Annu Rev Neurosci.* 2001;24:31–55.
15. Fredholm BB, Chern Y, Franco R, Sitkovsky M. Aspects of the general biology of adenosine A2A signaling. *Prog Neurobiol.* 2007;83:263–276.
16. Gödecke A. cAMP: fuel for extracellular adenosine formation? *Br J Pharmacol.* 2008;153:1087–1089.

17. Pedata F, Corsi C, Melani A, Bordoni F, Latini S. Adenosine extracellular brain concentrations and role of A2A receptors in ischemia. *Ann N Y Acad Sci*. 2001;939:74–84.
18. Cunha R. Adenosine as a neuromodulator and as a homeostatic regulator in the nervous system: different roles, different sources and different receptors. *Neurochem Int*. 2001;38:107–125.
19. Fredholm BB, Arslan G, Halldner L, Kull B, Schulte G, Wasserman W. Structure and function of adenosine receptors and their genes. *Naunyn Schmiedebergs Arch Pharmacol*. 2000;362:364–374.
20. Latini S, Pedata F. Adenosine in the central nervous system: release mechanisms and extracellular concentrations. *J Neurochem*. 2001;79:463–484.
21. Wei CJ, Lia W, Chen JF. Normal and abnormal functions of adenosine receptors in the central nervous system revealed by genetic knockout studies. *Biochim Biophys Acta*. 2011;1808:1358–1379.
22. Ralevic V, Burnstock G. Receptors for purines and pyrimidines. *Pharmacol Rev*. 1998;50:413–492.
23. Fredholm BB, IJzerman AP, Jacobson KA, Klotz KN, Linden J. International Union of Pharmacology. XXV. Nomenclature and classification of adenosine receptors. *Pharmacol Rev*. 2001;53:527–532.
24. Daly JW, Padgett WL. Agonist activity of 2- and 5′-substituted adenosine analogs and their N6-cycloalkyl derivatives at A1- and A2-adenosine receptors coupled to adenylate cyclase. *Biochem Pharmacol*. 1992;43:1089–1093.
25. van Calker D, Müller M, Hamprecht B. Adenosine regulates via two different types of receptors, the accumulation of cyclic AMP in cultured brain cells. *J Neurochem*. 1979;33:999–1005.
26. Ciruela F, Casadó V, Rodrigues RJ, Luján R, Burgueño J, Canals M, et al. Presynaptic control of striatal glutamatergic neurotransmission by adenosine A1-A2A receptor heteromers. *J Neurosci*. 2006;15:2080–2087.
27. Cunha R. Different cellular sources and different roles of adenosine: A1 receptor-mediated inhibition through astrocytic-driven volume transmission and synapse-restricted A2A receptor-mediated facilitation of plasticity. *Neurochem Int*. 2008;52:65–72.
28. Ciruela F, Albergaria C, Soriano A, Cuffí L, Carbonell L, Sánchez S, et al. Adenosine receptors interacting proteins (ARIPs): behind the biology of adenosine signaling. *Biochim Biophys Acta*. 2010;1798:9–20.
29. Boison D, Chen JF, Fredholm BB. Adenosine signaling and function in glial cells. *Cell Death Differ*. 2010;17:1071–1082.
30. Ribeiro J. What can adenosine neuromodulation do for neuroprotection? *Curr Drug Targets CNS Neurol Disord*. 2005;4:325–329.
31. Stone TW, Ceruti S, Abbracchio MP. Adenosine receptors and neurological disease: neuroprotection and neurodegeneration. *Handb Exp Pharmacol*. 2009;193:535–587.
32. Gouder N, Fritschy JM, Boison D. Seizure suppression by adenosine A1 receptor activation in a mouse model of pharmacoresistant epilepsy. *Epilepsia*. 2003;44:877–885.
33. Sebastião AM, Ribeiro JA. Tuning and fine-tuning of synapses with adenosine. *Curr Neuropharmacol*. 2009;7:180–194.
34. Shen HY, Coelho JE, Ohtsuka N, Canas PM, Day YJ, Huang QY, et al. A critical role of the adenosine A2A receptor in extrastriatal neurons in modulating psychomotor activity as revealed by opposite phenotypes of striatum and forebrain A2A receptor knock-outs. *J Neurosci*. 2008;28:2970–2975.
35. Palchykova S, Winsky-Sommerer R, Shen HY, Boison D, Gerling A, Tobler I. Manipulation of adenosine kinase affects sleep regulation in mice. *J Neurosci*. 2010;30:13157–13165.
36. Hanson E. Can adenosine substitute for SAM-e as an adjunct in the treatment of depression? *Med Hypotheses*. 2009;73:857–859.

37. Boison D, Singer P, Shen HY, Feldon J, Yee BK. Adenosine hypothesis of schizophrenia–opportunities for pharmacotherapy. *Neuropharmacology*. 2012;62:1527–1543.
38. Boison D, Scheurer L, Zumsteg V, Rülicke T, Litynski P, Fowler B, et al. Neonatal hepatic steatosis by disruption of the adenosine kinase gene. *Proc Natl Acad Sci U S A*. 2002;99:6985–6990.
39. Kobow K, Blümcke I. The methylation hypothesis: do epigenetic chromatin modifications play a role in epileptogenesis? *Epilepsia*. 2011;52:15–19.
40. Weaver DR. A2a adenosine receptor gene expression in developing rat brain. *Mol Bram Res*. 1993;20:313–327.
41. Ribeiro JA, Sebastião AM, de Mendonça A. Adenosine receptors in the nervous system: pathophysiological implications. *Prog Neurobiol*. 2002;68:377–392.
42. Fredholm BB, Bättig K, Holmén J, Nehlig A, Zvartau EE. Actions of caffeine in the brain with special reference to factors that contribute to its widespread use. *Pharmacol Rev*. 1999;51:83–133.
43. Nawrot P, Jordan S, Eastwood J, Rotstein J, Hugenholtz A, Feeley M. Effects of caffeine on human health. *Food Addit Contam*. 2003;20:1–30.
44. http://www.acog.org/~/media/Committee Opinions/Committee on Obstetric%20Practice/co462.pdf?dmc=1&ts=20130313T0649414568; 2010.
45. Salisbury AL, Ponder KL, Padbury JF, Lester BM. Fetal effects of psychoactive drugs. *Clin Perinatol*. 2009;36:595–619.
46. Mandel H. Update on caffeine consumption, disposition and action. *Food Chem Toxicol*. 2002;40:1231–1234.
47. Greenwood DC, Thatcher NJ, Ye J, Garrard L, Keogh G, King LG, et al. Caffeine intake during pregnancy and adverse birth outcomes: a systematic review and dose-response meta-analysis. *Eur J Epidemiol*. 2014;29:725–734.
48. Miller M. Effects of alcohol on the generation and migration of cerebral cortical neurons. *Science*. 1986;233:1308–1311.
49. Crandall JE, Hackett HE, Tobet SA, Kosofsky BE, Bhide PG. Cocaine exposure decreases GABA neuron migration from the ganglionic eminence to the cerebral cortex in embryonic mice. *Cereb Cortex*. 2004;14:665–675.
50. Anavi-Goffer S, Mulder J. The polarised life of the endocannabinoid system in CNS development. *Chembiochem*. 2009;10:1591–1598.
51. Berghuis P, Rajnicek AM, Morozov YM, Ross RA, Mulder J, Urbán GM, et al. Hardwiring the brain: endocannabinoids shape neuronal connectivity. *Science*. 2007;316:1212–1216.
52. Thompson BL, Levitt P, Stanwood GD. Prenatal exposure to drugs: effects on brain development and implications for policy and education. *Nat Rev Neurosci*. 2009;10:303–312.
53. Weaver D. A2a adenosine receptor gene expression in developing rat brain. *Brain Res Mol Brain Res*. 1993;20:313–327.
54. Doriat JF, Koziel V, Humbert AC, Daval JL. Medium- and long-term alterations of brain A1 and A2A adenosine receptor characteristics following repeated seizures in developing rats. *Epilepsy Res*. 1999;35:219–228.
55. Brito R, Pereira MR, Paes-de-Carvalho R, Calaza Kda C. Expression of A1 adenosine receptors in the developing avian retina: in vivo modulation by A(2A) receptors and endogenous adenosine. *J Neurochem*. 2012;123:239–249.
56. Stettler O, Joshi RL, Wizenmann A, Reingruber J, Holcman D, Bouillot C, et al. Engrailed homeoprotein recruits the adenosine A1 receptor to potentiate ephrin A5 function in retinal growth cones. *Development*. 2012;139:215–224.
57. Silva CG, Métin C, Fazeli W, Machado NJ, Darmopil S, Launay PS, et al. Adenosine receptor antagonists including caffeine alter fetal brain development in mice. *Sci Transl Med*. 2013;7:197ra104.

58. Haskó G, Pacher P, Vizi ES, Illes P. Adenosine receptor signaling in the brain immune system. *Trends Pharmacol Sci.* 2005;26:511–516.
59. Galéra C, Bernard JY, van der Waerden J, Bouvard MP, Lioret S, Forhan A, et al. EDEN Mother-Child Cohort Study Group. Prenatal caffeine exposure and child IQ at age 5.5 years: the EDEN mother-child cohort. *Biol Psych.* 2015;S 0006-3223:00733–00737.
60. Guerrini R, Parrini E. Neuronal migration disorders. *Neurobiol Dis.* 2010;38:154–166.
61. Marín O. Interneuron dysfunction in psychiatric disorders. *Nat Rev Neurosci.* 2012;13:107–120.
62. Dubé CM, Brewster AL, Baram TZ. Febrile seizures: mechanisms and relationship to epilepsy. *Brain Dev.* 2009;31:366–371.
63. Linnet KM, Wisborg K, Secher NJ, Thomsen PH, Obel C, Dalsgaard S, et al. Coffee consumption during pregnancy and the risk of hyperkinetic disorder and ADHD: a prospective cohort study. *Acta Paediatr.* 2009;98:173–179.
64. Loomans EM, Hofland L, van der Stelt O, van der Wal MF, Koot HM, Van den Bergh BR, et al. Caffeine intake during pregnancy and risk of problem behavior in 5- to 6-year-old children. *Pediatrics.* 2012;130:305–313.
65. Dzhala V, Desfreres L, Melyan Z, Ben-Ari Y, Khazipov R. Epileptogenic action of caffeine during anoxia in the neonatal rat hippocampus. *Ann Neurol.* 1999;46:95–102.
66. Porciúncula LO, Sallaberry C, Mioranzza S, Botton PH, Rosemberg DB. The Janus face of caffeine. *Neurochem Int.* 2013;63:594–609.
67. Mishra SK, Braun N, Shukla V, Füllgrabe M, Schomerus C, Korf HW, et al. Extracellular nucleotide signaling in adult neural stem cells: synergism with growth factor-mediated cellular proliferation. *Development.* 2006;133:675–684.
68. Migita H, Kominami K, Higashida M, Maruyama R, Tuchida N, McDonald F, et al. Activation of adenosine A1 receptor-induced neural stem cell proliferation via MEK/ERK and Akt signaling pathways. *J Neurosci Res.* 2008;86:2820–2828.
69. Stevens B, Porta S, Haak LL, Gallo V, Fields RD. Adenosine: a neuron-glial transmitter promoting myelination in the CNS in response to action potentials. *Neuron.* 2002;36:855–868.
70. Huang X, Saint-Jeannet JP. Induction of the neural crest and the opportunities of life on the edge. *Dev Biol.* 2004;275:1–11.
71. Bilodeau ML, Ji M, Paris M, Andrisani OM. Adenosine signaling promotes neuronal, catecholaminergic differentiation of primary neural crest cells and CNS-derived CAD cells. *Mol Cell Neurosci.* 2005;29:394–404.
72. Burnstock G, Fredholm BB, Verkhratsky A. Adenosine and ATP receptors in the brain. *Curr Top Med Chem.* 2011;11:973–1011.
73. Coppi E, Cellai L, Maraula G, Pugliese AM, Pedata F. Adenosine A2A receptors inhibit delayed rectifier potassium currents and cell differentiation in primary purified oligodendrocyte cultures. *Neuropharmacology.* 2013;73:301–310.
74. González-Fernández E, Sánchez-Gómez MV, Pérez-Samartín A, Arellano RO, Matute C. A3 Adenosine receptors mediate oligodendrocyte death and ischemic damage to optic nerve. *Glia.* 2014;62:199–216.
75. Konduri GG, Forman K, Mital S. Characterization of purine receptors in fetal lamb pulmonary circulation. *Pediatr Res.* 2000;47:114–120.
76. Mason BA, Ogunyemi D, Punla O, Koos BJ. Maternal and fetal cardiorespiratory responses to adenosine in sheep. *Am J Obstet Gynecol.* 1993;168:1558–1561.
77. Ghatpande SK, Billington Jr CJ, Rivkees SA, Wendler CC. Hypoxia induces cardiac malformations via a1 adenosine receptor activation in chicken embryos. *Birth Defects Res A Clin Mol Teratol.* 2008;82:121–130.
78. Wendler CC, Amatya S, McClaskey C, Ghatpande S, Fredholm BB, Rivkees SA. A1 adenosine receptors play an essential role in protecting the embryo against hypoxia. *Proc Natl Acad Sci U S A.* 2007;104:9697–9702.

79. Wendler CC, Poulsen RR, Ghatpande S, Greene RW, Rivkees SA. Identification of the heart as the critical site of adenosine mediated embryo protection. *BMC Dev Biol.* 2010;10:57.
80. Rivkees SA, Wendler CC. Regulation of cardiovascular development by adenosine and adenosine-mediated embryo protection. *Arterioscler Thromb Vasc Biol.* 2012;32: 851–855.
81. Rivkees S. The ontogeny of cardiac and neural A1 adenosine receptor expression in rats. *Brain Res Dev Brain Res.* 1995;89:202–213.
82. Liang BT, Haltiwanger B. Adenosine A2a and A2b receptors in cultured fetal chick heart cells. High- and low affinity coupling to stimulation of myocyte contractility and cAMP accumulation. *Circ Res.* 1995;76:242–251.
83. Browne M. Maternal exposure to caffeine and risk of congenital anomalies: a systematic review. *Epidemiology.* 2006;17:324–331.
84. Fredholm B. Astra award lecture. Adenosine, adenosine receptors and the actions of caffeine. *Pharmacol Toxicol.* 1995;76:93–101.
85. Christian MS, Brent RL. Teratogen update: evaluation of the reproductive and developmental risks of caffeine. *Teratology.* 2001;64:51–78.
86. Ross CP, Persaud TV. Cardiovascular primordium of the rat embryo following in utero exposure to alcohol and caffeine. *Can J Cardiol.* 1986;2:160–163.
87. Ross CP, Persaud TV. Early embryonic development in the rat following in utero exposure to alcohol and caffeine. *Histol Histopathol.* 1986;1:13–17.
88. Cothran DL, Lloyd TR, Taylor H, Linden J, Matherne GP. Ontogeny of rat myocardial A1 adenosine receptors. *Biol Neonate.* 1995;68:111–118.
89. Matherne GP, Byford AM, Gilrain JT, Dalkin AC. Changes in myocardial A1 adenosine receptor and message levels during fetal development and postnatal maturation. *Biol Neonate.* 1996;70:199–205.
90. Kumar R, Akita T, Joyner RW. Adenosine and carbachol are not equivalent in their effects on L-type calcium current in rabbit ventricular cells. *J Mol Cell Cardiol.* 1996;28:403–415.
91. Takano M, Noma A. Development of muscarinic potassium current in fetal and neonatal rat heart. *Am J Physiol.* 1997;272:H1188–H1195.
92. Blackburn MR, Vance CO, Morschl E, Wilson CN. Adenosine receptors and inflammation. *Handb Exp Pharmacol.* 2009;(193):215–269.
93. Dobson JJ. Mechanism of adenosine inhibition of catecholamine-induced responses in heart. *Circ Res.* 1983;52:151–160.
94. Romano FD, Dobson Jr JG. Adenosine modulates beta-adrenergic signal transduction in guinea-pig heart ventricular membranes. *J Mol Cell Cardiol.* 1990;22:1359–1370.
95. Fenton RA, Dobson Jr JG. Adenosine A1 and A2A receptor effects on G-protein cycling in beta-adrenergic stimulated ventricular membranes. *J Cell Physiol.* 2007;213: 785–792.
96. Lorbar M, Chung ES, Nabi A, Skalova K, Fenton RA, Dobson Jr JG, et al. Receptors subtypes involved in adenosine-mediated modulation of norepinephrine release from cardiac nerve terminals. *Can J Physiol Pharmacol.* 2004;82:1026–1031.
97. Richardt G, Waas W, Kranzhöfer R, Mayer E, Schömig A. Adenosine inhibits exocytotic release of endogenous noradrenaline in rat heart: a protective mechanism in early myocardial ischemia. *Circ Res.* 1987;61:117–123.
98. Burgdorf C, Richardt D, Kurz T, Seyfarth M, Jain D, Katus HA, et al. Adenosine inhibits norepinephrine release in the postischemic rat heart: the mechanism of neuronal stunning. *Cardiovasc Res.* 2001;49:713–720.
99. Headrick JP, Peart JN, Reichelt ME, Haseler LJ. Adenosine and its receptors in the heart: regulation, retaliation and adaptation. *Biochim Biophys Acta.* 2011;1808: 1413–1428.

100. Belardinelli L, Shryock JC, Song Y, Wang D, Srinivas M. Ionic basis of the electrophysiological actions of adenosine on cardiomyocytes. *FASEB J*. 1995;9:359–365.
101. Sato A, Terata K, Miura H, Toyama K, Loberiza Jr FR, Hatoum OA, et al. Mechanism of vasodilation to adenosine in coronary arterioles from patients with heart disease. *Am J Physiol Heart Circ Physiol*. 2005;288:H1633–H1640.
102. Hein TW, Belardinelli L, Kuo L. Adenosine A2A receptors mediate coronary microvascular dilation to adenosine: role of nitric oxide and ATP-sensitive potassium channels. *J Pharmacol Exp Ther*. 1999;291:655–664.
103. Belardinelli L, Shryock JC, Snowdy S, Zhang Y, Monopoli A, Lozza G, et al. The A2A adenosine receptor mediates coronary vasodilation. *J Pharmacol Exp Ther*. 1998;284:1066–1073.
104. Maddock HL, Broadley KJ, Bril A, Khandoudi N. Effects of adenosine receptor agonists on guinea-pig isolated working hearts and the role of endothelium and NO. *J Pharm Pharmacol*. 2002;54:859–867.
105. Flood A, Headrick JP. Functional characterization of coronary vascular adenosine receptors in the mouse. *Br J Pharmacol*. 2001;133:1063–1072.
106. Kemp BK, Cocks TM. Adenosine mediates relaxation of human small resistance-like coronary arteries via A2B receptors. *Br J Pharmacol*. 1999;126:1796–1800.
107. Hinschen AK, Rose'Meyer RB, Headrick JP. Adenosine receptor subtypes mediating coronary vasodilation in rat hearts. *J Cardiovasc Pharmacol*. 2003;41:73–80.
108. Ledent C, Vaugeois JM, Schiffmann SN, Pedrazzini T, El Yacoubi M, Vanderhaeghen JJ, et al. Aggressiveness, hypoalgesia and high blood pressure in mice lacking the adenosine A2A receptor. *Nature*. 1997;388:674–678.
109. Scislo J, O'Leary DS. Purinergic mechanisms of the nucleus of the solitary tract and neural cardiovascular control. *Neurol Res*. 2005;27:182–194.
110. Burnstock G. Physiology and pathophysiology of purinergic neurotransmission. *Physiol Rev*. 2007;87:659–797.
111. Koos B. Adenosine A2A receptors and O2 sensing in development. *Am J Physiol*. 2011;301:R601–R622.
112. Bairam A, Niane LM, Joseph V. Role of ATP and adenosine on carotid body function during development. *Respir Physiol Neurobiol*. 2013;185:57–66.
113. Johansson B, Halldner L, Dunwiddie TV, Masino SA, Poelchen W, Gimenez-Llort L, et al. Hyperalgesia, anxiety, and decreased hypoxic neuroprotection in mice lacking the adenosine A1 receptor. *Proc Natl Acad Sci U S A*. 2001;98:9407–9412.
114. Runold M, Lagercrantz H, Fredholm BB. Ventilatory effect of an adenosine analogue in unanesthetized rabbits during development. *J Appl Physiol*. 1986;61:255–259.
115. Elnazir B, Marshall JM, Kumar P. Postnatal development of the pattern of respiratory and cardiovascular response to systemic hypoxia in the piglet: the roles of adenosine. *J Physiol*. 1996;492:573–585.
116. Koos BJ, Kawasaki Y, Kim YH, Bohorquez F. Adenosine A2A-receptor blockade abolishes the roll-off respiratory response to hypoxia in awake lambs. *Am J Physiol*. 2005;288:R1185–R1194.
117. Schmidt B, Roberts RS, Davis P, Doyle LW, Barrington KJ, Ohlsson A, et al. Caffeine therapy for apnea of prematurity. *N Engl J Med*. 2006;354:2112–2121.
118. Davis PG, Schmidt B, Roberts RS, Doyle LW, Asztalos E, Haslam R, et al. Caffeine for Apnea of Prematurity trial: benefits may vary in subgroups. *J Pediatr*. 2010;156:382–387.
119. Thompson LF, Eltzschig HK, Ibla JC, Van De Wiele CJ, Resta R, Morote-Garcia JC, et al. Crucial role for ecto-5′-nucleotidase (CD73) in vascular leakage during hypoxia. *J Exp Med*. 2004;200:1395–1405.
120. Eckle T, Faigle M, Grenz A, Laucher S, Thompson LF, Eltzschig HK. A2B adenosine receptor dampens hypoxia-induced vascular leak. *Blood*. 2008;111:2024–2035.

121. Watt AH, Buss DC, Routledge PA. Effect of aminophylline on the respiratory depressant action of intravenous adenosine in neonatal rabbits. *Life Sci.* 1987;40:29–34.
122. Wilson CG, Martin RJ, Jaber M, Abu-Shaweesh J, Jafri A, Haxhiu MA, et al. Adenosine A2A receptors interact with GABAergic pathways to modulate respirationin neonatal piglets. *Respir Physiol Neurobiol.* 2004;141:201–211.
123. Gauda EB, Cooper RZ, Donnelly DF, Mason A, McLemore GL. The effect of development on the pattern of A1 and A2a-adenosine receptor gene and protein expression in rat peripheral arterial chemoreceptors. *Adv Exp Med Biol.* 2006;580:121–129.
124. Herlenius E, Lagercrantz H, Yamamoto Y. Adenosine modulates inspiratory neurons and the respiratory pattern in the brainstem of neonatal rats. *Pediatr Res.* 1997;42: 46–53.
125. Barros RC, Branco LG. Role of central adenosine in the respiratory and thermoregulatory responses to hypoxia. *NeuroReport.* 2000;11:193–197.
126. Brooke RE, Deuchars J, Deuchars SA. Input-specific modulation of neurotransmitter release in the lateral horn of the spinal cord via adenosine receptors. *J Neurosci.* 2004;24:127–137.
127. Henderson-Smart DJ, Steer P. Methylxanthine treatment for apnea in preterm infants. *Cochrane Database Syst Rev.* 2001;(3):CD000140.
128. Aranda JV, Grondin D, Sasyniuk BI. Pharmacologic considerations in the therapy of neonatal apnea. *Pediatr Clin North Am.* 1981;28:113–133.
129. Harris MC, Baugart S, Rooklin AR. Successful extubation of infants with respiratory distress syndrome using aminophylline. *J Pediatr.* 1983;103:303–305.
130. Carnielli VP, Verlato G, Benini F, Rossi K, Cavedagni M, Filippone M, et al. Metabolic and respiratory eVects of theophylline in the preterm infant. *Arch Dis Child Fetal Neonatal.* 2000;83:F39–F43.
131. Fredholm B. On the mechanism of action of theophylline and caVeine. *Acta Med Scand.* 1985;217:149–153.
132. Bucher HU, Wolf M, Keel M, von Siebenthal K, Duc G. EVect of aminophylline on cerebral haemodynamics and oxidative metabolism in premature infants. *Eur J Pediatr.* 1994;153:123–128.
133. Heyman E, Ohlsson A, Heyman Z, Fong K. The eVect of aminophylline on the excursion of the diaphragm in preterm neonates. *Acta Paediatr Scand.* 1991;80:308–315.
134. Eichenwald EC, Howell II RG, Leszczynski LE, Stark AR. Theophylline improves coordination of laryngeal abduction and inspiratory eVort in premature infants [abstract]. *Pediatr Res.* 1989;25:308A.
135. Henderson-Smart DJ, Davis PG. Prophylactic methylxanthines for endotracheal extubation in preterm infants. *Cochrane Database Syst Rev.* 2010;CD000139. [Medline]
136. Henderson-Smart DJ, Davis PG. Prophylactic methylxanthines for extubation in preterm infants. *Cochrane Database Syst Rev.* 2003;(1):CD000139.
137. Burnstock G. Noradrenaline and ATP: costrasmitters and neuromodulators. *J Physiol Pharmacol.* 1995;46:365–384.
138. Nicholls J, Hourani SM, Kitchen I. Degradation of extracellular adenosine and ATP by adult and neonate rat duodenum and urinary bladder. *Pharmacol Commun.* 1992;2:203–210.
139. Burnstock G, Ulrich H. Purinergic signaling in embryonic and stem cell development. *Cell Mol Life Sci.* 2011;68:1369–1394.
140. Peachey JA, Brownhill VR, Hourani SM, Kitchen I. The ontogenetic profiles of the pre- and postjunctional adenosine receptors in the rat vas deferens. *Br J Pharmacol.* 1996;117:1105–1110.
141. Wakade TD, Palmer KC, McCauley R, Przywara DA, Wakade AR. Adenosineinduced apoptosis in chick embryonic sympathetic neurons: a new physiological role for adenosine. *J Physiol.* 1995;488:123–138.

142. Furukawa K NT. Postnatal changes in response to adenosine and adenine nucleotides in the rat duodenum. *Br J Pharmacol.* 1989;97:1111–1118.
143. Nicholls J, Hourani SMO, Kitchen I. The ontogeny of purinoceptors in rat urinary bladder and duodenum. *Br J Pharmacol.* 1990;100:874–878.
144. Hourani SMO, Shaw DA, Kitchen I. Ontogeny of purinoceptors in the rat colon muscularis mucosae. *Pharmacol Commun.* 1993;2:317–322.
145. Giaroni C, Knight GE, Zanetti E, Chiaravalli AM, Lecchini S, Frigo G, et al. Postnatal development of P2 receptors in the murine gastrointestinal tract. *Neuropharmacology.* 2006;50:690–704.
146. Chinsky JM, Ramamurthy V, Fanslow WC, Ingolia DE, Blackburn MR, Shaffer KT, et al. Developmental expression of adenosine deaminase in the upper alimentary tract of mice. *Differentiation.* 1990;42:172–183.
147. Giaroni C. Purinergic signalling and development of the autonomic nervous system. *Auton Neurosci Basic Clin.* 2015;191:67–77.
148. Rolband GC, Furth ED, Staddon JM, Rogus EM, Goldberg AP. Effects of age and adenosine in the modulation of insulin action on rat adipocyte metabolism. *J Gerontol.* 1990;45:B174–B178.
149. Chapal J, Loubatières-Mariani MM, Roye M. Effect of adenosine and phosphate derivatives on insulin release from the newborn dog pancreas. *J Physiol Paris.* 1981;77:873–875.
150. Morton MJ, Sivaprasadarao A, Bowmer CJ, Yates MS. Adenosine receptor mRNA levels during postnatal renal maturation in the rat. *J Pharm Pharmacol.* 1998;50:649–654.
151. Cekic C, Sag D, Day YJ, Linden J. Extracellular adenosine regulates naive T cell development and peripheral maintenance. *J Exp Med.* 2013;210:2693–2706.
152. Schwindinger WF, Mihalcik LJ, Giger KE, Betz KS, Stauffer AM, Linden J, et al. Adenosine A2A receptor signaling and golf assembly show a specific requirement for the gamma7 subtype in the striatum. *J Biol Chem.* 2010;285:29787–29796.
153. Lou L, Urbani J, Ribeiro-Neto F, Altschuler DL. cAMP inhibition of Akt is mediated by activated and phosphorylated Rap1b. *J Biol Chem.* 2002;277:32799–32806.
154. Hand TW, Cui W, Jung YW, Sefik E, Joshi NS, Chandele A, et al. Differential effects of STAT5 and PI3K/AKT signaling on effector and memory CD8 T-cell survival. *Proc Natl Acad Sci U S A.* 2010;107:16601–16606.
155. Cekic C, Casella CR, Sag D, Antignano F, Kolb J, Suttles J, et al. MyD88-dependent SHIP1 regulates proinflammatory signaling pathways in dendritic cells after monophosphoryl lipid A stimulation of TLR4. *J Immunol.* 2011;186:3858–3865.
156. Coppi E, Pugliese AM, Urbani S, Melani A, Cerbai E, Mazzanti B, et al. ATP modulates cell proliferation and elicits two different electrophysiological responses in human mesenchymal stem cells. *Stem Cells.* 2007;25:1840–1849.
157. Carroll SH, Ravid K. Differentiation of mesenchymal stem cells to osteoblasts and chondrocytes: a focus on adenosine receptors. *Expert Rev Mol Med.* 2013;15:e1.
158. Carroll SH, Wigner NA, Kulkarni N, Johnston-Cox H, Gerstenfeld LC, Ravid K. A2B adenosine receptor promotes mesenchymal stem cell differentiation to osteoblasts and bone formation in vivo. *J Biol Chem.* 2012;287:15718–15727.
159. Gharibi B, Abraham AA, Ham J, Evans BA. Adenosine receptor subtype expression and activation influence the differentiation of mesenchymal stem cells to osteoblasts and adipocytes. *J Bone Miner Res.* 2011;26:2112–2124.
160. Mohamadnejad M, Sohail MA, Watanabe A, Krause DS, Swenson ES, Mehal WZ. Adenosine inhibits chemotaxis and induces hepatocyte-specific genes in bone marrow mesenchymal stem cells. *Hepatology.* 2010;51:963–973.
161. Montesinos MC, Shaw JP, Yee H, Shamamian P, Cronstein BN. Adenosine A(2A) receptor activation promotes wound neovascularization by stimulating angiogenesis and vasculogenesis. *Am J Pathol.* 2004;164:1887–1892.

CHAPTER

3

Influence of Adenosine on Synaptic Excitability

Detlev Boison[1,*], *Raquel B. Dias*[2], *Traci Plumb*[1], *Sofia Cristóvão-Ferreira*[2] *and Ana M. Sebastião*[2,*]

[1]Legacy Research Institute, Portland, OR, United States

[2]University of Lisbon, Lisbon, Portugal

OUTLINE

*Equal contribution

Adenosine Receptors in Neurodegenerative Diseases.
DOI: http://dx.doi.org/10.1016/B978-0-12-803724-9.00003-X

INTRODUCTION

Homeostatic plasticity in the brain is of fundamental importance to fine-tune synaptic activity, which in turn determines cognitive processes and susceptibility to seizures. Whereas plasticity mechanisms mediated by conventional plasticity-promoting molecules, such as brain derived neurotrophic factor (BDNF), are well characterized, new evidence suggests that adenosine acts as a metaregulator of plasticity.[1] Adenosine is derived from the degradation of adenosine triphosphate (ATP) and as a reaction product of transmethylation, but in the brain it can also be directly released by neurons and glia. Adenosine functions as an intracellular metabolite and as an extracellular signaling molecule that modulates synaptic transmission and passive membrane properties predominantly through activation of high-affinity inhibitory A_1 (A_1R) and excitatory A_{2A} ($A_{2A}R$) G protein–coupled receptors (GPCRs),[2] though A_{2B} and A_3 receptors ($A_{2B}R$ and A_3R) also affect neuronal functioning and excitability. Adenosine's primary and evolutionary conserved function is the maintenance of bioenergetic homeostasis.[3–5] In this chapter, we will first discuss fundamental mechanisms through which adenosine affects neuronal excitability and plasticity. In the second part of this chapter, we will discuss the implications of those mechanisms for epilepsy, a common and serious neurological condition, which is characterized by maladaptive changes in plasticity.

MOLECULAR & PHYSIOLOGICAL MECHANISMS

The ability of adenosine receptors to fine-tune neuronal communication in the central nervous system has been studied best in the hippocampus, where pre-, post- and nonsynaptic actions converge to modulate glutamatergic transmission and synaptic plasticity.[1] This action of adenosine is also known to occur at the level of inhibitory GABAergic transmission,[6] and has even been found at the level of metamodulation due to its ability to control the action of other neuromodulators.[7] How adenosine affects the availability (release and uptake) and action of neurotransmitters and neuromodulators will be briefly summarized in this section. We will predominantly focus on brain areas where synaptic plasticity phenomena are better known—i.e., those located in the forebrain.

Influence on Neurotransmitter Availability and Basal Synaptic Transmission

Adenosine was first described to negatively affect hippocampal glutamate release over 3 decades ago,[8] an effect later shown to require A_1 receptor activation.[9] Since then, studies have revealed that A_1 receptors regulate glutamate release at many other central nervous system synapses, such as the amygdala,[10] laterodorsal tegmentum,[11] cerebellum,[12,13] striatum,[14] hypothalamus,[15] and thalamus.[16]

Excitatory A_{2A} receptors afford opposite, facilitatory regulation of glutamate release.[17] In fact, their activation is required for endogenous astrocyte-mediated reinforcement of basal synaptic transmission induced by stimulation of single hippocampal synapses.[18] Adenosine modulates synaptic transmission by directly regulating glutamate transporters, thereby influencing synaptic glutamate levels. Excitatory A_{2A} receptors can diminish glutamate uptake by inhibiting glutamate transporter (GLT-1 and GLAST) activity and expression in astrocytes.[19] This leads to more glutamate being available to activate its receptors and as such, the overall role of A_{2A}Rs seems to be a facilitation of excitatory synaptic transmission, which in some pathophysiological conditions may lead to exacerbated excitotoxicity or even aberrant synaptic plasticity.[7]

Adenosine A_{2A}Rs, acting upstream of both protein kinase C- and protein kinase A-linked pathways, lead to facilitation of GABA release from hippocampal nerve terminals,[20] an action now known to selectively occur in nerve terminals from parvalbumin (PV)-positive interneurons that project onto other interneurons.[21] PV-positive interneurons are mostly responsible for stable and time-controlled inhibition onto their target cells,[22,23] thus being involved in the generation of burst-like hippocampal output. The facilitatory action of A_{2A}Rs at nerve terminals of PV-positive neurons that project to other interneurons may thus lead to disinhibition of pyramidal cells, which together with a facilitatory action upon glutamatergic synapses, likely contributes to the facilitatory action of A_{2A}Rs upon epileptiform hippocampal activity.[21]

Adenosine also directly regulates GABA transporters in nerve terminals. Activation of A_{2A}Rs promotes GABA uptake through GABA transporter 1 (GAT-1), a mechanism that likely involves activation of the adenylate cyclase/protein kinase A signaling cascade that in turn relieves the inhibition of GAT-1 mediated by protein kinase C.[24] A_1Rs were ineffective to modulate GAT-1 in hippocampal nerve terminals.[24] GABA uptake into glial cells is also regulated by adenosine through activation of A_1R-$A_{2A}R$ heteromers,[25] where the shift between A_1R- or $A_{2A}R$-dependent effects requires only a 10-fold change in concentration. At lower concentrations, adenosine activates the A_1R protomer of the heteromeric association and inhibits GABA uptake. At higher concentrations, adenosine activates the $A_{2A}R$ protomer of the heteromer, promoting GABA uptake into astrocytes.[25] Therefore, at the tripartite synapse, adenosine has the ability to drive GABA uptake into presynaptic terminals or into glial cells, which in turn may affect excitability. At low levels, by inhibiting uptake through astrocytes, adenosine may drive GABA uptake into presynaptic terminals (where A_1Rs are ineffective to control GAT-1), thus prompting the replenishment of intracellular neurotransmitter vesicles and facilitating phasic inhibitory transmission. At higher adenosine levels, GABA may be taken up by astrocytes, which will shunt GABA from presynaptic terminals and delay the recycling of intracellular GABA pools. Glial GABA uptake also reduces ambient GABA levels, decreasing tonic inhibition. Both actions converge to facilitate excitatory transmission.

Sasaki and colleagues[26] recently uncovered another mechanism through which endogenous adenosine signaling can influence the fidelity of signal transmission at tripartite synapses. By performing dual patch-clamp recordings from the soma and axon portions of CA3 pyramidal neurons under different experimental settings, it was shown that action potentials (APs), once initiated near the soma, can undergo rapid and reversible waveform modulation in response to local manipulation of glutamatergic and adenosinergic signaling. Application of glutamate (prompting AMPA receptor activation and depolarization-induced inactivation of potassium channels) or of an A_1R antagonist to axon shafts resulted in localized AP broadening. In turn, AP waveform modulation enabled larger calcium transients in presynaptic boutons, thus facilitating signal transmission onto corresponding postsynaptic contacts. Notably, astrocytes seem to provide the endogenous source of glutamate, since calcium uncaging onto peri-axonal astrocytes prompting glutamate release comparably broadened APs. Hence, at least in the hippocampus, APs may convey information in a graded analog manner, depending on local availability of glutamate and adenosine.

Adenosine Tunes the Responses of Postsynaptic Glutamate and GABA Receptors

The mechanisms operated by adenosine to tune basal synaptic transmission are not restricted to the regulation of neurotransmitter release and signal transmission fidelity. Adding to its actions at pre- and nonsynaptic sites, adenosine can operate postsynaptically to regulate the response to neurotransmitters.

Through A_1Rs, adenosine can directly inhibit tonic inhibition of hippocampal neurons, an action independent of GAT-1 and most probably exerted at the level of extrasynaptic $GABA_A$ receptors.[27] This action is exerted both over pyramidal neurons as well as over the cannabinoid receptor-1 containing (CB1R) interneurons,[27] thus affecting the interneuron subtype that mostly determines slow oscillatory output patterns at the hippocampus. Since tonic inhibition of interneurons is more pronounced than tonic inhibition of pyramidal neurons,[28] the overall consequence of this action of A_1Rs is likely a disinhibition of interneurons and a subsequent decrease in excitatory transmission by pyramidal cells.

Glutamatergic AMPA receptors mediate the bulk of excitatory transmission in the central nervous system. Adenosine A_1 and $A_{2A}Rs$ regulate AMPA receptor-mediated responses in opposite directions, respectively inhibiting or facilitating agonist-evoked postsynaptic currents recorded from hippocampal neurons,[29,30] with implications for synaptic plasticity modulation.[29,31] Activation of $A_{2A}Rs$ is required for ischemia-induced sustained facilitation of synaptic transmission, an action that involves synaptic recruitment of calcium-permeable AMPA receptors.[32] Calcium-permeable AMPA receptors play a relevant role in several forms of synapse modification and their membrane delivery primes synapses for plasticity in response to relevant stimuli such as acute stress,[33] prolonged wakefulness, cocaine exposure, and fear conditioning.[34]

Inhibition of postsynaptic NMDA receptors by A_1R activation was first demonstrated in isolated CA1 pyramidal neurons[35] and later shown to be maintained under hypoxic/ischemic conditions.[36] On the other hand, endogenous $A_{2A}R$ activation is required for mGluR5-mediated facilitation of NMDA receptor effects on hippocampal synaptic transmission,[37] as well as for a postsynaptic form of long-term potentiation (LTP) of NMDA receptor-mediated synaptic responses that is induced by short bursts of mossy fiber stimulation.[38] Although the underlying signaling pathways still require further clarification, mGluR5-mediated enhancement of NMDAR-dependent responses in the hippocampus may arise from GluN2B subunit phosphorylation by Src kinases, which is gated by $A_{2A}Rs$.[39]

Astrocyte-Derived Adenosine Controls Basal Synaptic Transmission and the Available Range for Plasticity Control

Astrocytes are increasingly regarded as active participants in neuronal communication and neuromodulation, by controlling neurotransmitter and neuromodulator levels in and around the synaptic cleft. This is achieved not only through regulated neurotransmitter uptake but also by the release of several gliotransmitters in response to sensed neuronal activity. The first evidence for activity-dependent release of glutamate dates back to the early 1990s.[40] Data favoring the regulation of excitatory and inhibitory neuronal synaptic transmission by astrocyte-released glutamate soon followed,[41] as did the characterization of other gliotransmitters such as D-serine and the rapid hydrolyzation of ATP to adenosine.

Using dn-SNARE mice with selectively impaired astrocytic exocytosis, Pascual et al.[42] demonstrated that astrocytes contribute to the major source of adenosine that is required to maintain endogenous basal synaptic transmission. Interestingly, transgenic mice with reduced gliotransmission display reduced expression of synaptic NMDA receptors, which can be rescued by activation of A_1Rs.[43] This finding indicates that extracellular adenosine, formed as a consequence of gliotransmission, is required to homeostatically adjust surface expression of NMDA receptors.[43] The relevance of astrocyte-derived purines for the homeostatic control of synaptic transmission in the hippocampus was also highlighted in a study showing that manipulation of extracellular adenosine levels by driving under- or over-expression of astrocytic adenosine kinase (ADK) markedly influences glutamatergic and GABAergic transmission, impacting synaptic plasticity in the hippocampus.[44]

It is now clear that many of the mechanisms by which adenosine receptors orchestrate endogenous regulation of synaptic transmission and plasticity actually rely on glial-derived adenosine, be it in the brain,[18,26,42,44] the spinal cord,[45] or the peripheral nervous system.[46] It should also be noted that although glia-derived adenosine can potently tune down basal synaptic transmission[42,44] and expand the range for plasticity,[42] smaller increases in extracellular adenosine may actually lower the threshold to induce LTP by affording selective suppression of inhibitory transmission.[47] The influence of adenosine on the astrocyte-mediated control of extracellular GABA[25] and glutamate[19] clearance also needs to be taken into account while considering the role of purines in the control of excitability. In this context, it is important to keep in mind that there are also other relevant sources of purines, such as microglia. Indeed, the presence or absence of microglia in astrocytic cultures seems to determine the switch between predominant tonic A_1R-mediated inhibition and tonic $A_{2A}R$-mediated facilitation of GABA uptake into astrocytes.[25,48]

Metamodulation by Adenosine: Impact on Synaptic Plasticity

The role of adenosine as a metamodulator has been reviewed recently[7] and therefore, we herein only briefly address the influence of adenosine upon the action of the neurotrophin BDNF, well-known for its pivotal role in plasticity and learning.[49,50] Evidence has been accumulating that the synaptic actions of BDNF in the hippocampus are fully lost upon $A_{2A}R$ blockade or upon removal of endogenous extracellular adenosine.[51–53] This ability of $A_{2A}Rs$ to gate the action of BDNF at synapses involves the cyclic AMP signaling cascade [50,51,54,55] and facilitation of TrkB receptor translocation to lipid rafts.[56] The actions of $A_{2A}Rs$ modulate the facilitatory actions of BDNF on LTP[54] and long-term depression (LTD),[50] as well as determine the age-dependent effects of BDNF on hippocampal synapses.[44] Genetic deletion of $A_{2A}Rs$ leads to reduced BDNF levels,[52] and chronic blockade of $A_{2A}Rs$ in vivo decreases both mRNA and protein levels of the TrkB-FL receptor in the hippocampus.[57] Both manipulations of A_{2A} receptor activity have a negative impact upon synaptic plasticity. $A_{2A}R$ activity is also required for BDNF-induced facilitation of GABA transport into astrocytes,[58] in contrast to nerve terminals, where BDNF inhibits GABA transport even when activation of adenosine $A_{2A}Rs$ is blocked.[59] However, in spite of the $A_{2A}Rs$ not being absolutely required for the modulatory effects of BDNF on GABA uptake in nerve terminals, this action of BDNF is enhanced when $A_{2A}Rs$ are coactivated.[59]

Metaplasticity refers to the propensity for synapses to undergo activity-dependent modifications, which is thought to reflect their past history.[60] Among other factors, changes in AMPA and NMDA receptors are considered substrates for metaplasticity, by directly affecting the ability of synapses to undergo plasticity (e.g., Refs. 34,61). As mentioned above, adenosine can regulate both AMPA and NMDA receptors with consequences for plasticity.[29,31,35,38,43] In line with this, several studies have supported that adenosine signaling provides essential regulation of several synaptic plasticity forms, as we will detail in the next section.

Adenosine Shapes Synapses for Plasticity

The ability of adenosine and its analogs to inhibit LTP was first established in the CA1 area of the rat hippocampus by two independent groups, almost simultaneously.[62,63] Interestingly, adenosine was only able to exert this inhibitory effect when applied within a critical peri-induction period,[63] suggesting the occurrence of a vulnerability window during which the released adenosine could "short-circuit" plasticity induction—as later shown to be the case.[64] Endogenous regulation of LTP by adenosine was subsequently shown to depend on A_1R activation,[65,66] also required for activity-dependent LTD of synapses undergoing prolonged, low-frequency stimulation.[67,68]

Regulation of activity-dependent plasticity by adenosine was not exclusive to A_1Rs, however; A_{2A}Rs would soon enough be reported to share the ability to shape synaptic plasticity[65,69] and their activation has been associated with the expression of noncanonical forms of LTP (e.g., Refs. 38,69). In addition to more direct mechanisms, the activation of A_{2A}Rs leads to the metamodulation of synaptic plasticity (e.g., [3,7]) by gating the action of other regulatory molecules such as BDNF.[44,50,51,57]

Adenosine not only shapes synaptic plasticity in the hippocampus—the most extensively studied brain area in what concerns plasticity modulation—but also shapes activity-dependent synaptic plasticity outside this brain structure. Indeed, and as summarized in Tables 3.1 and 3.2, adenosine receptors convey regulation of bidirectional synapse modification through multiple mechanisms at different synapses within the brain. This is the case in the auditory receptive fields of the adult neocortex, where adenosine has recently been shown to adjust the requirements for plasticity to development.[70] Perceived information reaches the auditory thalamus and is then relayed to thalamorecipient neurons within the sensory cortices. The authors show that, with development, the threshold required to drive reorganization of this neuronal network is heightened through acquisition of presynaptic gating by adenosine. Indeed, as a result of tonic A_1R-mediated inhibition of glutamate release, low-frequency stimulation of glutamatergic thalamic inputs alone is not sufficient to trigger LTD of mature thalamocortical connections. However, combining delivery of low-frequency afferent stimulation with bath application of a cholinergic agonist (simulating incoming cholinergic information), it becomes possible to overcome A_1R-mediated negative priming of plasticity, ultimately enabling postsynaptic LTD to take place.[70]

A_1R-mediated gating of synaptic plasticity can also be triggered by stimuli that negatively impact memory formation. Indeed, three independent studies have shown that prior exposure to relevant experience, such as acute stress,[74] chronic morphine,[75] and sleep deprivation,[76] negatively affect subsequent synaptic plasticity through activation of A_1Rs. Hence, changes in adenosinergic signaling may account for negative priming of LTP following the integration of relevant stimuli. It should be noted that in each case, synaptic plasticity was assessed using ex vivo LTP protocols after rodents had been exposed to relevant experiences. These studies thus provide evidence for the involvement of adenosine in metaplasticity, since they evaluate how prior experience is able to condition subsequent plasticity events through adenosine A_1 receptor-mediated modulation of neuronal activity.

The manipulation of adenosine receptor signaling in vivo has been shown to impact plasticity, behavior, and learning. One of the first in vivo studies implicating adenosine in synaptic plasticity addressed the effects of a locally perfused adenosine analog upon LTP of the

TABLE 3.1 Adenosine Receptors Regulate Bidirectional Synaptic Plasticity in the Central Nervous System

	Central nervous system synapse	Experimental protocol	Pharmacological and/or genetic manipulation	Major findings	References
Ex vivo Studies	Thalamocortical synapses	Low-frequency stimulation (LFS) of thalamic afferents; Glutamate release onto single spines	Bath application (BA) of cholinergic agonists (carbachol); BA of A_1R antagonists	A_1R tonically inhibit glutamate release; Cholinergic activation overcomes adenosine-mediated presynaptic gating of LTD	Blundon et al.[70]
	Nucleus accumbens	High-frequency stimulation (HFS) of cortical afferents	BA of $A_{2A}R$ antagonists; $A_{2A}R$ KO mice	Genetic or pharmacological $A_{2A}R$ blockade inhibits HFS-induced LTP	d'Alcantara et al.[201]
	Cerebellum (cultured Purkinje cells)	Repetitive pulses of somatic depolarization coupled to glutamate iontophoresis to induce mGluR1-mediated LTD	Application of adenosine at near-physiological concentrations or application of A_1 receptor agonists	A_1 and mGluR1 receptors from heteromeric complexes; A_1 receptor activation precludes mGluR1-induced LTD by decreasing ligand sensitivity	Kamikubo et al.[202]
	Striatum (corticostriatal synapses)	Afferent stimulation followed or preceded by postsynaptic spiking at theta frequency to induce LTP or LTD, respectively	BA of dopamine D_2 and adenosine A_{2A} agonists and antagonists; Transgenic mice expressing GFP-tagged dopamine receptors	$A_{2A}R$ blockade disrupts LTP in D_2-expressing neurons; $A_{2A}R$ activation restores LTP when postsynaptic spiking precedes afferent stimulation; Balanced A_{2A}/D_2 signaling enables bidirectional striatal plasticity	Shen et al.[71]
	Hypothalamus	Low- and high-frequency afferent stimulation onto hypocretin neurons	BA of A_1 and A_{2A} receptor antagonists	Blockade of A_1Rs enhances HFS-induced LTP	Xia et al.[203]

Data from ex vivo studies.

TABLE 3.2 Adenosine Receptors Regulate Bidirectional Synaptic Plasticity in the Central Nervous System

	Central nervous system synapse	Experimental protocol	Pharmacological and/or genetic manipulation	Major findings	References
In vivo Studies	Hippocampus (CA3-CA1 synapses)	Eye-blink conditioning (EBC); HFS of CA3 afferents to CA1 neurons	Intraperitoneal delivery of an $A_{2A}R$ antagonist	$A_{2A}R$ blockade prevented associative memory acquisition and in vivo HFS-induced LTP	Fontinha et al.[72]
	Hippocampus (CA3-CA1 synapses)	Morris Water maze spatial memory test; HFS of CA3 afferents to CA1 neurons	Maternal separation stress model; oral delivery of an $A_{2A}R$ antagonist in the animal's drinking water	$A_{2A}R$ blockade reverted both stress-induced spatial memory deficits and HFS-induced ex-vivo LTP	Batalha et al.[73]
	Somatosensory cortex	Eye-blink conditioning; Transcranial direct current stimulation (tDCS)	Local microinjection of an A_1R antagonist	EBC acquisition is either potentiated or depressed by anodal or cathodal tDCS; Blockade of A_1R activation impairs cathodal tDCS-induced LTD	Márquez-Ruiz et al.[204]
	Superior colicullus (SC)	A_1 and $A_{2A}R$ expression in the SC, through development; Anatomical studies using intraocular injection of horseradish peroxidase	In vivo blockade of adenosine deaminase (ADA), thereby increasing adenosine availability	A_1 and A_{2A} receptor expression in the visual layer of the SC varies with development; Short-term ADA blockade during the 2nd postnatal week induced an expansion of the retinotectal terminal fields in the rostrocaudal axis of the tectum	Tavares Gomes et al.[205]

Data from in vivo studies.

perforant path and showed that adenosine markedly diminishes the ability of this hippocampal region to undergo long-term facilitation in response to high-frequency stimulation.[77] Later work has reinforced the idea that manipulation of adenosinergic signaling can directly affect learning and synaptic plasticity in vivo. Fontinha et al.[72] examined synaptic responses from chronically implanted electrodes during eye-blink conditioning, a classical model of associative learning. They reported that $A_{2A}R$ blockade (by intraperitoneal delivery of a selective antagonist) impaired synaptic reinforcement of CA3-CA1 synapses with trial progression. While $A_{2A}R$ blockade may perturb learning under physiological conditions, i.e., situations where animals have not been subject to any distress or pathology, it may not do so under pathological conditions. Batalha et al.[73] showed that the spatial memory deficits and impairments in synaptic plasticity elicited by chronic mild stress could be rescued by systemic administration of an $A_{2A}R$ antagonist. Another study showed that genetic or pharmacological inactivation of $A_{2A}Rs$ ameliorates cognitive impairment normally observed in those with Huntington's disease.[78] Taken together, the evidence indicates that $A_{2A}Rs$ may have quite distinct and even opposite roles under physiological or pathological conditions, as depicted in Fig. 3.1.

A comment on LTP and adenosine upon aging is deserved. Aging seems to differentially affect theta-burst LTP in basal and apical dendrites of pyramidal neurons, since aged animals have reduced LTP in basal dendrites[79] but enhanced LTP in apical dendrites.[80,81] A_1Rs reduce theta-burst LTP both in apical[80] and basal[79] dendrites of aged rats, and may even be responsible for the reduced LTP at basal dendrites seen with ageing.[79] Remarkably, the cause for enhanced LTP in aged rats is a higher adenosinergic tonus leading to A_{2A} receptor activation[81] that then triggers a facilitatory action of BDNF.[82] Theta-burst LTP in young animals is not affected by A_{2A} receptor blockade,[82,83] in contrast to what occurs with high-frequency induced LTP.[83] It is as yet unknown whether this relates to different levels of adenosine attained at synapses as a consequence of different patterns of stimulation.

Adenosine can also affect structural plasticity since A_{2A} receptors promote axonal elongation and dendritic branching.[84] Interestingly, A_{2A} receptors interact with BDNF to facilitate dendritic branching, but its promotion of axonal elongation utilizes a BDNF-independent mechanism, instead correlating with a decrease in microtubule stability and an increase in microtubule growth speed in axonal growth cones.[84]

In vivo manipulations of A_1R activity also affect cognitive function. It has been reported that upregulation of forebrain A_1 receptors induced by prolonged caffeine intake is associated with an exacerbation of memory deficits induced by cannabinoids.[85] Trusel and colleagues[86] recently showed that interfering with endocannabinoid and A_1 receptor-mediated

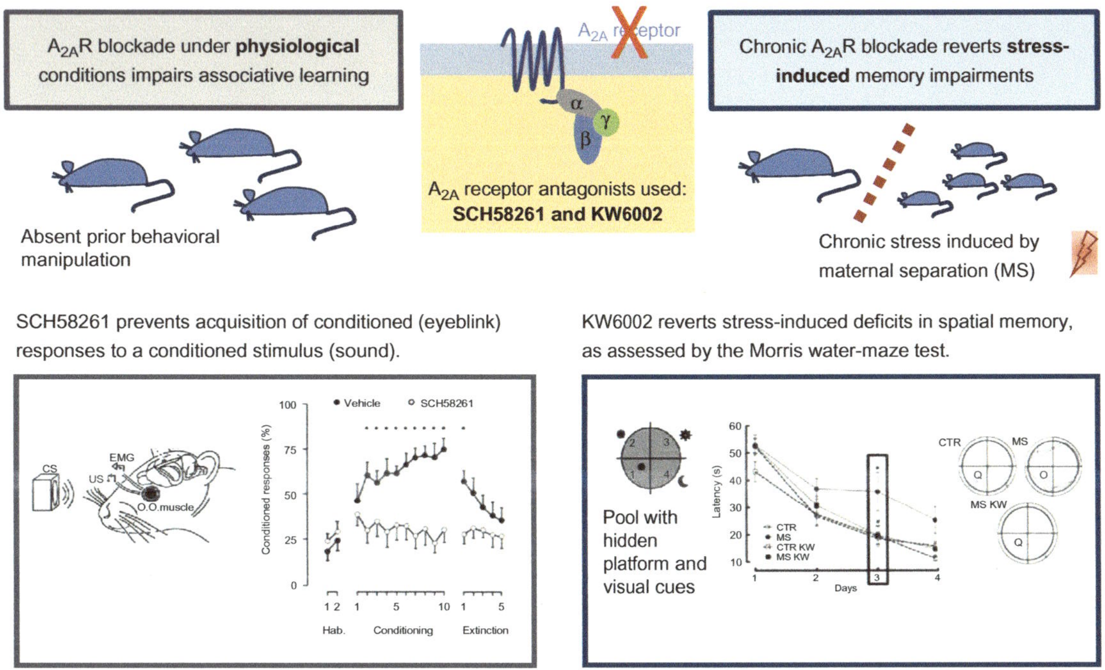

FIGURE 3.1 **Blockade of adenosine A_{2A} receptors differentially affects synaptic plasticity under physiological versus pathological conditions.** (left) In the absence of prior behavioral manipulation, intraperitoneal administration with a selective A_{2A} receptor antagonist (SCH58261) prevents acquisition of conditioned (eyeblink) responses to the conditioned sound (CS) stimulus. In the work by Fontinha et al. SCH58261 was injected before each conditioning session; conditioned responses were measured by electromiographic (EMG) recordings from the orbis orbicularis (OO) muscle. The unconditioned stimulus (US) consisted of a pulse applied to the supraorbital nerve, presented 500 ms after the end of the CS (right) oral chronic administration of KW6002 (KW), a second A_{2A} receptor antagonist, can rescue the deficits in spatial memory associated with chronic stress from a maternal separation (MS) model. The study by Batalha et al. evaluated hippocampal-dependent memory performance by applying the Morris water maze test and measuring the average latency that it took rats from each experimental group to find the hidden platform. Source: *(Left) Fontinha BM, Delgado-García JM, Madroñal N, Ribeiro JA, Sebastião AM, Gruart A. Adenosine A2A receptor modulation of hippocampal CA3-CA1 synapse plasticity during associative learning in behaving mice. Neuropharmacology. 2009;34 (7):1865–1874; (Right) Batalha VL, Pego JM, Fontinha BM, et al. Adenosine A2A receptor blockade reverts hippocampal stress-induced deficits and restores corticosterone circadian oscillation. Mol Psychiatry. 2013;18:320–331.*

signaling at the basal ganglia alleviates motor abnormalities in animal models of Parkinson's disease. Purinergic gliotransmission may furthermore contribute to memory impairment following sleep loss, since deficits in novel object recognition usually observed in sleep-deprived mice are largely reverted by systemic blockade of adenosine A_1 receptors.[87]

In summary, adenosine signaling seems to not only be required for physiological synaptic plasticity, but may also contribute to maladaptive forms of plasticity under pathological conditions.

IMPLICATIONS FOR EPILEPSY

Epilepsy is a heterogeneous syndrome characterized by spontaneous recurrent seizures. As one of the most frequent neurological conditions, it affects around 1% of the population. Seizures can manifest as brief periods of altered consciousness or absences, but they can also generalize and spread to the motor cortex leading to clonic and/or tonic convulsions. Seizures are generated by excessive electrical discharges in neuronal networks; however, the cellular and molecular basis of epilepsy is still largely unknown. Today epilepsy is more and more considered to be a spectrum disorder, which is far more complex than merely being a condition of increased neuronal excitation and decreased neuronal inhibition. Accordingly, the contribution of inflammatory, glial, bioenergetic, and epigenetic changes to the pathophysiology of epilepsy has gained a great deal of attention.[88–94] Therefore, epilepsy presents a disorder of complex network dysfunction, which is also reflected by distinct histopathological changes commonly found in the epileptic brain. Temporal lobe epilepsy (TLE), the most common form of epilepsy in adulthood, is characterized by neural cell loss in select areas of the hippocampal formation, granule cell dispersion, mossy fiber sprouting, ectopic neurons and aberrant or recurrent circuitry. Importantly, glial changes that control homeostatic functions of brain activity also play an important role. Seizures can be suppressed by antiepileptic drugs (AEDs), but it is estimated that one third of all persons with epilepsy, or over 20 million persons worldwide, cannot be treated with AEDs. Findings from the past 15 years suggest that adenosine-dependent homeostatic plasticity changes contribute to the epileptic phenotype. A compelling theory suggests that homeostatic plasticity mechanisms in the epileptic brain remain active, but their continuing attempts to reset neuronal activity become maladaptive and finally result in disruption of homeostasis with neurobehavioral consequences.[95] Based on this rationale, homeostasis-restoring adenosine augmentation therapies (AATs) are considered a novel and promising approach for the suppression of seizures and the prevention of epilepsy.

Pathophysiological Mechanisms

In the following, we will focus on TLE, because (1) it is the most common form of adulthood epilepsy, (2) it is highly refractory to treatment, and (3) and maladaptive changes in adenosine homeostasis and signaling are well established in this form of epilepsy. It is now widely accepted that TLE is an acquired condition, in which an initial trigger such as brain injury, status epilepticus (SE), a febrile seizure, or an infection triggers a cascade of events that turn a healthy brain into an epileptic brain.[96–99] This period of "epileptogenesis" is characterized by inflammatory processes involving first microglial and later astroglial activation that lead to permanent alterations in neuronal circuitry and excitability.[100–102] New findings suggest that epigenetic alterations play an intrinsic role in epileptogenesis; those epigenetic changes have the potential to reprogram entire gene networks.[94,103,104] Overall, epileptogenesis is likely a maladaptive attempt to "repair" an initiating injurious event, thereby prompting the brain to switch into a "repair mode," which is characterized by enhanced plasticity and physiological and cellular features that are more characteristic of an immature brain. For example, an epileptogenesis-triggering SE induces profound neurogenesis; however, whether newborn neurons constitute a beneficial repair attempt to replace lost neurons or whether they contribute to epileptogenesis through the formation of aberrant networks is still a matter of debate.[105–109] As a result of the initial injury and subsequent inflammatory processes and neurogenesis, the epileptic hippocampus exhibits a very characteristic histopathology comprised of distinct neuronal cell loss (primarily in area CA1 and within the dentate), dispersed granular neurons, and sprouting of mossy fibers.[110] In particular, mossy fiber sprouting is closely associated with the epileptogenic process, and the prevention of mossy fiber sprouting has been associated with the prevention of epileptogenesis.[111–113]

Apart from those well-documented changes in neuronal organization and excitability, there is growing evidence that glial pathology plays a major role in the pathogenesis and pathophysiology of TLE.[102,114–116] Hippocampal sclerosis is a pathological hallmark of TLE[117] and the surgical excision of epileptogenic brain tissue—a routine practice to treat AED refractory TLE—has demonstrated that onset zones for chronic temporal lobe-derived and posttraumatic seizures correlate with gliotic scarring. Astrocytes influence pathogenesis and pathophysiology of epilepsy via the release of gliotransmitters such as glutamate, ATP, and

D-serine,[118] in addition to the reuptake of neurotransmitters such as glutamate[119] or neuromodulators such as adenosine.[120] An "astrocytic basis of epilepsy" was proposed because prolonged episodes of neuronal depolarization evoked by the astrocytic release of glutamate contributed to epileptiform discharges.[121] Astrocytes play important homeostatic roles in the "upstream control" of uptake, degradation, and recycling of neurotransmitters. In addition, disruption of the blood brain barrier, as well as neuroimmunological functions governed by glia, have been implicated not only in seizure generation (i.e., ictogenesis) but most importantly in the pathophysiological processes that lead to the development of epilepsy (i.e., epileptogenesis). Because glia communicate with each other and assume a role that is upstream of neuronal function, perturbations of glial homeostasis can affect entire neuronal networks. Those network effects of glia might indeed be a reason why neuronal networks in epilepsy synchronize; similarly, fluctuations in homeostatic functions of glia might explain why seizures are sporadic.

Adenosine Receptor-Dependent Effects

As mentioned above, adenosine exerts its homeostatic action through the activation of four GPCRs: A_1, A_{2A}, A_{2B}, and A_3.[122] The A_1 and A_{2A} subtypes are the most widely studied, and have been implicated in a number of physiological and psychological disorders, including epilepsy. Adenosine A_1 receptors are found in large numbers in the hippocampus, a common locus of epileptic activity.[123,124] Activation of A_1 receptors by adenosine or adenosine agonists have been shown to suppress seizure activity in the hippocampus in rodent models of TLE.[125,126] Deletion of the A_1R in mice causes lethal SE following traumatic brain injury or exposure to an excitotoxic drug,[127,128] whereas therapeutic adenosine augmentation effectively stops seizures through enhanced activation of A_1Rs.[129] Several lines of evidence suggest that maladaptive changes in adenosine receptor signaling contribute to the pathophysiology of epilepsy. It is conceivable that any shift in the ratio of inhibitory A_1R versus stimulatory $A_{2A}Rs$ directly affects neuronal excitability. However, it is currently unknown whether changes in adenosine receptor expression are cause for or consequence of epilepsy.

Experimentally, a decrease in A_1R densities and A_1R-mediated neuromodulation have been described in the rat kindling model of epilepsy, suggesting the failure of endogenous seizure control mechanisms in epilepsy.[130] Receptor knockout studies have shown that mice lacking the A_1R have spontaneous electrographic seizures[131] and develop lethal SE following an intrahippocampal injection of kainic acid or a traumatic brain injury.[127,128] These studies directly show that A_1R activation is also needed to prevent seizure spread. Histopathological and biochemical analyses from specimens surgically resected from patients with intractable epilepsy show decreased expression levels of A_1 receptors, suggesting that decreased A_1R expression may contribute to seizure generation in human chronic epilepsy.[132] Experimentally, dynamic changes in A_1R signaling or expression have been described as a direct consequence of acute seizures. Desensitization of A_1R responses, but normal receptor levels, have been described in the hippocampus of rats after SE elicited by performant path stimulation,[133] whereas upregulation of the A_1R in the entorhinal cortex has been described as a response to spontaneous seizures induced by electrical stimulation.[134] In a human genomic study, variants in the A_1R gene have been associated with the development of posttraumatic seizures after a severe traumatic brain injury, suggesting that deficiency in A_1R signaling might be associated with posttraumatic epileptogenesis.[135] Together, these data suggest that dysregulation of A_1R signaling is intricately linked to the pathophysiology of epilepsy.

The synaptic fraction of $A_{2A}Rs$ can mediate synaptotoxic effects of adenosine,[136–138] which is largely dependent on the neuronal release of adenosine or its precursor ATP.[139] Thereby, neuronal hyperexcitability in epilepsy likely leads to enhanced synaptic $A_{2A}R$ activation, which could aggravate synaptotoxicity and thereby further the degeneration of normal circuitry contributing to the progressive course of epilepsy. Interestingly, genetic variants of the $A_{2A}R$ gene have been associated with acute encephalopathy with biphasic seizures and late reduced diffusion in children, suggesting that $A_{2A}R$ dysregulation promotes seizures and excitotoxic brain damage in those patients.[140] In Wistar Albino Glaxo/Rijswijk (WAG/Rij) rats, a model of human absence epilepsy, increased expression of $A_{2A}Rs$ in epileptic rats, but not in presymptomatic animals, has been described, suggesting that increased $A_{2A}R$ expression in this model supports the epileptic phenotype.[141] In line with a potentially pro-convulsive role of the $A_{2A}R$, $A_{2A}R$ knockout mice were partially resistant to limbic seizures.

The pro-convulsive action of A_{2A} receptors may be due to its ability to modulate A_1 receptor densities and level of activation. High-frequency stimulation of nerve terminals results in the release of ATP into the synapse.[142] The extracellular breakdown of ATP to adenosine preferentially binds to A_{2A} receptors in hippocampal slices, resulting in presynaptic excitation.[143] A_{2A} receptor activation can subsequently lead to downregulation of A_1 receptors.[144–146] As described above, studies of chronic epilepsy have confirmed a reduced number of A_1 receptors in the hippocampus.[130,132,147] Thus, high-frequency conditions in epileptic circuits may preferentially stimulate A_{2A} receptors and lead to downregulation of A_1 receptors. In conclusion, increased $A_{2A}R$ activation may promote the epileptic state.

Little is known about the contribution of $A_{2B}Rs$ and A_3Rs in epilepsy. Using an in vitro system of tissue from human resected epileptogenic foci, which was microtransplanted into Xenopus oocytes, it was shown that the $A_{2B}R$ selective antagonist MRS1706 as well as the A_3R selective antagonist MRS1334 reduced the run-down of GABA currents.[148] These findings suggest that cortical adenosine A_{2B} and A_3 receptors alter the stability of $GABA_A$ receptors and thereby fine-tune neuronal excitability.

Adenosine Receptor-Independent Effects

In addition to receptor-dependent effects, adenosine also regulates the epigenome through biochemical enzyme reactions independent of adenosine receptor activity. Adenosine tone regulates DNA methylation through interference of the transmethylation pathway. Adenosine is formed as a product of transmethylation reactions that are dependent on *S*-adenosyl methionine (SAM), and these reactions can include the transfer of methyl groups to DNA through DNA methyltransferases,[149,150] a process that regulates gene expression. A slight change in adenosine homeostasis can trigger a change in DNA methylation and can turn "on" genes that are normally silenced or turn "off" genes that should be expressed. For example, a reduction in adenosine levels through overexpression of ADK, an astrocyte-based enzyme that converts adenosine to adenosine monophosphate, directly leads to DNA hypermethylation[149] and reduced gene expression. Conversely, an increase in adenosine blocks DNA methylation and results in more gene transcription.[5,149,151,152]

Importantly, epigenetic changes in DNA methylation states have been linked to epileptogenesis and disease progression.[104] Human and animal studies of TLE frequently exhibit DNA hypermethylation in epileptogenic areas of the brain.[92,149,153] Adenosine delivered directly to the brain of rats after the onset of epilepsy reverses DNA hypermethylation and prevents progression of the disease.[149] These adenosine receptor-independent effects further support the critical role of adenosine in TLE and indicate that AATs should reduce seizure progression through changes in DNA methylation.

Disruption of Adenosine Homeostasis in Epilepsy

Based on the evidence above that adenosine plays a key role in regulating epileptic states in both humans and animals, it is becoming exceedingly clear that dysregulation of adenosine homeostasis is associated with or perhaps even responsible for producing epileptogenesis. Several mechanisms contribute to dysregulation of adenosine homeostasis in epilepsy, including impairments in adenosine transporter activity and the conversion of ATP to adenosine (see Ref. 154 for a review). Perhaps the most compelling is the profound astroglial dysfunction and subsequent increase in the astrocyte-based enzyme ADK that characterizes epileptogenic regions of the brain.

Adenosine homeostasis in the brain is largely under the control of metabolic clearance through astrocytes and ADK.[154–157] Adenosine transported into astrocytes is quickly converted to adenosine monophosphate through the action of ADK. Because extracellular and intracellular adenosine levels can freely equilibrate through equilibrative nucleoside transporters,[158] the expression levels and activity levels of the intracellular enzyme ADK determine the flux of adenosine from the synaptic compartment into astrocytes. Consequentially, overexpression of ADK in astrocytes, a pathological hallmark of TLE, leads to reduced extracellular levels of adenosine and reduced adenosine receptor activation.[5]

Research on epileptogenesis has suggested that the disease occurs in two phases. The acute phase of epileptogenesis is characterized by a profound surge of adenosine and concurrent reduction of ADK within 2–3 h of an initiating injury,[159,160] such as stroke,[160] traumatic brain injury,[161] or seizures.[159,162] This increase in adenosine likely results in reduced DNA methylation,[149] and may lead to the transcription and expression of genes that initiate epileptogenesis. The latent phase of epileptogenesis, anywhere from days to months following the injury, is characterized by an inflammatory response that results in the activation of microglia and astrogliosis.[115,163,164] Adenosine deficiency soon follows as a result of this astrogliosis and the consequent upregulation of ADK. A number of studies have shown a link between seizure activity in epileptogenesis and a change in ADK levels. For example, astrogliosis and ADK overexpression are routinely found at the origination point of seizure activity,[165,166] and coincide with seizure onset.[131] Further, ADK overexpression is associated with DNA hypermethylation[149] and is sufficient to trigger electrographic seizures.[131,165] Therefore, it is likely that an increase in ADK following injury and the subsequent increase in DNA methylation are responsible for the progression and maintenance of epileptogenesis.

It should be noted that astrogliosis is associated not only with increases in ADK, but also with the upregulation of A_{2A} receptors on astrocytes expressing glial fibrillary acidic protein (GFAP),[136,167] a marker of reactive astrocytes. This overexpression of A_{2A}Rs on reactive astrocytes has been found as early as 3–7 days following kainate-induced neurotoxicity.[167] Importantly, A_{2A}Rs play a role in the regulation of astrocyte physiology, where activation or inhibition of A_{2A}Rs can trigger[168] or prevent[169] astrogliosis, respectively. The purpose of this upregulation of A_{2A}Rs on GFAP-positive astrocytes is still unknown. As ADK and A_{2A}Rs are both overexpressed in reactive astrocytes, it is possible that the maladaptive changes of these two systems are related during epileptogenesis.[154] Perhaps the ADK-induced reduction of adenosine tone during the latent phase results in a compensatory increase in A_{2A}Rs. Alternatively, neuronal hyperexcitability may trigger increased A_{2A}R expression and the compensatory increase in ADK may serve as a negative feedback loop to reduce adenosine and prevent further A_{2A}R-induced astrogliosis. Consequently, further studies are needed to elucidate the nature of the relationship between A_{2A}Rs and ADK in reactive astrogliosis.

Comorbidities of Epilepsy

Patients with epilepsy commonly suffer from additional physiological comorbidities, such as cardiac, respiratory, and gastrointestinal disorders.[170] Equally troublesome are the major psychiatric disorders that are often comorbid with epilepsy, including obsessive-compulsive disorder, cognitive impairments, anxiety, major depression, bipolar-affective disorder and schizophrenia.[171,172] Evidence suggests that those with poorly controlled seizures or those with refractory epilepsy show a higher prevalence to these comorbidities.[173] In pediatric epilepsy, comorbidities are also common. When pediatric epilepsy cooccurs with other common isolated disorders, such as reading or math disorders and even autism, the characteristic cognitive pattern of the isolated disorder is altered.[174]

Apart from mediating seizure control, adenosine is a crucial regulator of behavior. In particular, the disruption of adenosine metabolism has been linked with cognitive and psychiatric phenotypes.[71,175–177] The brain-wide transgenic overexpression of ADK in mice was sufficient to trigger adenosine-deficiency[178] and impairment of cognitive function,[175,177] in particular severe learning deficits in the Morris water maze task and in Pavlovian conditioning.[177] Conversely, reconstruction of adenosine homeostasis emerges as a rational and innovative approach to restore cognitive function under conditions of increased metabolic adenosine clearance (as found in TLE). We recently demonstrated that therapeutic adenosine augmentation with the ADK inhibitor ABT-702 exerted antipsychotic-like activity in the prepulse inhibition paradigm for psychiatric gating deficits typical for schizophrenia, whereas adenosine-releasing cell grafts to the hippocampal formation restored cognitive performance in Adk-transgenic mice.[71] Together these findings suggest that maladaptive changes in adenosine homeostasis can not only give rise to epileptic seizures, but also give rise to comorbidities commonly associated with epilepsy.

Therapeutic Implications

There is a large body of evidence that suggests adenosine plays a key role in seizure suppression. However, the use of pharmacological agents to modulate adenosine levels presents unique challenges due to the widespread distribution of adenosine receptors. Systemic treatment with adenosine agonists can result in suppression of cardiovascular function,[179,180] decreased locomotion or catalepsy,[181] and behavioral suppression.[182,183] Due to risks associated with those potential side effects, alternative therapies must be explored.

It has been suggested that instead of trying to increase extracellular adenosine with adenosine analogs, pharmacological approaches should instead target the key regulators of adenosine metabolism or clearance.[155] One reasonable target is ADK, a key metabolic enzyme in astrocytes that regulates endogenous adenosine in the brain. Because of the profound overexpression of ADK seen in brain regions considered epileptogenic,[116,120] and the ability of ADK overexpression to trigger seizure activity,[165,166,184] pharmacological agents that inhibit ADK are promising treatments for TLE. Importantly, there is evidence that they may even suppress seizure activity when traditional AEDs fail.[159] Unfortunately, ADK inhibitors administered systemically have similar side effects as adenosine agonists, including cognitive changes and sedation.[5]

To avoid widespread systemic side effects of adenosine, local or brain-specific treatment strategies are required. Four successful approaches have been demonstrated: (1) silk-based brain implants engineered to release defined doses of adenosine with known kinetics prevented kindled seizures in the rat. Importantly, a transient dose of adenosine delivered during kindling robustly attenuated epilepsy development, indicating a potential antiepileptogenic effect of therapeutic adenosine augmentation,[185] (2) stem cells engineered to release adenosine were used as intrahippocampal cell grafts and shown to prevent acute and induced seizures, to attenuate development of astrogliosis, to prevent overexpression of ADK, and to prevent development of spontaneous recurrent epileptic seizures in mice and rats,[186,187] (3) a gene therapy designed to selectively target ADK in astrocytes based on RNA interference technology almost completely abrogated any spontaneous seizure activity in epileptic Adk-tg mice, and[188] (4) a high-fat low-carbohydrate ketogenic diet (see below) suppressed seizures in mice via increased activation of adenosine A_1 receptors.[189] Based on the potent antiictogenic and antiepileptogenic properties the translation into clinical practice is likely. Possible hurdles are the development of local AATs, which might best be achieved by gene therapy; however, dietary alternatives provide a readily available alternative.

A ketogenic diet has been used clinically for over 80 years to treat pediatric epilepsy and drug-resistant epilepsy.[190,191] It is a high-fat, low-carbohydrate metabolic regimen designed to force the brain to use ketones as the primary energy source instead of glucose. A large body of evidence indicates that the antiepileptic effect of a ketogenic diet is due to an increase in adenosine signaling.[192–195] In line with this, mice given a ketogenic diet show reduced ADK levels, and seizure suppression with this diet is dependent on functional A_1 receptors.[196]

CONCLUSIONS AND OUTLOOK

The excitability of the brain is determined by neurons, which are connected in networks and broadly controlled by the homeostatic environment of their surroundings. It becomes clear that complex neurological syndromes, such as epilepsy, which are not only defined by a dominant symptom (i.e., a seizure), but also by a growing number of associated comorbidities, can best be explained by the disruption of network homeostasis. Disruption of network homeostasis implies the simultaneous and concordant dysregulation of several molecular pathways, which in turn can affect each other and lead to a progression of the disease in the sense that "seizures beget seizures." This old concept has a lot of truth and can best be explained by the self-reinforcing interplay of several homeostatic systems that become progressively dysregulated during disease progression. Conventional AEDs with a target-centric mode of action are unlikely to affect network homeostasis and to prevent the progressive maladaptive changes occurring during epileptogenesis, which can be considered a lifelong process of disease progression. Novel therapeutic interventions based on adenosine, epigenetic mechanisms, or dietary and lifestyle interventions might hold promise to affect network homeostasis as a novel conceptual strategy to treat and prevent epilepsy on the network level. Adenosine emerges as a prototype homeostatic network regulator with the proven capability to control network activity both through receptor-dependent as well as through receptor-independent epigenetic and bioenergetic mechanisms.[5] Consequently, therapeutic adenosine augmentation has been demonstrated to suppress epileptic seizures,[186,187] to prevent disease progression and epileptogenesis,[186,197] but also to prevent psychosis and to improve cognition[71] without any known adverse effects. The robust antipsychotic and pro-cognitive effect of adenosine in mice[71] suggests that therapeutic adenosine augmentation might combine anticonvulsant with cognition-enhancing effects. In preclinical toxicity studies of intrathecal adenosine in dogs, no side effects were observed with intrathecal adenosine infused chronically for 26 days.[198] Likewise, intrathecal adenosine was tested in humans in escalating doses of up to 2 mg without any adverse effects.[199,200] Several therapeutic adenosine augmentation strategies ranging from gene therapy to dietary intervention are currently in preclinical development. It is within the scope of possibilities that adenosine augmenting therapies are introduced into the clinic as a novel class of "homeostatic network therapy" within the next 10 years. Challenges will be to develop strategies to confine adenosine's action to identifiable target areas and cells; the advent of cell-type specific gene therapy vectors might offer a promising strategy to manipulate adenosine homeostasis in a localized and cell-type selective manner.

Acknowledgments

The authors are indebted to outstanding experimental work from their research teams. The authors' work was funded through grants from the NIH (R01 NS084920, R01 MH083973, R21 088024), the Legacy Hospital Foundations, and Fundação para a Ciência e Tecnologia (EXPL/BIM-MEC/0009/2013; PTDC/DTP-FTO/3346/2014).

References

1. Dias RB, Rombo DM, Ribeiro JA, Henley JM, Sebastião AM. Adenosine: setting the stage for plasticity. *Trends Neurosci*. 2013;36(4):248–257.
2. Chen JF, Eltzschig HK, Fredholm BB. Adenosine receptors as drug targets–what are the challenges? *Nat Rev Drug Discov*. 2013;12(4):265–286.
3. Sebastião AM, Ribeiro JA. Tuning and fine-tuning of synapses with adenosine. *Curr Neuropharmacol*. 2009;7(3):180–194.
4. Boison D, Sandau US, Ruskin DN, Kawamura Jr M, Masino SA. Homeostatic control of brain function - new approaches to understand epileptogenesis. *Front Cell Neurosci*. 2013;7:109.
5. Boison D. Adenosine kinase: exploitation for therapeutic gain. *Pharmacol Rev*. 2013;65:906–943.
6. Sebastião AM, Rombo DM, Ribeiro JA. Adenosine receptor modulation of GABAergic transmission. In: Ramkumar V, Carvalho RPD, eds. *Adenosine Signaling Mechanisms: Pharmacology, Functions and Therapeutic Aspects*. Hauppauge, NY: Nova Science Publishers; 2015:193–220.
7. Sebastião AM, Ribeiro JA. Neuromodulation and metamodulation by adenosine: impact and subtleties upon synaptic plastic6ity regulation. *Brain Res*. 2015;1621: 102–113.
8. Dunwiddie TV, Fredholm BB. Adenosine receptors mediating inhibitory electrophysiological responses in rat hippocampus are different from receptors mediating cyclic AMP accumulation. *Naunyn-Schmiedeberg's Arch Pharmacol*. 1984;326(4):294–301.
9. Sebastião AM, Stone TW, Ribeiro JA. The inhibitory adenosine receptor at the neuromuscular junction and hippocampus of the rat: antagonism by 1,3,8-substituted xanthines. *Br J Pharmacol*. 1990;101(2):453–459.
10. Rau AR, Ariwodola OJ, Weiner JL. Presynaptic adenosine A_1 receptors modulate excitatory transmission in the rat basolateral amygdala. *Neuropharmacology*. 2014;77:465–474.

11. Arrigoni E, Rainnie DG, McCarley RW, Greene RW. Adenosine-mediated presynaptic modulation of glutamatergic transmission in the laterodorsal tegmentum. *J Neurosci.* 2001;21(3):1076–1085.
12. Wall MJ, Atterbury A, Dale N. Control of basal extracellular adenosine concentration in rat cerebellum. *J Physiol.* 2007;582:137–151.
13. Zhang W, Linden DJ. Neuromodulation at single presynaptic boutons of cerebellar parallel fibers is determined by bouton size and basal action potential-evoked Ca transient amplitude. *J Neurosci.* 2009;29(49):15586–15594.
14. Borycz J, Pereira MF, Melani A, et al. Differential glutamate-dependent and glutamate-independent adenosine A1 receptor-mediated modulation of dopamine release in different striatal compartments. *J Neurochem.* 2007;101(2):355–363.
15. Oliet SH, Poulain DA. Adenosine-induced presynaptic inhibition of IPSCs and EPSCs in rat hypothalamic supraoptic nucleus neurons. *J Physiol.* 1999;520:815–825.
16. Ulrich D, Huguenard JR. Purinergic inhibition of GABA and glutamate release in the thalamus: implications for thalamic network activity. *Neuron.* 1995;15(4):909–918.
17. Lopes LV, Cunha RA, Kull B, Fredholm BB, Ribeiro JA. Adenosine A_{2A} receptor facilitation of hippocampal synaptic transmission is dependent on tonic A_1 receptor inhibition. *Neuroscience.* 2002;112(2):319–329.
18. Panatier A, Vallée J, Haber M, Murai KK, Lacaille JC, Robitaille R. Astrocytes are endogenous regulators of basal transmission at central synapses. *Cell.* 2011;146(5):785–798.
19. Matos M, Augusto E, Santos-Rodrigues AD, et al. Adenosine A_{2A} receptors modulate glutamate uptake in cultured astrocytes and gliosomes. *Glia.* 2012;60(5):702–716.
20. Cunha RA, Ribeiro JA. Purinergic modulation of [^{3}H]GABA release from rat hippocampal nerve terminals. *Neuropharmacology.* 2000;39(7):1156–1167.
21. Rombo DM, Newton K, Nissen W, et al. Synaptic mechanisms of adenosine A_{2A} receptor-mediated hyperexcitability in the hippocampus. *Hippocampus.* 2015;25:566–580.
22. Doischer D, Hosp JA, Yanagawa Y, et al. Postnatal differentiation of basket cells from slow to fast signaling devices. *J Neurosci.* 2008;28:12956–12968.
23. Hefft S, Jonas P. Asynchronous GABA release generates long-lasting inhibition at a hippocampal interneuron-principal neuron synapse. *Nat Neurosci.* 2005;8:1319–1328.
24. Cristóvão-Ferreira S, Vaz SH, Ribeiro JA, Sebastião AM. Adenosine A2A receptors enhance GABA transport into nerve terminals by restraining PKC inhibition of GAT-1. *J Neurochem.* 2009;109(2):336–347.
25. Cristóvão-Ferreira S, Navarro G, Brugarolas M, et al. A_1R-$A_{2A}R$ heteromers coupled to G_s and $G_{i/0}$ proteins modulate GABA transport into astrocytes. *Purinergic Signal.* 2013;9(3):433–449.
26. Sasaki T, Matsuki N, Ikegaya Y. Action-potential modulation during axonal conduction. *Science.* 2011;331(6017):599–601.
27. Rombo DM, Dias RB, Duarte ST, Ribeiro JA, Lamsa KP, Sebastião AM. Adenosine A_1 receptor suppresses Tonic GABAA receptor currents in hippocampal pyramidal cells and in a defined subpopulation of interneurons. *Cereb Cortex.* 2016;26:1081–1095.
28. Semyanov A, Walker MC, Kullmann DM, Silver RA. Tonically active GABA A receptors: modulating gain and maintaining the tone. *Trends Neurosci.* 2004;27:262–269.
29. Dias RB, Ribeiro JA, Sebastião AM. Enhancement of AMPA currents and GluR1 membrane expression through PKA-coupled adenosine A_{2A} receptors. *Hippocampus.* 2012;22(2):276–291.
30. Di Angelantonio S, Bertollini C, Piccinin S, et al. Basal adenosine modulates the functional properties of AMPA receptors in mouse hippocampal neurons through the activation of A_1R $A_{2A}R$ and A_3R. *Front Cell Neurosci.* 2015;9:409.

31. Chen Z, Stockwell J, Cayabyab FS. Adenosine A_1 receptor-mediated endocytosis of AMPA receptors contributes to impairments in long-term potentiation (LTP) in the middle-aged rat hippocampus. *Neurochem Res*. 2016;41:1085–1097.
32. Dias RB, Rombo DM, Ribeiro JA, Sebastião AM. Ischemia-induced synaptic plasticity drives sustained expression of calcium-permeable AMPA receptors in the hippocampus. *Neuropharmacology*. 2013;65:114–122.
33. Whitehead G, Jo J, Hogg EL, et al. Acute stress causes rapid synaptic insertion of Ca2 + -permeable AMPA receptors to facilitate long-term potentiation in the hippocampus. *Brain*. 2013;136:3753–3765.
34. Shepherd JD. Memory, plasticity and sleep - a role for calcium permeable AMPA receptors? *Front Mol Neurosci*. 2012;11(5):49.
35. de Mendonça A, Sebastião AM, Ribeiro JA. Inhibition of NMDA receptor-mediated currents in isolated rat hippocampal neurones by adenosine A1 receptor activation. *Neuroreport*. 1995;6(8):1097–1100.
36. Sebastião AM, de Mendonca A, Moreira T, Ribeiro JA. Activation of synaptic NMDA receptors by action potential-dependent release of transmitter during hypoxia impairs recovery of synaptic transmission on reoxygenation. *J Neurosci*. 2001;21(21):8564–8571.
37. Tebano MT, Martire A, Rebola N, et al. Adenosine A_{2A} receptors and metabotropic glutamate 5 receptors are co-localized and functionally interact in the hippocampus: a possible key mechanism in the modulation of N-methyl-D-aspartate effects. *J Neurochem*. 2005;95(4):1188–1200.
38. Rebola N, Luján R, Cunha RA, Mulle C. Adenosine A_{2A} receptors are essential for long-term potentiation of NMDA-EPSCs at hippocampal mossy fiber synapses. *Neuron*. 2008;57(1):121–134.
39. Sarantis K, Tsiamaki E, Kouvaros S, Papatheodoropoulos C, Angelatou F. Adenosine A_{2A} receptors permit mGluR5-evoked tyrosine phosphorylation of NR2B (Tyr1472) in rat hippocampus: a possible key mechanism in NMDA receptor modulation. *J Neurochem*. 2015;135(4):714–726.
40. Parpura V, Basarsky TA, Liu F, Jeftinija K, Jeftinija S, Haydon PG. Glutamate-mediated astrocyte-neuron signalling. *Nature*. 1994;369:744–747.
41. Araque A, Parpura V, Sanzgiri RP, Haydon PG. Glutamate-dependent astrocyte modulation of synaptic transmission between cultured hippocampal neurons. *Eur J Neurosci*. 1998;10:2129–2142.
42. Pascual O, Casper KB, Kubera C, et al. Astrocytic purinergic signaling coordinates synaptic networks. *Science*. 2005;310(5745):113–116.
43. Deng Q, Terunuma M, Fellin T, Moss SJ, Haydon PG. Astrocytic activation of A1 receptors regulates the surface expression of NMDA receptors through a Src kinase dependent pathway. *Glia*. 2011;59(7):1084–1093.
44. Diógenes MJ, Neves-Tomé R, Fucile S, et al. Homeostatic control of synaptic activity by endogenous adenosine is mediated by adenosine kinase. *Cereb Cortex*. 2014;24(1):67–80.
45. Carlsen EM, Perrier JF. Purines released from astrocytes inhibit excitatory synaptic transmission in the ventral horn of the spinal cord. *Front Neural Circuits*. 2014;4(8):60.
46. Todd KJ, Darabid H, Robitaille R. Perisynaptic glia discriminate patterns of motor nerve activity and influence plasticity at the neuromuscular junction. *J Neurosci*. 2010;30(35):11870–11882.
47. Lee HU, Yamazaki Y, Tanaka KF, et al. Increased astrocytic ATP release results in enhanced excitability of the hippocampus. *Glia*. 2013;61(2):210–224.
48. Jacob JP, Vaz SH, Ribeiro JA, Sebastião AM. P2Y1 receptor inhibits GABA transport through a calcium signalling-dependent mechanism in rat cortical astrocytes. *Glia*. 2014;62:1211–1226.

49. Lu B, Nagappan G, Lu Y. BDNF and synaptic plasticity, cognitive function, and dysfunction. In: Lewin RG, Carter DB, eds. *Neurotrophic Factors*. Berlin, Heidelberg: Springer; 2014:223–250.
50. Rodrigues TM, Jerónimo-Santos A, Sebastião AM, Diógenes MJ. Adenosine A_{2A} receptors as novel upstream regulators of BDNF-mediated attenuation of hippocampal Long-Term Depression (LTD). *Neuropharmacology*. 2014;79:389–398.
51. Diógenes MJ, Fernandes CC, Sebastião AM, Ribeiro JA. Activation of adenosine A_{2A} receptor facilitates brain-derived neurotrophic factor modulation of synaptic transmission in hippocampal slices. *J Neurosci*. 2004;24:2905–2913.
52. Tebano MT, Martire A, Potenza RL, et al. Adenosine A_{2A} receptors are required for normal BDNF levels and BDNF-induced potentiation of synaptic transmission in the mouse hippocampus. *J Neurochem*. 2008;104:279–286.
53. Fernandes CC, Pinto-Duarte A, Ribeiro JA, Sebastião AM. Postsynaptic action of brain-derived neurotrophic factor attenuates alpha7 nicotinic acetylcholine receptor-mediated responses in hippocampal interneurons. *J Neurosci*. 2008;28: 5611–5618.
54. Fontinha BM, Diógenes MJ, Ribeiro JA, Sebastião AM. Enhancement of long-term potentiation by brain-derived neurotrophic factor requires adenosine A_{2A} receptor activation by endogenous adenosine. *Neuropharmacology*. 2008;54:924–933.
55. Pousinha PA, Diógenes MJ, Ribeiro JA, Sebastião AM. Triggering of BDNF facilitatory action on neuromuscular transmission by adenosine A_{2A} receptors. *Neurosci Lett*. 2006;404:143–147.
56. Assaife-Lopes N, Sousa VC, Pereira DB, Ribeiro JA, Sebastião AM. Regulation of TrkB receptor translocation to lipid rafts by adenosine A2A receptors and its functional implications for BDNF-induced regulation of synaptic plasticity. *Purinergic Signal*. 2014;10:251–267.
57. Jerónimo-Santos A, Batalha VL, Müller CE, et al. Impact of in vivo chronic blockade of adenosine A_{2A} receptors on the BDNF-mediated facilitation of LTP. *Neuropharmacology*. 2014;83:99–106.
58. Vaz SH, Jørgensen TN, Cristóvão-Ferreira S, et al. Brain-derived neurotrophic factor (BDNF) enhances GABA transport by modulating the trafficking of GABA transporter-1 (GAT-1) from the plasma membrane of rat cortical astrocytes. *J Biol Chem*. 2011;286:40464–40476.
59. Vaz SH, Cristóvão-Ferreira S, Ribeiro JA, Sebastião AM. Brain-derived neurotrophic factor inhibits GABA uptake by the rat hippocampal nerve terminals. *Brain Res*. 2008;1219:19–25.
60. Abraham WC. Metaplasticity: tuning synapses and networks for plasticity. *Nat Rev Neurosci*. 2008;9:387.
61. Lee MC, Yasuda R, Ehlers MD. Metaplasticity at single glutamatergic synapses. *Neuron*. 2010;66(6):859–870.
62. de Mendonça A, Ribeiro JA. 2-Chloroadenosine decreases long-term potentiation in the hippocampal CA1 area of the rat. *Neurosci Lett*. 1990;118(1):107–111.
63. Arai A, Kessler M, Lynch G. The effects of adenosine on the development of long-term potentiation. *Neurosci Lett*. 1990;119(1):41–44.
64. zur Nedden S, Hawley S, Pentland N, Hardie DG, Doney AS, Frenguelli BG. Intracellular ATP influences synaptic plasticity in area CA1 of rat hippocampus via metabolism to adenosine and activity-dependent activation of adenosine A_1 receptors. *J Neurosci*. 2011;31(16):6221–6234.
65. de Mendonça A, Ribeiro JA. Endogenous adenosine modulates long-term potentiation in the hippocampus. *Neuroscience*. 1994;62(2):385–390.
66. Forghani R, Krnjević K. Adenosine antagonists have differential effects on induction of long-term potentiation in hippocampal slices. *Hippocampus*. 1995;5(1):71–77.

67. de Mendonça A, Almeida T, Bashir ZI, Ribeiro JA. Endogenous adenosine attenuates long-term depression and depotentiation in the CA1 region of the rat hippocampus. *Neuropharmacology*. 1997;36(2):161–167.
68. Kemp N, Bashir ZI. A role for adenosine in the regulation of long-term depression in the adult rat hippocampus in vitro. *Neurosci Lett*. 1997;225(3):189–192.
69. Kessey K, Trommer BL, Overstreet LS, Ji T, Mogul DJ. A role for adenosine A2 receptors in the induction of long-term potentiation in the CA1 region of rat hippocampus. *Brain Res*. 1997;756(1–2):184–190.
70. Blundon JA, Bayazitov IT, Zakharenko SS. Presynaptic gating of postsynaptically expressed plasticity at mature thalamocortical synapses. *J Neurosci*. 2011;31 (44):16012–16025.
71. Shen HY, Singer P, Lytle N, et al. Adenosine augmentation ameliorates psychotic and cognitive endophenotypes of schizophrenia. *J Clin Inv*. 2012;122 (7):2567–2577.
72. Fontinha BM, Delgado-García JM, Madroñal N, Ribeiro JA, Sebastião AM, Gruart A. Adenosine A_{2A} receptor modulation of hippocampal CA3-CA1 synapse plasticity during associative learning in behaving mice. *Neuropharmacology*. 2009;34(7):1865–1874.
73. Batalha VL, Pego JM, Fontinha BM, et al. Adenosine A_{2A} receptor blockade reverts hippocampal stress-induced deficits and restores corticosterone circadian oscillation. *Mol Psychiatry*. 2013;18:320–331.
74. Chen CC, Yang CH, Huang CC, Hsu KS. Acute stress impairs hippocampal mossy fiber-CA3 long-term potentiation by enhancing cAMP-specific phosphodiesterase 4 activity. *Neuropsychopharmacology*. 2010;35(7):1605–1617.
75. Lu G, Zhou QX, Kang S, et al. Chronic morphine treatment impaired hippocampal long-term potentiation and spatial memory via accumulation of extracellular adenosine acting on adenosine A_1 receptors. *J Neurosci*. 2010;30(14):5058–5070.
76. Florian C, Vecsey CG, Halassa MM, Haydon PG, Abel T. Astrocyte-derived adenosine and A_1 receptor activity contribute to sleep loss-induced deficits in hippocampal synaptic plasticity and memory in mice. *J Neurosci*. 2011;31(19):6956–6962.
77. Dolphin AC. The adenosine agonist 2-chloroadenosine inhibits the induction of long-term potentiation of the perforant path. *Neurosci Lett*. 1983;39(1):83–89.
78. Li W, Silva HB, Real J, et al. Inactivation of adenosine A_{2A} receptors reverses working memory deficits at early stages of Huntington's disease models. *Neurobiol Dis*. 2015;79:70–80.
79. Rex CS, Kramár EA, Colgin LL, Lin B, Gall CM, Lynch G. Long-term potentiation is impaired in middle-aged rats: regional specificity and reversal by adenosine receptor antagonists. *J Neurosci*. 2005;25:5956–5966.
80. Costenla AR, de Mendonça A, Ribeiro JA. Adenosine modulates synaptic plasticity in hippocampal slices from aged rats. *Brain Res*. 1999;985:228–234.
81. Costenla AR, Diógenes MJ, Canas PM, et al. Enhanced role of adenosine A_{2A} receptors in the modulation of LTP in the rat hippocampus upon ageing. *Eur J Neurosci*. 2011;34:12–21.
82. Diógenes MJ, Costenla AR, Lopes LV, et al. Enhancement of LTP in aged rats is dependent on endogenous BDNF. *Neuropsychopharmacology*. 2011;36:1823–1836.
83. Zhu G, Liu Y, Wang Y, Bi X, Baudry M. Different patterns of electrical activity lead to long-term potentiation by activating different intracellular pathways. *J Neurosci*. 2015;35:621–633.
84. Ribiero FF, Neves-Tomé R, Assaife-Lopes N, et al. Axonal elongation and dendritic branching is enhanced by adenosine A_{2A} receptors activation in cerebral cortical neurons. *Brain Struct Funct*. 2016;221:2777–2799.
85. Sousa VC, Assaife-Lopes N, Ribeiro JA, Pratt JA, Brett RR, Sebastião AM. Regulation of hippocampal cannabinoid CB1 receptor actions by adenosine A_1 receptors and chronic caffeine administration: implications for the effects of Δ9-tetrahydrocannabinol on spatial memory. *Neuropsychopharmacology*. 2011;36(2):472–487.

86. Trusel M, Cavaccini A, Gritti M, et al. Coordinated regulation of synaptic plasticity at striatopallidal and striatonigral neurons orchestrates motor control. *Cell Reports*. 2015;13(7):1353–1365.
87. Halassa MM, Florian C, Fellin T, et al. Astrocytic modulation of sleep homeostasis and cognitive consequences of sleep loss. *Neuron*. 2009;61(2):213–219.
88. Seifert G, Carmignoto G, Steinhauser C. Astrocyte dysfunction in epilepsy. *Brain Res Rev*. 2010;63(1–2):212–221.
89. Vezzani A, Ravizza T, Balosso S, Aronica E. Glia as a source of cytokines: implications for neuronal excitability and survival. *Epilepsia*. 2008;49(suppl 2):24–32.
90. Ravizza T, Gagliardi B, Noe F, Boer K, Aronica E, Vezzani A. Innate and adaptive immunity during epileptogenesis and spontaneous seizures: evidence from experimental models and human temporal lobe epilepsy. *Neurobiol Dis*. 2008;29(1):142–160.
91. Pan JW, Kim JH, Cohen-Gadol A, Pan C, Spencer DD, Hetherington HP. Regional energetic dysfunction in hippocampal epilepsy. *Acta Neurol Scand*. 2005;111(4):218–224.
92. Williamson A, Patrylo PR, Pan J, Spencer DD, Hetherington H. Correlations between granule cell physiology and bioenergetics in human temporal lobe epilepsy. *Brain*. 2005;128(Pt 5):1199–1208.
93. Boison D, Masino MA, Geiger JD. Homeostatic bioenergetic network regulation – a novel concept to avoid pharmacoresistance in epilepsy. *Expert Opin Drug Discov*. 2011;6(7):713–724.
94. Qureshi IA, Mehler MF. Epigenetic mechanisms underlying human epileptic disorders and the process of epileptogenesis. *Neurobiol Dis*. 2010;39(1):53–60.
95. Swann JW, Rho JM. How is homeostatic plasticity important in epilepsy? *Adv Exp Med Biol*. 2014;813:123–131.
96. Dudek FE, Staley KJ. The time course of acquired epilepsy: implications for therapeutic intervention to suppress epileptogenesis. *Neurosci Lett*. 2011;497(3):240–246.
97. Graef JD, Godwin DW. Intrinsic plasticity in acquired epilepsy: too much of a good thing? *Neuroscientist*. 2010;16(5):487–495.
98. Ben-Ari Y, Dudek FE. Primary and secondary mechanisms of epileptogenesis in the temporal lobe: there is a before and an after. *Epilepsy Curr*. 2010;10(5):118–125.
99. Bender RA, Dube C, Baram TZ. Febrile seizures and mechanisms of epileptogenesis: insights from an animal model. *Adv Exp Med Biol*. 2004;548:213–225.
100. van Vliet EA, Aronica E, Gorter JA. Role of blood-brain barrier in temporal lobe epilepsy and pharmacoresistance. *Neuroscience*. 2014;277:455–473.
101. Pitkanen A, Engel Jr J. Past and present definitions of epileptogenesis and its biomarkers. *Neurotherapeutics*. 2014;11(2):231–241.
102. Vezzani A, Aronica E, Mazarati A, Pittman QJ. Epilepsy and brain inflammation. *Exp Neurol*. 2013;244:11–21.
103. Henshall DC, Kobow K. Epigenetics and epilepsy. *Cold Spring Harb Perspect Med*. 2015;5(12):pii:a022731.
104. Kobow K, Blumcke I. The methylation hypothesis: do epigenetic chromatin modifications play a role in epileptogenesis? *Epilepsia*. 2011;52(suppl 4):15–19.
105. Scharfman HE, McCloskey DP. Postnatal neurogenesis as a therapeutic target in temporal lobe epilepsy. *Epilepsy Res*. 2009;85(2–3):150–161.
106. Siebzehnrubl FA, Blumcke I. Neurogenesis in the human hippocampus and its relevance to temporal lobe epilepsies. *Epilepsia*. 2008;49:55–65.
107. Sanabria Ydel C, Arganaraz GA, Lima E, et al. Neurogenesis induced by seizures in the dentate gyrus is not related to mossy fiber sprouting but is age dependent in developing rats. *Arq Neuropsiquiatr*. 2008;66(4):853–860.
108. Parent JM, Murphy GG. Mechanisms and functional significance of aberrant seizure-induced hippocampal neurogenesis. *Epilepsia*. 2008;49(suppl 5):19–25.

109. Parent JM. Persistent hippocampal neurogenesis and epilepsy. *Epilepsia.* 2008;49 (suppl 5):1–2.
110. Blumcke I, Thom M, Aronica E, et al. International consensus classification of hippocampal sclerosis in temporal lobe epilepsy: a Task Force report from the ILAE Commission on Diagnostic Methods. *Epilepsia.* 2013;54(7):1315–1329.
111. Thomas EA, Reid CA, Petrou S. Mossy fiber sprouting interacts with sodium channel mutations to increase dentate gyrus excitability. *Epilepsia.* 2010;51 (1):136–145.
112. Buckmaster PS, Ingram EA, Wen X. Inhibition of the mammalian target of rapamycin signaling pathway suppresses dentate granule cell axon sprouting in a rodent model of temporal lobe epilepsy. *J Neurosci.* 2009;29(25):8259–8269.
113. Dudek FE, Sutula TP. Epileptogenesis in the dentate gyrus: a critical perspective. *Prog Brain Res.* 2007;163:755–773.
114. Seifert G, Steinhauser C. Neuron-astrocyte signaling and epilepsy. *Exp Neurol.* 2013;244:4–10.
115. Devinsky O, Vezzani A, Najjar S, De Lanerolle NC, Rogawski MA. Glia and epilepsy: excitability and inflammation. *Trends Neurosci.* 2013;36(3):174–184.
116. Aronica E, Sandau US, Iyer A, Boison D. Glial adenosine kinase – a neuropathological marker of the epileptic brain. *Neurochem Int.* 2013;63:688–695.
117. Malmgren K, Thom M. Hippocampal sclerosis–origins and imaging. *Epilepsia.* 2012;53(suppl 4):19–33.
118. Haydon PG, Carmignoto G. Astrocyte control of synaptic transmission and neurovascular coupling. *Physiol Rev.* 2006;86(3):1009–1031.
119. Coulter DA, Eid T. Astrocytic regulation of glutamate homeostasis in epilepsy. *Glia.* 2012;60(8):1215–1226.
120. Boison D. Adenosine dysfunction in epilepsy. *Glia.* 2012;60(8):1234–1243.
121. Tian GF, Azmi H, Takano T, et al. An astrocytic basis of epilepsy. *Nat Med.* 2005;11 (9):973–981.
122. Hass HL, Selbach O. Functions of neuronal adenosine receptors. *Naunyn-Schmiedeberg's Arch Pharmacol.* 2000;362:375–381.
123. Fredholm BB, Ijzerman AP, Jacobson KA, Klotz K, Linden J. International Union of Pharmacology. XXV. Nomenclature and classification of adenosine receptors. *Pharmacol Rev.* 2001;53(4):527–552.
124. Cunha RA. Neuroprotection by adenosine in the brain: from A_1 receptor activation to A_{2A} receptor blockade. *Purinergic Signal.* 2005;1(2):111–134.
125. Boison D. Cell and gene therapies for refractory epilepsy. *Curr Neuropharmacol.* 2007;5(2):115–125.
126. Gouder N, Fritschy J, Boison D. Seizure suppression by adenosine A_1 receptor activation in a mouse model of pharmacoresistant epilepsy. *Epilepsia.* 2003;44 (7):877–885.
127. Kochanek PM, Vagni VA, Janesko KL, et al. Adenosine A1 receptor knockout mice develop lethal status epilepticus after experimental traumatic brain injury. *J Cereb Blood Flow Metab.* 2006;26:565–575.
128. Fedele DE, Li T, Lan JQ, Fredholm BB, Boison D. Adenosine A_1 receptors are crucial in keeping an epileptic focus localized. *Exp Neurol.* 2006;200(1):184–190.
129. Boison D. Adenosine augmentation therapy for epilepsy. In: Noebels JL, Avoli M, Rogawski MA, Olsen RW, Delgado-Escueta AV, eds. *Jasper's Basic Mechanisms of the Epilepsies.* 4th ed. Oxford: Oxford University Press; 2012:1150–1160.
130. Rebola N, Coelho JE, Costenla AR, et al. Decrease of adenosine A_1 receptor density and of adenosine neuromodulation in the hippocampus of kindled rats. *Eur J Neurosci.* 2003;18(4):820–828.

131. Li T, Lan JQ, Fredholm BB, Simon RP, Boison D. Adenosine dysfunction in astrogliosis: cause for seizure generation? *Neuron Glia Biol.* 2007;3:353–366.
132. Glass M, Faull RL, Bullock JY, et al. Loss of A1 adenosine receptors in human temporal lobe epilepsy. *Brain Res.* 1996;710(1–2):56–68.
133. Hamil NE, Cock HR, Walker MC. Acute down-regulation of adenosine A(1) receptor activity in status epilepticus. *Epilepsia.* 2012;53(1):177–188.
134. Hargus NJ, Jennings C, Perez-Reyes E, Bertram EH, Patel MK. Enhanced actions of adenosine in medial entorhinal cortex layer II stellate neurons in temporal lobe epilepsy are mediated via A(1)-receptor activation. *Epilepsia.* 2012;53(1):168–176.
135. Wagner AK, Miller MA, Scanlon J, Ren D, Kochanek PM, Conley YP. Adenosine A1 receptor gene variants associated with post-traumatic seizures after severe TBI. *Epilepsy Res.* 2010;90(3):259–272.
136. Matos M, Augusto E, Machado NJ, dos Santos-Rodrigues A, Cunha RA, Agostinho P. Astrocytic adenosine A2A receptors control the amyloid-beta peptide-induced decrease of glutamate uptake. *J Alzheimers Dis.* 2012;31(3):555–567.
137. Popoli P, Frank C, Tebano MT, et al. Modulation of glutamate release and excitotoxicity by adenosine A2A receptors. *Neurology.* 2003;61(11 suppl 6):S69–S71.
138. Silva CG, Porciuncula LO, Canas PM, Oliveira CR, Cunha RA. Blockade of adenosine A(2A) receptors prevents staurosporine-induced apoptosis of rat hippocampal neurons. *Neurobiol Dis.* 2007;27(2):182–189.
139. Lovatt D, Xu Q, Liu W, et al. Neuronal adenosine release, and not astrocytic ATP release, mediates feedback inhibition of excitatory activity. *Proc Natl Acad Sci U S A.* 2012;109(16):6265–6270.
140. Shinohara M, Saitoh M, Nishizawa D, et al. ADORA2A polymorphism predisposes children to encephalopathy with febrile status epilepticus. *Neurology.* 2013;80(17):1571–1576.
141. D'Alimonte I, D'Auro M, Citraro R, et al. Altered distribution and function of A2A adenosine receptors in the brain of WAG/Rij rats with genetic absence epilepsy, before and after appearance of the disease. *Eur J Neurosci.* 2009;30(6):1023–1035.
142. Cunha RA, Vizi ES, Ribeiro JA, Sebastião AM. Preferential release of ATP and its extracellular catabolism as a source of adenosine upon high- but not low-frequency stimulation of rat hippocampal slices. *J Neurochem.* 1996;67(5):2180–2187.
143. Cunha RA, Correia-de-Sa P, Sebastião AM, Ribeiro JA. Preferential activation of excitatory adenosine receptors at rat hippocampal and neuromuscular synapses by adenosine formed from released adenine nucleotides. *Br J Pharmacol.* 1996;119(2):253–260.
144. Ciruela F, Casadó V, Rodrigues RJ, et al. Presynaptic control of striatal glutamatergic neurotransmission by adenosine A_1-A_{2A} receptor heteromers. *J Neurosci.* 2006;26(7):2080–2087.
145. Lopes LV, Cunha RA, Ribeiro JA. Cross talk between A_1 and A_{2A} adenosine receptors in the hippocampus and cortex of young adult and old rats. *J Neurophysiol.* 1999;82(6):3196–3203.
146. Cunha RA, Johansson B, van der Ploeg I, Sebastião AM, Ribeiro JA, Fredholm BB. Evidence for functionally important adenosine A_{2A} receptors in the rat hippocampus. *Brain Res.* 1994;649(1–2):208–216.
147. Ekonomou A, Sperk G, Kostopoulos G, Angelatou F. Reduction of A1 adenosine receptors in rat hippocampus after kainic acid-induced limbic seizures. *Neurosci Lett.* 2000;284(1–2):49–52.
148. Roseti C, Martinello K, Fucile S, et al. Adenosine receptor antagonists alter the stability of human epileptic $GABA_A$ receptors. *Proc Natl Acad Sci U S A.* 2008;105(39):15118–15123.

149. Williams-Karnesky RL, Sandau US, Lusardi TA, et al. Epigenetic changes induced by adenosine augmentation therapy prevent epileptogenesis. *J Clin Invest*. 2013;123 (8):3552–3563.
150. Mato JM, Martínez-Chantar ML, Lu SC. Methionine metabolism and liver disease. *Annu Rev Nutr*. 2008;28:273–293.
151. James SJ, Melnyk S, Pogribna M, Pogribny IP, Caudill MA. Elevation in S-Adenosylhomocysteine and DNA hypomethylation: potential epigenetic mechanism for homocysteine-related pathology. *J Nutr*. 2002;132(8):2361S–2366S.
152. Mandaviya PR, Stolk L, Heil SG. Homocysteine and DNA methylation: a review of animal and human literature. *Mol Genet Metab*. 2014;113(4):243–252.
153. Kobow K, Kaspi A, Harikrishnan KN, et al. Deep sequencing reveals increased DNA methylation in chronic rat epilepsy. *Acta Neuropathol*. 2013;126(5):741–756.
154. Boison D. Adenosinergic signaling in epilepsy. *Neuropharmacology*. 2016;104:131–139.
155. Boison D. Astrogliosis and adenosine kinase: a glial basis of epilepsy. *Future Neurol*. 2008;3(3):221–224.
156. Boison D. The biochemistry and epigenetics of epilepsy: focus on adenosine and glycine. *Front Mol Neurosci*. 2016;9:26.
157. Diamond ML, Ritter AC, Jackson EK, et al. Genetic variation in the adenosine regulatory cycle is associated with posttraumatic epilepsy development. *Epilepsia*. 2015;56 (8):1198–1206.
158. Peng L, Huang R, Yu AC, Fung KY, Rathbone MP, Hertz L. Nucleoside transporter expression and function in cultured mouse astrocytes. *Glia*. 2005;52(1):25–35.
159. Gouder N, Scheurer L, Fritschy J, Boison D. Overexpression of adenosine kinase in epileptic hippocampus contributes to epileptogenesis. *J Neurosci*. 2004;24(3):692–701.
160. Pignataro G, Meller R, Inoue K, et al. In vivo and in vitro characterization of a novel neuroprotective strategy for stroke: ischemic postconditioning. *J Cereb Blood Flow Metab*. 2008;28(2):232–241.
161. Clark RS, Chen J, Watkins SC, et al. Apoptosis-suppressor gene *bcl-2* expression after traumatic brain injury in rats. *J Neurosci*. 1997;17(23):9172–9182.
162. During MJ, Spencer DD. Adenosine: a potential mediator of seizure arrest and postictal refractoriness. *Ann Neurol*. 2004;32(5):618–624.
163. Vezzani A, French J, Bartfai T, Baram TZ. The role of inflammation in epilepsy. *Nat Rev Neurol*. 2011;7:31–40.
164. Nabbout R, Vezzani A, Dulac O, Chiron C. Acute encephalopathy with inflammation-mediated status epilepticus. *Lancet Neurol*. 2011;10:99–108.
165. Li T, Lan JQ, Boison D. Uncoupling of astrogliosis from epileptogenesis in adenosine kinase (ADK) transgenic mice. *Neuron Glia Biol*. 2008;4(2):91–99.
166. Li T, Lytle NK, Lan JQ, Sandau US, Boison D. Local disruption of glial adenosine homeostasis in mice associates with focal electrographic seizures: a first step in epileptogenesis? *Glia*. 2012;60(1):83–95.
167. Orr AG, Hsiao EC, Wang MM, et al. Astrocytic adenosine receptor A_{2A} and Gs-coupled signaling regulate memory. *Nat Neurosci*. 2015;18:423–434.
168. Ke RH, Xiong J, Liu Y, Ye ZR. Adenosine A2A receptor induced gliosis via Akt/NF-kappaB pathway in vitro. *Neurosci Res*. 2009;65:280–285.
169. Brambilla R, Cottini L, Fumagalli M, Ceruti S, Abbracchio MP. Blockade of A_{2A} adenosine receptors prevents basic fibroblast growth factor-induced reactive astrogliosis in rat striatal primary astrocytes. *Glia*. 2003;43:190–194.
170. Gaitatzis A, Sisodiya SM, Sander JW. The somatic comorbidity of epilepsy: a weighty but often unrecognized burden. *Epilepsia*. 2012;53(8):1282–1293.
171. Krishnamoorthy ES, Trimble MR, Blumer D. The classification of neuropsychiatric disorders in epilepsy: a proposal by the ILAE Commission on Psychobiology of Epilepsy. *Epilepsy Behavior*. 2007;10(3):349–353.

172. Gaitatzis A, Trimble MR, Sander JW. The psychiatric comorbidity of epilepsy. *Acta Neurol Scand*. 2004;110(4):207–220.
173. LaFrance Jr WC, Kanner AM, Hermann B. Chapter 20 psychiatric comorbities in epilepsy. *Int Rev Neurobiol*. 2008;83:347–383.
174. van Iterson L, de Jong PF, Zijlstra BJH. Pediatric epilepsy and comorbid reading disorders, math disorders, or autism spectrum disorders: Impact of epilepsy on cognitive patterns. *Epilepsy Behavior*. 2015;44:159–168.
175. Boison D, Singer P, Shen HY, Feldon J, Yee BK. Adenosine hypothesis of schizophrenia - opportunities for pharmacotherapy. *Neuropharmacology*. 2012;62: 1527–1543.
176. Wei CJ, Singer P, Boison D, Yee BK, Chen JF. Selective inactivation of adenosine A2A receptors in striatal neurons enhances working memory and reversal learning. *Learn Mem*. 2011;18:459–474.
177. Yee BK, Singer P, Chen JF, Feldon J, Boison D. Transgenic overexpression of adenosine kinase in brain leads to multiple learning impairments and altered sensitivity to psychomimetic drugs. *Eur J Neurosci*. 2007;26:3237–3252.
178. Shen HY, Lusardi TA, Williams-Karnesky RL, Lan JQ, Poulsen DJ, Boison D. Adenosine kinase determines the degree of brain injury after ischemic stroke in mice. *J Cereb Blood Flow Metab*. 2011;31(7):1648–1659:PMCID: PMC3137468.
179. Dunwiddie TV, Masino SA. The role and regulation of adenosine in the central nervous system. *Annu Rev Neurosci*. 2001;24:31–55.
180. Monopoli A, Conti A, Zocchi C, et al. Pharmacology of the new selective A2a adenosine receptor agonist 2-hexynyl-5′-N-ethylcarboxamidoadenosine. *Arzneimittelforschung*. 1994;44(12):1296–1304.
181. Wardas J, Konieczny J, Pietraszek M. Influence of CGS 21680, a selective adenosine A_{2A} agonist, on the phencyclidine-induced sensorimotor gating deficit and motor behaviour in rats. *Psychopharmacology (Berl)*. 2003;168(3):299–306.
182. Minor TR, Winslow JL, Chang W. Stress and adenosine: II. Adenosine analogs mimic the effect of inescapable shock. *Behav Neurosci*. 1994;108(2):265–276.
183. Mingote S, Font L, Farrar AM, et al. Nucleus accumbens adenosine A_{2A} receptors regulate exertion of effort by acting on the ventral striatopallidal pathway. *J Neurosci*. 2008;28(36):9037–9046.
184. Etherington LV, Patterson GE, Meechan L, et al. Astrocytic adenosine kinase regulates basal synaptic adenosine levels and seizure activity but not activity-dependent adenosine release in the hippocampus. *Neuropharmacology*. 2009;56 (2):429–437.
185. Szybala C, Pritchard EM, Wilz A, Kaplan DL, Boison D. Antiepileptic effects of silk-polymer based adenosine release in kindled rats. *Exp Neurol*. 2009;219:126–135.
186. Li T, Ren G, Lusardi T, et al. Adenosine kinase is a target for the prediction and prevention of epileptogenesis in mice. *J Clin Inv*. 2008;118(2):571–582.
187. Li T, Steinbeck JA, Lusardi T, et al. Suppression of kindling epileptogenesis by adenosine releasing stem cell-derived brain implants. *Brain*. 2007;130(Pt 5):1276–1288.
188. Theofilas P, Brar S, Stewart K-A, et al. Adenosine kinase as a target for therapeutic antisense strategies in epilepsy. *Epilepsia*. 2011;52(3):589–601:PMCID: PMC3075862.
189. Masino SA, Li T, Theofilas P, et al. A ketogenic diet suppresses seizures in mice through adenosine A1 receptors. *J Clin Inv*. 2011;121(7):2679–2683.
190. Neal EG, Chaffe H, Schwartz RH, et al. The ketogenic diet for the treatment of childhood epilepsy: a randomised controlled trial. *Lancet Neurol*. 2008;7(6):500–506.
191. Kossoff EH, Rho JM. Ketogenic diets: evidence for short- and long-term efficacy. *Neurotherapeutics*. 2009;6(2):406–414.
192. Masino SA, Kawamura M, Ruskin DN, Geiger JD, Boison D. Purines and neuronal excitability: links to the ketogenic diet. *Epilepsy Res*. 2012;100:229–238.

193. Masino SA, Kawamura M, Wasser CA, Pomeroy LT, Ruskin DN. Adenosine, ketogenic diet and epilepsy: the emerging therapeutic relationship between metabolism and brain activity. *Curr Neuropharmacol*. 2009;7(3):257–268.
194. Masino SA, Geiger JD. The ketogenic diet and epilepsy: is adenosine the missing link? *Epilepsia*. 2009;50(2):332–333.
195. Masino SA, Geiger JD. Are purines mediators of the anticonvulsant/neuroprotective effects of ketogenic diets? *Trends Neurosci*. 2008;31(6):273–278.
196. Masino SA, Li T, Theofilas P, et al. A ketogenic diet suppresses seizures in mice through adenosine A_1 receptors. *J Clin Invest*. 2011;121(7):2679–2683.
197. Williams-Karnesky RL, Sandau US, Lusardi TA, et al. Epigenetic changes induced by adenosine augmentation therapy prevent epileptogenesis. *J Clin Inv*. 2013;123 (8):3552–3563.
198. Chiari A, Yaksh TL, Myers RR, et al. Preclinical toxicity screening of intrathecal adenosine in rats and dogs. *Anesthesiology*. 1999;91(3):824–832.
199. Eisenach JC, Hood DD, Curry R. Preliminary efficacy assessment of intrathecal injection of an American formulation of adenosine in humans. *Anesthesiology*. 2002;96 (1):29–34.
200. Eisenach JC, Hood DD, Curry R. Phase I safety assessment of intrathecal injection of an American formulation of adenosine in humans. *Anesthesiology*. 2002;96(1):24–28.
201. d'Alcantara P, Ledent C, Swillens S, Schiffmann SN. Inactivation of adenosine A2A receptor impairs long term potentiation in the accumbens nucleus without altering basal synaptic transmission. *Neuroscience*. 2001;107:455–464.
202. Kamikubo Y, Shimomura T, Fujita Y, Tabata T, Kashiyama T, Sakurai T, et al. Functional cooperation of metabotropic adenosine and glutamate receptors regulates postsynaptic plasticity in the cerebellum. *J Neurosci*. 2013;33:18661–18671.
203. Xia J, Chen F, Ye J, Yan J, Wang H, Duan S, et al. Activity-dependent release of adenosine inhibits the glutamatergic synaptic transmission and plasticity in the hypothalamic hypocretin/orexin neurons. *Neuroscience*. 2009;162:980–988.
204. Marquez-Ruiz J, Leal-Campanario R, Sanchez-Campusano R, Molaee-Ardekani B, Wendling F, Miranda PC, et al. Transcranial direct-current stimulation modulates synaptic mechanisms involved in associative learning in behaving rabbits. *Proc Natl Acad Sci U S A*. 2012;109:6710–6715.
205. Tavares Gomes AL, Maia FB, Oliveira-Silva P, Marques Ventura AL, Paes-De-Carvalho R, Serfaty CA, et al. Purinergic modulation in the development of the rat uncrossed retinotectal pathway. *Neuroscience*. 2009;163:1061–1068.

CHAPTER

4

Regulation of Synaptic Transmission by Adenosine at the Neuromuscular Junction

Paula Pousinha[1] *and Joaquim A. Ribeiro*[2]

[1]Institute of Molecular and Cellular Pharmacology, CNRS UMR 7275, France [2]University of Lisbon, Lisbon, Portugal

OUTLINE

Adenosine Receptors in Neurodegenerative Diseases.
DOI: http://dx.doi.org/10.1016/B978-0-12-803724-9.00004-1

INTRODUCTION

Most of the knowledge about chemical synapses derives from the functional contact (synapse) between a motor neuron terminal and a muscle fiber; the neuromuscular junction. It is probably the cholinergic synapse (the neurotransmitter is acetylcholine—ACh) that is the most studied, and it has been an inspiration in the studies of all other excitatory synapses in the brain. According to numerous studies, not only ACh is released at the vertebrate neuromuscular synapse, but also other synaptically active molecules modulating the action of the primary neurotransmitter. Adenosine is one of these modulators. In this review we propose to mention the main events and actors involved in the regulation of transmission by adenosine at the neuromuscular junction.

NEUROMUSCULAR JUNCTION

Structural Organization of the Neuromuscular Junction

Motor neurons in the ventral region of the spinal cord send axons out toward the periphery, reaching the muscle fibers, where a specialized synapse is originated; the neuromuscular junction. It transmits electrical impulses from the nerve terminal to the skeletal muscle via a chemical neurotransmitter, ACh. In mammals and many higher vertebrates, each muscle fiber typically has a single synaptic site innervated by a single motor axon branch. Because of the metabolism, physiology, and gene expression of muscle, fibers from different muscles are highly diverse. The size and shape of neuromuscular junctions also varies significantly.[1] The neuromuscular junction has three core structural and functional components, which enclose organelles and molecules specifically found or preferentially expressed in the synaptic area that contribute to the functioning of the neuromuscular junction. Namely: (1) the motor nerve terminal, presynaptic or prejunctional region, where synthesis, storage, and release of ACh occurs, (2) the synaptic or junctional cleft or intrasynaptic region (gap between the nerve and muscle) composed by a structural and functional matrix that promotes synaptic membranes adhesion, and where ACh is either diffused to the postsynaptic region or hydrolyzed, and (3) the postsynaptic or postjunctional region or endplate (muscle membrane), where ACh binds to specific ACh receptors, opening voltage gated sodium channels and generating a depolarization at the postsynaptic muscle fiber (e.g., Refs. 1–3). All three components enclose organelles and molecules specifically found or preferentially expressed in the synaptic area that contribute to the functioning of the neuromuscular junction.

Neuromuscular Transmission

ACh is the neurotransmitter used by motor neurons of the spinal cord and therefore is released at all vertebrate neuromuscular junctions. This neurotransmitter is synthesized in nerve terminals from acetyl coenzyme A (acetyl CoA, which is synthesized from glucose) and choline (derived from the diet), in a reaction catalyzed by choline acetyl transferase (CAT). The postsynaptic action of ACh at the neuromuscular junction is not terminated by reuptake but by a powerful hydrolytic enzyme, acetylcholinesterase (AChE). This enzyme is concentrated in the synaptic cleft, ensuring a rapid decrease in ACh concentration after its release from the presynaptic terminal.[4] The rapid hydrolysis of ACh limits the duration of the action of transmitter and also provides a local source of choline for reutilization within the nerve terminal. Two classes of cholinergic receptors have been identified: ionotropic receptors—nicotinic ACh receptors; and metabotropic receptors—muscarinic ACh receptors. All nicotinic ACh receptors produce excitatory postsynaptic effects because they are nonselective cation channels, giving rise to currents with a reversal potential close to 0 mV.[4] Muscarinic ACh receptors are metabotropic receptors and mediate most of the effects of ACh in nervous system.

The transmission of signals from nerve to muscle is an extremely reliable process. This is mainly because more transmitter is released from the motor nerve terminal by each nerve impulse than is required to excite the muscle fiber. The term "safety factor for neuromuscular transmission" is used to describe this excess. The release of transmitter is triggered by an increase in the concentration of free Ca^{2+} within the nerve terminal. Each vesicle appears to contain 5000–10,000 molecules of ACh.[5] The ACh contained in a single vesicle is often referred to as a 'quantum' of transmitter, though its amount and effect on the postsynaptic membrane shows considerable statistical variation. One of the Katz's seminal findings, in studies carried out with Paul Fatt,[6] was that spontaneous changes in the muscle cell membrane potential occur even in the absence of stimulation of the presynaptic motor neuron. Moreover, Del Castillo and Stark[7] reported that a reduction in calcium concentration has little effect on the spontaneous discharge. Spontaneous exocytosis of the content of individual vesicles occurs at most motor nerve terminals at rates of from several per minute to several per second. These give rise to small depolarizations (<1 mV) of the muscle fiber membrane, the miniature endplate potentials (MEPPs). Presynaptic action potential causes a synchronized release of many transmitter quanta, producing an endplate potential (EPP) on the postjunctional cell. The magnitude of the EPP provides a convenient electrical assay of neurotransmitter secretion from a motor neuron terminal because it is made up of individual units.[8] The number of quanta released by a nerve impulse at a given neuromuscular junction is known as the 'quantal content' of the EPP. During continued stimulation of a motor nerve, the quantal content of an EPP is not maintained unless the frequency is less than 1 Hz. At higher frequencies there is an initial fall, which is steepest during the first few impulses, after which EPP size is fairly well maintained at a level that is frequency-dependent: the higher the frequency the lower the quantal content.[9]

Coexistence of Transmitters and Neuromodulators

In the neurobiology field, the Dale's principle has dominated for decades: one neuron synthesizes, stores, and releases a single transmitter liberated from all axon terminals. In this regard, ACh was considered the only neurotransmitter at the neuromuscular synapse. The suggestion that some nerve cells store and release more than one transmitter was made in 1976,[10] largely on the basis of comparative studies of the evolution of the autonomic nervous system[11] and on the evidence for the coexistence of biologically active substances in certain invertebrate nerves.[12] Since then, considerable data has been accumulated in support of this possibility, being the modern theory of cotransmission accepted in the early 1990s.[13,14] According to this theory, one or several types of synaptically active molecules—cotransmitters—are released from the neuron together with the basic mediator. These cotransmitters are capable of exerting their own effects in the target cell, regulating the release of the primary neurotransmitter (presynaptic modulation), or modulating the physiological response in the postsynaptic cell (postsynaptic modulation). At present, it can be stated that the phenomenon of corelease of several neurotransmitters is the rule rather than the exception for the entire nervous system, including the peripheral part (see Ref. 15). Some signaling molecules that do not meet the definition of "cotransmitters" are involved in the functioning of the synapse apparatus too. They are released from either the presynaptic neuronal terminal, but independently of the primary neurotransmitter, or they have a glial origin, or they are released from the postsynaptic cell and, along with cotransmitters, exert their modulation on synaptic transmission. Adenosine is one of these neuromodulators.

ADENOSINE AS A NEUROMODULATOR AT THE NEUROMUSCULAR JUNCTION

Purines and purine nucleotides are essential constituents of all living cells (see Refs. 16–18 for review), adenosine is neither stored nor released as a classical neurotransmitter as it does not accumulate in synaptic vesicles, being released from the cytoplasm into the extracellular space through a nucleoside transporter. The first observations on the relevance of adenosine in the nervous system were made in the 1970s, when it was reported that adenosine decreased the release of ACh from motor nerve terminals[19,20] and affected cAMP accumulation in nerve cells.[21] Several studies have been conducted since then, which confirm and further detail the action of adenosine as neuromodulator of almost every neurotransmitter in the nervous system (see Refs. 18,22,23 for review). Purinergic function in the peripheral nervous system (PNS) and at the neuromuscular junction indicates roles during development,[24] at mature neuromuscular junctions,[20,25–27] in PNS myelination,[28] and in neuron–glia interactions.[26,29]

Pathways: Synthesis and Degradation of Adenosine at the Neuromuscular Junction

Adenosine is at a crossroads between several key metabolic pathways in cellular metabolism, it is a product of ATP degradation, and is also a substrate for its synthesis (see Ref. 30). Inside the cells, adenosine is formed by the action of endo-5′-nucleotidase over AMP[31] and also by hydrolysis of *S*-adenosylhomocysteine (SAM),[32] reaction catalyzed by SAM hydrolase. Judging by the kinetic constants of some of the major enzymes involved, the intracellular concentration of adenosine at equilibrium is probably around 100 nM.[33] This level is controlled by the rapid phosphorylation of adenosine by adenosine kinase and also by its conversion to inosine by adenosine deaminase.[34,35]

In the case of the neuromuscular junction, it is generally accepted that ATP is stored together with ACh in cholinergic synaptic vesicles.[36] ATP can be coreleased with ACh from motor nerve terminals[37] or released independently of ACh secretion.[38,39] ATP and ACh are released from rat phrenic-nerve diaphragm muscle preparations stimulated at physiological frequencies; ATP is rapidly hydrolyzed to ADP.[37] Because large amounts of ATP are found in cholinergic synaptic vesicles,[40,41] they are thought to be the source of adenosine. When ATPase is inhibited in preparations of Torpedo synaptosomes, the ACh/ATP released is almost identical to that in the synaptic vesicles,[42] which is an important evidence for vesicular release. When ATP is released at the neuromuscular junction, it is degraded into ADP, AMP, IMP, adenosine, inosine, and hypoxantine[43] through the ecto-nucleotidase pathway.[44]

The concentration of extracellular endogenous adenosine is around 120–200 nM.[45] The amount of adenosine released, upon electrical stimulation of innervated skeletal muscle preparations in the presence of a supramaximal concentration of tubocurarine, is approximately half the amount released in the absence of tubocurarine, suggesting that half of the adenosine being released comes from the nerve and the other half comes from contracting muscle fibers.[46] Inactivation of extracellular adenosine at the rat diaphragm, in contrast with what occurs in other preparations from the peripheral and central nervous systems, might occur through both uptake and deamination.[39] The nucleoside transporters constitute an extended family[47] with different kinetic and pharmacological properties. Adenosine transportation is bi-directional, results from facilitated diffusion process[48] and is kinetically independent of the intracellular metabolism of adenosine.[49] Thus, adenosine transporters also mediate adenosine reuptake, the direction of the transport being dependent upon the concentration gradient at both sides of the membrane.[50]

Adenosine Receptors

Nerve Terminals

Since the pioneering studies of Ribeiro et al.[20] in the early 1970s, it is known that adenosine and ATP released by nerve endings modulate the presynaptic function. A nomenclature on purinergic receptors was advanced by Burnstock in 1978: P1 receptors, from the P1 (adenosine selective) and P2 (ATP selective).[51] The field for identification of different subtypes of adenosine receptors was then opened, and a further breakthrough was attained by the end of the 1970s by van Calker and colleagues,[52] who were the first to propose adenosine receptors in brain cells as A_1 (inhibitory) and A_2 (stimulatory). Adenosine receptor cloning possibilities had to wait until the beginning of the 1990s. The first adenosine receptor to be cloned was the A_1 receptor from brain tissue.[53] All four G protein coupled adenosine receptors (A_1, A_{2A}, A_{2B}, and A_3) have since then been cloned. All of these receptors possess seven transmembrane domains and are linked to a variety of transducing mechanisms. Neuromodulation by adenosine is exerted through activation of high-affinity receptors (A_1 and A_{2A}), which are probably of physiological importance, and of low affinity receptors (A_{2B}), which might be relevant in pathological conditions. The A_3 receptor is a high-affinity receptor in humans, but has a low density in most tissues.

Both nerve terminal and muscular activity contribute to the build up of extracellular adenosine,[46] which strongly supports the concept of activity-associated generation of an important auto-control modulator at the neuromuscular junction.[54] Adenosine controls the release of neurotransmitters by activating inhibitory A_1 and facilitatory A_{2A} adenosine receptors, which coexist in the same nerve terminal.[25] A_1 receptors are coupled to pertussis toxin sensitive G-proteins (G_i/G_o) that lead to an inhibition of the activity of the enzyme adenylate cyclase (through subunits α and $\beta\gamma$) (see Ref. 55). The inhibitory effects of adenosine are attenuated by activators of protein kinase C or by Li^+, which is known to affect phosphoinositide metabolism. However, the effects are not attenuated by activators of protein kinase A.[56] Depression of ACh release via presynaptic A_1 receptors is by inhibition of N-type Ca^{2+} channels,[57] but is not the basis of tetanic fade at the rat neuromuscular junctions.[58] Tetanic depression is overcome by tonic adenosine A_{2A} receptor facilitation of Ca^{2+} influx through L-type channels at the rat motor nerve terminals.[59] Therefore, extracellular adenosine mediates two independent actions on neuromuscular transmission: a fast and easily washable inhibitory action, mediated by A_1 receptors[19,60,61] and a slower A_{2A} receptor-mediated enhancement of EPPs, which is hardly reversible.[25,62] The reason for this slower and sustained adenosine A_{2A} receptor-mediated action might be the transducing system operated by the receptor, which involves cyclic AMP formation and PKA activation[63] with subsequent protein phosphorylation, whereas in the case of adenosine A_1 receptors, in mammalian nerve terminals, it involves a fast inhibition of calcium entry.[64,65] It is therefore conceivable that the consequences of adenosine A_{2A} receptor activation extend far beyond receptor activation and adenosine removal from the synaptic cleft, whereas the consequences of A_1 receptor activation vanish shortly after dissociation of adenosine from the receptor.

The coexistence of both adenosine A_1 and A_{2A} receptors in the same nerve terminal opens the question: in which conditions adenosine activates A_1 or A_{2A} receptors? There is some uncertainty about how P1 receptors are distributed in the cells of the paradigmatic neuromuscular junction. Lynge and Hellsten[66] showed, by immunohistochemistry and western-blot analysis, the presence of A_{2A} receptors but the absence of A_1 receptors in the sarcolemma and cytosol of skeletal muscle cells. On the other hand, Baxter and colleagues[67] showed, by immunohistochemistry, that A_{2A} receptors are present in the nerve terminals but not in intramuscular axons, muscle cells or Schwann cells. More recently, Garcia[68] showed, by western-blot analysis, that A_1 and A_{2A} receptors are abundantly expressed in muscle. A_1 receptor is more abundant in the adult whereas A_{2A} receptor is more abundant in the newborn. By immunochemistry in semithin sections of Levator auris longus (LAL) mouse neuromuscular junctions, they also show that these receptors are differently localized in the three cells (muscle, nerve, and Schwann cells) that configure the neuromuscular junction. A_1 receptor localizes in the terminal Schwann cell and nerve terminal whereas A_{2A} receptor localizes in the postsynaptic muscle, in the axon, and in the nerve terminal. These apparent contradictory results make the interpretation of functional studies difficult. Indeed, most of the functional experiments have been done in recording conditions, where the prevention of muscle contraction is needed. The different strategies to prevent muscle contraction alter differently the synapse physiology. For example, high Mg^{2+} reduces ACh release and curare reduces the postsynaptic response.

After directly measuring the activity-dependent efflux of radiolabeled ACh incorporated in nerve terminals, Correia-de-Sá and colleagues[25] conclude that adenosine receptors control the nerve-evoked release of ACh. It has been reported that micromolar concentrations of the P1 receptor endogenous agonist adenosine reduced evoked quantal content and/or spontaneous ACh release in frog neuromuscular junctions.[69–71] In rats, however, submicromolar adenosine can have the opposite effect.[62] In other studies in mice, only very high adenosine concentrations (0.5–1 mM) reduce perineural calcium currents and affect neurotransmission.[65] In more recent studies, in the LAL mouse neuromuscular junctions,[68] micromolar concentrations of the P1 receptor endogenous agonist, when applied in resting conditions, reduced the evoked quantal content only when applied in the presence of high Mg^{2+} (ACh release reduction) or D-tubocurarine (reduction in the postsynaptic response), whereas in the presence of μ-CgTx-GIIIB,: A specific blocker of muscle sodium channels. (paralyzed muscles but fully operative molecular signaling within the neuromuscular synapse), the inhibitory effect disappears. In the same study, the authors also studied the involvement of the P1 receptors in modulating transmitter release, in the presence of μ-CgTx-GIIIB, during intense synaptic activity, stimulating the nerve with a frequency of 40 Hz, 2 min. In these conditions micromolar concentrations of adenosine reduced the magnitude of the depression observed in the last EPPs of the train. By the interpretation of these results Garcia and colleagues[68] point out an interesting hypothesis, suggesting that P1 receptors might have no real effect on neuromuscular transmission in resting conditions, however these receptors can conserve resources by limiting spontaneous quantal leak of acetylcholine, and may protect synaptic function by reducing the magnitude of depression during repetitive activity.

Schwann Cells

In the PNS there are three types of Schwann cells: the myelinating and nonmyelinating Schwann cells associated with axons, and the perisynaptic Schwann cells located at the neuromuscular junction. Perisynaptic Schwann cells perform similar functions to central nervous system astrocytes, where they participate in synapse-glia signaling, modulate synaptic function, and are, in all senses of the term, active synaptic partners.[72] Purinergic signaling to perisynaptic Schwann cells was first described in 1992 and was observed in acutely isolated muscles using calcium imaging.[73] Since the initial description of ATP signaling to perisynaptic Schwann cells, information was provided regarding similar signaling pathways in amphibian[26] and mouse perisynaptic Schwann cells.[29] ATP can be released by perisynaptic Schwann cells[74] and as mentioned before P1 receptors are present at perisynaptic glial cells of both frog[26] and mammalian[29] neuromuscular junctions. The presence of P1 receptors in the perisynaptic Schwann cells appear to be represented by only A_1 receptor types, since there is no evidence for the presence of A_{2A} receptors[68] in these cells.

The first studies of neuron–glia interactions at the neuromuscular junction demonstrated that perisynaptic Schwann cells could respond to nerve activity with Ca^{2+} elevations[73,75] and these elevations were frequency-dependent.[73] A later study, performed by the same group, demonstrated a further role for perisynaptic Schwann cells in the modulation of neuromuscular junction function. This was done by direct and specific Ca^{2+}-mediated activation or inactivation of glial cells. Direct activation of the IP3 cascade caused a potentiation of neurotransmission at 0.2 Hz.[76] Furthermore, during 10 Hz stimulation, injection of perisynaptic Schwann cells with the Ca^{2+} chelator BAPTA induced greater depression. Thus, perisynaptic Schwann cells can regulate differently the neuromuscular function, depending on neuronal activity.

Adenosine Receptors and Their Interactions With Other Receptors at the Neuromuscular Junction

Crosstalk Between A_1 and A_{2A} Receptors

Pousinha and colleagues[62] showed that by manipulating the endogenous concentration of adenosine with an adenosine kinase blocker, or by applying submicromolar concentrations of adenosine or CADO to the bath perfusion, the amplitude of EPPs increased, revealing an expected facilitatory effect mediated by adenosine A_{2A} receptors. These effects were antagonized not only by an A_{2A} receptor selective antagonist but also by an A_1 receptor selective antagonist. The inhibitory effect of the A_{2A} receptor agonist, CGS21680, in micromolar concentrations at both the hippocampus[77] and neuromuscular junction,[25] is probably due to A_1 receptor activation. However, one would not expect the observed excitatory effects seen with low nanomolar concentrations of CGS21680, which are well below its Ki for A_1 receptor.[78] These results open the discussion of the existence of adenosine receptor heteromers at the neuromuscular junction, as reported in the central nervous system. For decades in the field of adenosine neuromodulation A_{2A} receptors were considered as isolated entities. However, the ability of these receptors to form heteromers with other GPCRs, such as the dopamine and glutamate receptors, and the fact that heteromers are constituted of different adenosine receptor subtypes,[79] is now established at the anatomic, biochemical, and functional levels. Coimmunoprecipitation and bioluminescence resonance energy transfer techniques have shown the existence of $A_1 - A_{2A}$ receptor heteromers in cotransfected HEK cells, as well as the existence of an intermolecular crosstalk, and radioligand-binding techniques have allowed the identification of an intramembrane receptor–receptor interaction in the $A_1 - A_{2A}$ receptor heteromer.[80] Indeed, heteromerization of G-protein coupled receptors may affect receptor functioning,[81] since monomers and heteromers can have distinct pharmacological and signaling properties. The presence of A_1–A_{2A} heteromers at the neuromuscular junction and their functional implications remain to be clarified, although presynaptic interactions between A_1 and A_{2A} receptors were observed at motor nerve terminals.[27,62,82]

Crosstalk Between Adenosine Receptors and Acetylcholine Receptors

At the neuromuscular junction ACh modulates its own release by activating presynaptic nicotinic and muscarinic receptors. The overall outcome of the muscarinic ACh receptors is to stabilize the conditions of the spontaneous quantal output of ACh (a facilitatory M_1 and inhibitory M_2 role), and to save the synapse function by decreasing the extent of evoked release (mainly an M_2 action) and reducing tetanic depression.[59,83,84] Adenosine receptors and muscarinic ACh receptors intervene differently in spontaneous and evoked ACh release modulation, but seem to share some functional control over synaptic depression. In basal conditions adenosine and muscarinic receptors are mutually dependent on the modulation of evoked and spontaneous ACh release.[85–87] In these conditions the muscarinic M_1 facilitation predominates over tonic M_2 inhibition.[87] When the muscarinic M_1 positive feedback loop is operational, the function of both M_2 and A_{2A} receptors is suppressed and ACh overflow is controlled by endogenous formation of small amounts of adenosine acting via inhibitory A_1 receptors, which also prevents the muscarinic M_2 negative feedback mechanism becoming active.[85] During high-frequency stimulation ($>$50 Hz), activation of A_{2A} receptors by high concentrations of adenosine favors a shift in muscarinic M_1/M_2 receptor balance towards the activation of inhibitory M_2 receptors.[87] The autofacilitatory nicotinic receptors function is also modulated by adenosine receptors. Thus, in basal conditions, A_1 and A_{2A} receptors are tonically involved and have opposite actions with respect to their ability to facilitate (A_{2A}) or attenuate (A_1) their desensitization.[88] During repetitive neuronal firing, the endogenous produced adenosine shifts tonic A_1/A_{2A} receptor balance towards the activation of adenosine A_{2A} receptors, favoring the desensitization of the autofacilitatory nicotinic receptors.[89] Thus, activation of A_{2A} receptors prevents the posttetanic fade by modulating both muscarinic M_2 and autofacilitatory nicotinic receptors function.

Adenosine Receptors and BDNF Actions

Presynaptic depolarization,[90] which is known to increase extracellular adenosine levels,[91] as well as the enhancement of intracellular cAMP,[92] the most frequent A_{2A} receptor transducing pathway, triggers synaptic actions of BDNF. Also, A_{2A} receptors are known to transactivate TrkB receptors in the absence of the neurotrophin.[93] This transactivation requires long-term incubation with G-protein coupled receptor agonists and receptor internalization.[94] Indeed, it has been recognized that adenosine A_{2A} receptor activation is a crucial prerequisite for the functioning of neurotrophic receptors at synapses. This has been shown for the facilitatory actions of BDNF on synaptic transmission[95,96] and on LTP[97] at the CA1 area of the hippocampus. The mechanism associated with this interaction between A_{2A} receptors and BDNF actions was described for the first time at the neuromuscular junction,[98] where the activation of adenosine A_{2A} receptors leads to activation of the cyclic AMP/PKA transducing mechanism, favoring of the BDNF-TrkB-PLCγ signaling pathway.

Adenosine Signaling in Ageing and Disease

The neuromuscular junction has been reported to undergo significant detrimental changes during ageing. Early studies found that ageing was associated with morphological changes confined primarily to nerve endings with little or no degeneration or loss of primary axons, suggesting that ageing was associated with a functional denervation.[99,100] Subsequent studies revealed that age-related changes in neuromuscular junction morphology vary among different muscle types and could potentially be related to muscle activity levels.[101] Ageing skeletal muscle is characterized by a progressive loss of muscle mass as well as a decrease in function.[102] This age-acquired deficit, known as sarcopenia, contributes profoundly to a decrease in quality of life in elderly people, and predisposes them to an increased risk of morbidity, disability, and mortality.[103] Whether this loss of muscle mass is a cause or a consequence of nerve terminals degeneration remains to be clarified. Several authors reported morphological, molecular, and functional age-related changes at the nerve cells[104–107] which were dependent on the muscle type, with the diaphragm being the most preserved muscle.[108]

There are not many studies on adenosine in aged animals at the neuromuscular junction, although several studies performed in the central nervous system showed that the neuromodulatory action of adenosine changes with age. Ageing decreases the ability of adenosine A_1 receptors to inhibit neuronal activity.[109] This may be a function of an age-related decrease in the density of A_1 receptors in the brain, which has been shown in both mice[110] and humans.[111] Low A_1 receptor density and function, however, can be compensated for by higher levels of extracellular adenosine, which keep tonic inhibition high in aged animals.[112] In contrast to A_1 receptors, A_{2A} receptors have a different pattern of change with age which depends on the brain area, as indicated: there is a significant increase in the density of A_{2A} receptors in the cortex[113] and hippocampus[114] of aged rats, which correlates with their enhanced ability to facilitate glutamatergic synaptic transmission[115] and ACh release[116] in the hippocampus. In the striatum there is a tendency for a decrease in A_{2A} receptor density in aged rats,[113] and within the striatum, age may influence the A_{2A} receptors in glutamatergic, dopaminergic, or GABAergic nerve terminals in different ways.[117–119] At the neuromuscular junction, one of the few preparations where it is experimentally possible to separate the neuromodulatory and homeostatic roles of adenosine, the tonic A_1 receptor inhibition of ACh release is not modified in aged rats.[27,120] The excitatory action of A_{2A} receptors, however, decreases with ageing, even disappearing in very old animals ($\sim$2 years old).[27] Age-related changes in adenosine receptors at the neuromuscular junction seem to be more similar to what has been observed in the striatum from aged animals, rather than in the cortex or hippocampus. The reduction in the effects of A_{2A} receptors in very old animals might be one of the causes of the age-related changes in the neuromuscular transmission, because these receptors are known to influence the action of several neurotransmitters/neuromodulators/receptors at the nerve terminal.[121] Accordingly, studies performed on adenosine receptors knockout (KO) mice revealed that A_{2A} receptors play much more important roles than A_1 receptors in regulating locomotor activity,[122] because both A_{2A} receptor KO and double A_1–A_{2A} receptors KO mice present lower locomotion activity and consume less oxygen than their wild type controls. These parameters are unaltered in A1 receptor KO mice.

Amyotrophic Lateral Sclerosis (ALS), the most common adult-onset motor neuron disease, is characterized by the degeneration of motor neurons in the corticospinal tract. This progressive loss leads to widespread muscle weakness, atrophy, and paralysis, and ultimately to death after the involvement of the respiratory muscles. Although most cases have no known cause, 10% are inherited, with 20% of these familial cases being caused by a mutation in the copper/zinc superoxide dismutase gene (Cu/Zn, SOD1). Both familial and sporadic forms of ALS present similar pathological and clinical features, suggesting a common pathogenesis.[123,124] A_{2A} receptors are known to have a neuroprotective role in some pathological conditions,[125] and have been considered as a potential therapeutic target for ALS.[126,127] Nascimento and colleagues showed that the role of adenosine A_{2A} receptors at the neuromuscular junction of the ALS SOD1(G93A) mouse model changes with disease progression. In the presymptomatic phase, the magnitude of the excitatory effects on neuromuscular transmission, caused by the A_{2A} receptor, is enhanced compared to age-matched controls. In contrast, in the symptomatic SOD1(G93A) mice, the A_{2A} receptor-mediated facilitation is absent.[128] These alterations in the function of A_{2A} receptors, observed with disease progression, changes the fine-tuned neuromodulation exerted by adenosine, as the A_1/A_{2A} receptors functional crosstalk is lost in ALS. Therefore, symptomatic mice display a higher A_1 receptor tonic activation than age-matched controls.[129] Interestingly, one can observe similarities between the symptomatic SOD1(G93A) mice[128–130] and aged rats[27,107] in what relates to the neuromuscular transmission physiological properties and adenosine receptors' neuromodulatory roles.

CONCLUSIONS

The neuromodulation exerted at the neuromuscular junction by adenosine seems to go beyond simply controlling the release of the neurotransmitter. Adenosine receptors have a crucial function in regulating the activation of multiple receptors that affect neurotransmitter release, synaptic transmission, neuron-glia communication and trophic actions. Adenosine actions can take place at the motor nerve terminal, but also at a more complex level, involving the Schwann cells and/or the muscle fiber. Adenosine actions can take place at the motor nerve terminal, but also at a more complex level, involving the Schwann cells and/or the muscle fiber. By the literature herein reviewed, the A_{2A} receptor seems to be the key adenosine receptor in the modulation of neuromuscular transmission, whereas the A_1 receptors behave mainly as targets instead of a trigger. In addition to providing a review on the role of adenosine at the neuromuscular junction, one could verify that most of the adenosine receptors' actions and interactions herein reported, observed at the neuromuscular junction, have also been described at the CNS, making the neuromuscular junction an excellent model of synapse for investigating the cellular and molecular details of synaptic transmission.

References

1. Punga AR, Ruegg MA. Signaling and aging at the neuromuscular synapse: lessons learnt from neuromuscular diseases. *Curr Opin Pharmacol*. 2012;12(3):340–346.

2. Fagerlund MJ, Eriksson LI. Current concepts in neuromuscular transmission. *Br J Anaesth*. 2009;103(1):108–114.
3. Martyn JA, Fagerlund MJ, Eriksson LI. Basic principles of neuromuscular transmission. *Anaesthesia*. 2009;64(suppl 1):1–9.
4. Kandel ER, Schwartz JH, Jessel TM, Siegelbaum SA, Hudspeth AJ. *Principles of Neural Science*. 5th ed., New York: McGraw-Hill; 2012.
5. Kuffler SW, Yoshikami D. The number of transmitter molecules in a quantum: an estimate from iontophoretic application of acetylcholine at the neuromuscular synapse. *J Physiol*. 1975;251(2):465–482.
6. Fatt P, Katz B. Spontaneous subthreshold activity at motor nerve endings. *J Physiol*. 1952;117(1):109–128.
7. Del Castillo J, Stark L. The effect of calcium ions on the motor end-plate potentials. *J Physiol*. 1952;116(4):507–515.
8. Katz B, Miledi R. Spontaneous and evoked activity of motor nerve endings in calcium Ringer. *J Physiol*. 1969;203(3):689–706.
9. Brooks VB, Thies RE. Reduction of quantum content during neuromuscular transmission. *J Physiol*. 1962;162:298–310.
10. Burnstock G. Do some nerve cells release more than one transmitter? *Neuroscience*. 1976;1(4):239–248.
11. Burnstock G. Evolution of the autonomic innervation of visceral and cardiovascular systems in vertebrates. *Pharmacol Rev*. 1969;21(4):247–324.
12. Brownstein MJ, Saavedra JM, Axelrod J, Zeman GH, Carpenter DO. Coexistence of several putative neurotransmitters in single identified neurons of Aplysia. *Proc Natl Acad Sci U S A*. 1974;71(12):4662–4665.
13. Kupfermann I. Functional studies of cotransmission. *Physiol Rev*. 1991;71(3):683–732.
14. Burnstock G. Purinergic cotransmission. *F1000 Biol Rep*. 2009;1:46.
15. Trudeau LE, Gutierrez R. On cotransmission & neurotransmitter phenotype plasticity. *Mol Interv*. 2007;7(3):138–146.
16. Arch JR, Newsholme EA. The control of the metabolism and the hormonal role of adenosine. *Essays Biochem*. 1978;14:82–123.
17. Van den Berghe G, Bontemps F, Vincent MF, Van den Bergh F. The purine nucleotide cycle and its molecular defects. *Prog Neurobiol*. 1992;39(5):547–561.
18. Burnstock G. Physiology and pathophysiology of purinergic neurotransmission. *Physiol Rev*. 2007;87(2):659–797.
19. Ginsborg BL, Hirst GD. The effect of adenosine on the release of the transmitter from the phrenic nerve of the rat. *J Physiol*. 1972;224(3):629–645.
20. Ribeiro JA, Walker J. Action of adenosine triphosphate on endplate potentials recorded from muscle fibres of the rat-diaphragm and frog sartorius. *Br J Pharmacol*. 1973;49(4):724–725.
21. Sattin A, Rall TW. The effect of adenosine and adenine nucleotides on the cyclic adenosine 3′, 5′-phosphate content of guinea pig cerebral cortex slices. *Mol Pharmacol*. 1970;6(1):13–23.
22. Ribeiro JA, Sebastião AM. Modulation and metamodulation of synapses by adenosine. *Acta Physiol (Oxf)*. 2010;199(2):161–169.
23. Todd KJ, Robitaille R. Purinergic modulation of synaptic signalling at the neuromuscular junction. *Pflugers Arch*. 2006;452(5):608–614.
24. Fu WM, Poo MM. ATP potentiates spontaneous transmitter release at developing neuromuscular synapses. *Neuron*. 1991;6(5):837–843.
25. Correia-de-Sá P, Sebastião AM, Ribeiro JA. Inhibitory and excitatory effects of adenosine receptor agonists on evoked transmitter release from phrenic nerve ending of the rat. *Br J Pharmacol*. 1991;103(2):1614–1620.
26. Robitaille R. Purinergic receptors and their activation by endogenous purines at perisynaptic glial cells of the frog neuromuscular junction. *J Neurosci*. 1995;15(11):7121–7131.

27. Pousinha PA, Correia AM, Sebastião AM, Ribeiro JA. Neuromuscular transmission modulation by adenosine upon aging. *Neurobiol Aging*. 2012;33(12):2869–2880.
28. Stevens B, Ishibashi T, Chen JF, Fields RD. Adenosine: an activity-dependent axonal signal regulating MAP kinase and proliferation in developing Schwann cells. *Neuron Glia Biol*. 2004;1(1):23–34.
29. Rochon D, Rousse I, Robitaille R. Synapse-glia interactions at the mammalian neuromuscular junction. *J Neurosci*. 2001;21(11):3819–3829.
30. Fredholm BB. Adenosine and neuroprotection. *Int Rev Neurobiol*. 1997;40:259–280.
31. Phillips E, Newsholme EA. Maximum activities, properties and distribution of 5′ nucleotidase, adenosine kinase and adenosine deaminase in rat and human brain. *J Neurochem*. 1979;33(2):553–558.
32. Nagata H, Mimori Y, Nakamura S, Kameyama M. Regional and subcellular distribution in mammalian brain of the enzymes producing adenosine. *J Neurochem*. 1984;42 (4):1001–1007.
33. Meghji P, Pearson JD, Slakey LL. Regulation of extracellular adenosine production by ectonucleotidases of adult rat ventricular myocytes. *Am J Physiol*. 1992;263(1 Pt 2):H40–H47.
34. Wu PH, Phillis JW. Uptake by central nervous tissues as a mechanism for the regulation of extracellular adenosine concentrations. *Neurochem Int*. 1984;6(5):613–632.
35. Nagy JI, Geiger JD, Staines WA. Adenosine deaminase and purinergic neuroregulation. *Neurochem Int*. 1990;16(3):211–221.
36. Dowdall MJ, Boyne AF, Whittaker VP. Adenosine triphosphate. A constituent of cholinergic synaptic vesicles. *Biochem J*. 1974;140(1):1–12.
37. Silinsky EM. On the association between transmitter secretion and the release of adenine nucleotides from mammalian motor nerve terminals. *J Physiol*. 1975;247 (1):145–162.
38. Rabasseda X, Solsona C, Marsal J, Egea G, Bizzini B. ATP release from pure cholinergic synaptosomes is not blocked by tetanus toxin. *FEBS Lett*. 1987;213(2):337–340.
39. Sebastião AM, Ribeiro JA. On the adenosine receptor and adenosine inactivation at the rat diaphragm neuromuscular junction. *Br J Pharmacol*. 1988;94(1):109–120.
40. Dunant Y, Muller D. Quantal release of acetylcholine evoked by focal depolarization at the Torpedo nerve-electroplaque junction. *J Physiol*. 1986;379:461–478.
41. Whittaker VP. The structure and function of cholinergic synaptic vesicles. The Third Thudichum Lecture. *Biochem Soc Trans*. 1984;12(4):561–576.
42. Yu SP, Van der Kloot W. Non-quantal acetylcholine release at mouse neuromuscular junction: effects of elevated quantal release and aconitine. *Neurosci Lett*. 1990;117(1–2): 111–116.
43. Cunha RA, Sebastião AM. Extracellular metabolism of adenine nucleotides and adenosine in the innervated skeletal muscle of the frog. *Eur J Pharmacol*. 1991;197(1):83–92.
44. Cunha RA, Vizi ES, Ribeiro JA, Sebastião AM. Preferential release of ATP and its extracellular catabolism as a source of adenosine upon high- but not low-frequency stimulation of rat hippocampal slices. *J Neurochem*. 1996;67(5):2180–2187.
45. Dunwiddie TV, Diao L. Extracellular adenosine concentrations in hippocampal brain slices and the tonic inhibitory modulation of evoked excitatory responses. *J Pharmacol Exp Ther*. 1994;268(2):537–545.
46. Cunha RA, Sebastião AM. Adenosine and adenine nucleotides are independently released from both the nerve terminals and the muscle fibres upon electrical stimulation of the innervated skeletal muscle of the frog. *Pflugers Arch*. 1993;424 (5–6):503–510.
47. Agbanyo FR, Vijayalakshmi D, Craik JD, et al. 5′-S-(2-aminoethyl)-N6-(4-nitrobenzyl)-5′-thioadenosine (SAENTA), a novel ligand with high affinity for polypeptides associated with nucleoside transport. Partial purification of the nitrobenzylthioinosine-binding protein of pig erythrocytes by affinity chromatography. *Biochem J*. 1990;270(3):605–614.

48. Plagemann PG, Wohlhueter RM. Nucleoside transport in cultured mammalian cells. Multiple forms with different sensitivity to inhibition by nitrobenzylthioinosine or hypoxanthine. *Biochim Biophys Acta*. 1984;773(1):39–52.
49. Gu JG, Kala G, Geiger JD. [3H]adenosine transport in synaptoneurosomes of postmortem human brain. *J Neurochem*. 1993;60(6):2232–2237.
50. Gu JG, Foga IO, Parkinson FE, Geiger JD. Involvement of bidirectional adenosine transporters in the release of L-[3H]adenosine from rat brain synaptosomal preparations. *J Neurochem*. 1995;64(5):2105–2110.
51. Burnstock G. *A Basis for Distinguishing Two Types of Purinergic Receptor*. New York: Raven Press; 1978.
52. van Calker D, Muller M, Hamprecht B. Adenosine regulates via two different types of receptors, the accumulation of cyclic AMP in cultured brain cells. *J Neurochem*. 1979;33 (5):999–1005.
53. Mahan LC, McVittie LD, Smyk-Randall EM, et al. Cloning and expression of an A1 adenosine receptor from rat brain. *Mol Pharmacol*. 1991;40(1):1–7.
54. Ribeiro JA, Cunha RA, Correia-de-Sá P, Sebastião AM. Purinergic regulation of acetylcholine release. *Prog Brain Res*. 1996;109:231–241.
55. Cunha RA. Adenosine as a neuromodulator and as a homeostatic regulator in the nervous system: different roles, different sources and different receptors. *Neurochem Int*. 2001;38(2):107–125.
56. Sebastião AM, Ribeiro JA. Interactions between adenosine and phorbol esters or lithium at the frog neuromuscular junction. *Br J Pharmacol*. 1990;100(1):55–62.
57. Schwartz AD, Whitacre CL, Lin Y, Wilson DF. Adenosine inhibits N-type calcium channels at the rat neuromuscular junction. *Clin Exp Pharmacol Physiol*. 2003;30 (3):174–177.
58. Malinowski MN, Cannady SB, Schmit KV, Barr PM, Schrock JW, Wilson DF. Adenosine depresses transmitter release but is not the basis for 'tetanic fade' at the neuromuscular junction of the rat. *Neurosci Lett*. 1997;230(2):81–84.
59. Oliveira L, Timoteo MA, Correia-de-Sa P. Tetanic depression is overcome by tonic adenosine A(2A) receptor facilitation of L-type Ca(2+) influx into rat motor nerve terminals. *J Physiol*. 2004;560(Pt 1):157–168.
60. Ribeiro JA, Sebastião AM. On the role, inactivation and origin of endogenous adenosine at the frog neuromuscular junction. *J Physiol*. 1987;384:571–585.
61. Sebastião AM, Stone TW, Ribeiro JA. The inhibitory adenosine receptor at the neuromuscular junction and hippocampus of the rat: antagonism by 1,3,8-substituted xanthines. *Br J Pharmacol*. 1990;101(2):453–459.
62. Pousinha PA, Correia AM, Sebastião AM, Ribeiro JA. Predominance of adenosine excitatory over inhibitory effects on transmission at the neuromuscular junction of infant rats. *J Pharmacol Exp Ther*. 2010;332(1):153–163.
63. Correia-de-Sá P, Ribeiro JA. Evidence that the presynaptic A2a-adenosine receptor of the rat motor nerve endings is positively coupled to adenylate cyclase. *Naunyn Schmiedebergs Arch Pharmacol*. 1994;350(5):514–522.
64. Ribeiro JA, Sa-Almeida AM, Namorado JM. Adenosine and adenosine triphosphate decrease 45Ca uptake by synaptosomes stimulated by potassium. *Biochem Pharmacol*. 1979;28(8):1297–1300.
65. Silinsky EM. Adenosine decreases both presynaptic calcium currents and neurotransmitter release at the mouse neuromuscular junction. *J Physiol*. 2004;558(Pt 2):389–401.
66. Lynge J, Hellsten Y. Distribution of adenosine A1, A2A and A2B receptors in human skeletal muscle. *Acta Physiol Scand*. 2000;169(4):283–290.
67. Baxter RL, Vega-Riveroll LJ, Deuchars J, Parson SH. A2A adenosine receptors are located on presynaptic motor nerve terminals in the mouse. *Synapse*. 2005;57 (4):229–234.

68. Garcia N, Priego M, Obis T, et al. Adenosine A(1) and A(2)A receptor-mediated modulation of acetylcholine release in the mice neuromuscular junction. *Eur J Neurosci.* 2013;38(2):2229–2241.
69. Searl TJ, Silinsky EM. Modulation of Ca(2+)-dependent and Ca(2+)-independent miniature endplate potentials by phorbol ester and adenosine in frog. *Br J Pharmacol.* 2005;145(7):954–962.
70. Shakirzyanova AV, Bukharaeva EA, Nikolsky EE, Giniatullin RA. Negative cross-talk between presynaptic adenosine and acetylcholine receptors. *Eur J Neurosci.* 2006;24 (1):105–115.
71. Adamek S, Shakirzyanova AV, Malomouzh AI, Naumenko NV, Vyskocil F. Interaction of glutamate- and adenosine-induced decrease of acetylcholine quantal release at frog neuromuscular junction. *Physiol Res.* 2010;59(5):803–810.
72. Auld DS, Robitaille R. Perisynaptic Schwann cells at the neuromuscular junction: nerve- and activity-dependent contributions to synaptic efficacy, plasticity, and reinnervation. *Neuroscientist.* 2003;9(2):144–157.
73. Jahromi BS, Robitaille R, Charlton MP. Transmitter release increases intracellular calcium in perisynaptic Schwann cells in situ. *Neuron.* 1992;8(6):1069–1077.
74. Liu GJ, Werry EL, Bennett MR. Secretion of ATP from Schwann cells in response to uridine triphosphate. *Eur J Neurosci.* 2005;21(1):151–160.
75. Reist NE, Smith SJ. Neurally evoked calcium transients in terminal Schwann cells at the neuromuscular junction. *Proc Natl Acad Sci U S A.* 1992;89 (16):7625–7629.
76. Castonguay A, Robitaille R. Differential regulation of transmitter release by presynaptic and glial Ca2+ internal stores at the neuromuscular synapse. *J Neurosci.* 2001;21 (6):1911–1922.
77. Lupica CR, Cass WA, Zahniser NR, Dunwiddie TV. Effects of the selective adenosine A2 receptor agonist CGS 21680 on in vitro electrophysiology, cAMP formation and dopamine release in rat hippocampus and striatum. *J Pharmacol Exp Ther.* 1990;252 (3):1134–1141.
78. Jacobson KA, Stiles GL, Ji XD. Chemical modification and irreversible inhibition of striatal A2a adenosine receptors. *Mol Pharmacol.* 1992;42(1):123–133.
79. Ferre S, Ciruela F, Quiroz C, et al. Adenosine receptor heteromers and their integrative role in striatal function. *ScientificWorldJournal.* 2007;7:74–85.
80. Ciruela F, Casado V, Rodrigues RJ, et al. Presynaptic control of striatal glutamatergic neurotransmission by adenosine A1-A2A receptor heteromers. *J Neurosci.* 2006;26 (7):2080–2087.
81. Franco R, Casado V, Cortes A, et al. G-protein-coupled receptor heteromers: function and ligand pharmacology. *Br J Pharmacol.* 2008;153(suppl 1):S90–S98.
82. Correia-de-Sá P, Timoteo MA, Ribeiro JA. Presynaptic A1 inhibitory/A2A facilitatory adenosine receptor activation balance depends on motor nerve stimulation paradigm at the rat hemidiaphragm. *J Neurophysiol.* 1996;76(6):3910–3919.
83. Tomas J, Santafe MM, Garcia N, et al. Presynaptic membrane receptors in acetylcholine release modulation in the neuromuscular synapse. *J Neurosci Res.* 2014;92 (5):543–554.
84. Wessler I. Control of transmitter release from the motor nerve by presynaptic nicotinic and muscarinic autoreceptors. *Trends Pharmacol Sci.* 1989;10(3):110–114.
85. Oliveira L, Timoteo MA, Correia-de-Sá P. Negative crosstalk between M1 and M2 muscarinic autoreceptors involves endogenous adenosine activating A1 receptors at the rat motor endplate. *Neurosci Lett.* 2009;459(3):127–131.
86. Santafe MM, Priego M, Obis T, et al. Adenosine receptors and muscarinic receptors cooperate in acetylcholine release modulation in the neuromuscular synapse. *Eur J Neurosci.* 2015;42(2):1775–1787.

87. Oliveira L, Timoteo MA, Correia-de-Sá P. Modulation by adenosine of both muscarinic M1-facilitation and M2-inhibition of [3H]-acetylcholine release from the rat motor nerve terminals. *Eur J Neurosci.* 2002;15(11):1728–1736.
88. Correia-de-Sá P, Ribeiro JA. Tonic adenosine A2A receptor activation modulates nicotinic autoreceptor function at the rat neuromuscular junction. *Eur J Pharmacol.* 1994;271(2–3):349–355.
89. Timoteo MA, Faria M, Correia-de-Sá P. Endogenous adenosine prevents post-tetanic release facilitation mediated by alpha3beta2 nicotinic autoreceptors. *Eur J Pharmacol.* 2003;464(2–3):115–125.
90. Boulanger LM, Poo MM. Presynaptic depolarization facilitates neurotrophin-induced synaptic potentiation. *Nat Neurosci.* 1999;2(4):346–351.
91. Latini S, Pedata F. Adenosine in the central nervous system: release mechanisms and extracellular concentrations. *J Neurochem.* 2001;79(3):463–484.
92. Boulanger L, Poo MM. Gating of BDNF-induced synaptic potentiation by cAMP. *Science.* 1999;284(5422):1982–1984.
93. Lee FS, Chao MV. Activation of Trk neurotrophin receptors in the absence of neurotrophins. *Proc Natl Acad Sci U S A.* 2001;98(6):3555–3560.
94. Rajagopal R, Chen ZY, Lee FS, Chao MV. Transactivation of Trk neurotrophin receptors by G-protein-coupled receptor ligands occurs on intracellular membranes. *J Neurosci.* 2004;24(30):6650–6658.
95. Diógenes MJ, Fernandes CC, Sebastião AM, Ribeiro JA. Activation of adenosine A2A receptor facilitates brain-derived neurotrophic factor modulation of synaptic transmission in hippocampal slices. *J Neurosci.* 2004;24(12):2905–2913.
96. Tebano MT, Martire A, Potenza RL, et al. Adenosine A(2A) receptors are required for normal BDNF levels and BDNF-induced potentiation of synaptic transmission in the mouse hippocampus. *J Neurochem.* 2008;104(1):279–286.
97. Fontinha BM, Diógenes MJ, Ribeiro JA, Sebastião AM. Enhancement of long-term potentiation by brain-derived neurotrophic factor requires adenosine A2A receptor activation by endogenous adenosine. *Neuropharmacology.* 2008;54(6):924–933.
98. Pousinha PA, Diógenes MJ, Ribeiro JA, Sebastião AM. Triggering of BDNF facilitatory action on neuromuscular transmission by adenosine A2A receptors. *Neurosci Lett.* 2006;404(1–2):143–147.
99. Fujisawa K. Some observations on the skeletal musculature of aged rats-III. Abnormalities of terminal axons found in motor end-plates. *Exp Gerontol.* 1976;11(1–2):43–47.
100. Gutmann E, Hanzlikova V. Basic mechanisms of aging in the neuromuscular system. *Mech Ageing Dev.* 1973;1(5):327–349.
101. Prakash YS, Sieck GC. Age-related remodeling of neuromuscular junctions on type-identified diaphragm fibers. *Muscle Nerve.* 1998;21(7):887–895.
102. Narici MV, Maffulli N. Sarcopenia: characteristics, mechanisms and functional significance. *Br Med Bull.* 2010;95:139–159.
103. Janssen I, Baumgartner RN, Ross R, Rosenberg IH, Roubenoff R. Skeletal muscle cutpoints associated with elevated physical disability risk in older men and women. *Am J Epidemiol.* 2004;159(4):413–421.
104. Smith DO, Rosenheimer JL. Decreased sprouting and degeneration of nerve terminals of active muscles in aged rats. *J Neurophysiol.* 1982;48(1):100–109.
105. Balice-Gordon RJ. Age-related changes in neuromuscular innervation. *Muscle Nerve Suppl.* 1997;5:S83–S87.
106. Ceballos D, Cuadras J, Verdu E, Navarro X. Morphometric and ultrastructural changes with ageing in mouse peripheral nerve. *J Anat.* 1999;195(Pt 4):563–576.
107. Pousinha PA, Correia AM, Sebastião AM, Ribeiro JA. The giant miniature endplate potentials frequency is increased in aged rats. *Neurosci Lett.* 2015;584:224–229.

108. Banker BQ, Kelly SS, Robbins N. Neuromuscular transmission and correlative morphology in young and old mice. *J Physiol*. 1983;339:355–377.
109. Sebastião AM, Cunha RA, de Mendonça A, Ribeiro JA. Modification of adenosine modulation of synaptic transmission in the hippocampus of aged rats. *Br J Pharmacol*. 2000;131(8):1629–1634.
110. Pagonopoulou O, Angelatou F. Reduction of A1 adenosine receptors in cortex, hippocampus and cerebellum in ageing mouse brain. *Neuroreport*. 1992;3 (9):735–737.
111. Meyer PT, Elmenhorst D, Boy C, et al. Effect of aging on cerebral A1 adenosine receptors: a [18F]CPFPX PET study in humans. *Neurobiol Aging*. 2007;28 (12):1914–1924.
112. Bauman LA, Mahle CD, Boissard CG, Gribkoff VK. Age-dependence of effects of A1 adenosine receptor antagonism in rat hippocampal slices. *J Neurophysiol*. 1992;68 (2):629–638.
113. Cunha RA, Constantino MC, Sebastião AM, Ribeiro JA. Modification of A1 and A2a adenosine receptor binding in aged striatum, hippocampus and cortex of the rat. *Neuroreport*. 1995;6(11):1583–1588.
114. Diógenes MJ, Assaife-Lopes N, Pinto-Duarte A, Ribeiro JA, Sebastião AM. Influence of age on BDNF modulation of hippocampal synaptic transmission: interplay with adenosine A2A receptors. *Hippocampus*. 2007;17(7):577–585.
115. Rebola N, Sebastião AM, de Mendonça A, Oliveira CR, Ribeiro JA, Cunha RA. Enhanced adenosine A2A receptor facilitation of synaptic transmission in the hippocampus of aged rats. *J Neurophysiol*. 2003;90(2):1295–1303.
116. Lopes LV, Cunha RA, Ribeiro JA. Cross talk between A(1) and A(2A) adenosine receptors in the hippocampus and cortex of young adult and old rats. *J Neurophysiol*. 1999;82(6):3196–3203.
117. Corsi C, Melani A, Bianchi L, Pedata F. Striatal A2A adenosine receptor antagonism differentially modifies striatal glutamate outflow in vivo in young and aged rats. *Neuroreport*. 2000;11(11):2591–2595.
118. Corsi C, Melani A, Bianchi L, Pepeu G, Pedata F. Effect of adenosine A2A receptor stimulation on GABA release from the striatum of young and aged rats in vivo. *Neuroreport*. 1999;10(18):3933–3937.
119. Corsi C, Melani A, Bianchi L, Pepeu G, Pedata F. Striatal A2A adenosine receptors differentially regulate spontaneous and K+-evoked glutamate release in vivo in young and aged rats. *Neuroreport*. 1999;10(4):687–691.
120. Pereira MF, Cunha RA, Ribeiro JA. Tonic adenosine neuromodulation is preserved in motor nerve endings of aged rats. *Neurochem Int*. 2000;36(7):563–566.
121. Sebastião AM, Ribeiro JA. Fine-tuning neuromodulation by adenosine. *Trends Pharmacol Sci*. 2000;21(9):341–346.
122. Huang ZL, Qu WM, Eguchi N, et al. Adenosine A2A, but not A1, receptors mediate the arousal effect of caffeine. *Nat Neurosci*. 2005;8(7):858–859.
123. Boillee S, Vande Velde C, Cleveland DW. ALS: a disease of motor neurons and their nonneuronal neighbors. *Neuron*. 2006;52(1):39–59.
124. Ferraiuolo L, Kirby J, Grierson AJ, Sendtner M, Shaw PJ. Molecular pathways of motor neuron injury in amyotrophic lateral sclerosis. *Nat Rev Neurol*. 2011;7 (11):616–630.
125. Sebastião AM, Ribeiro JA. Adenosine receptors and the central nervous system. *Handb Exp Pharmacol*. 2009;193:471–534.
126. Beghi E, Pupillo E, Messina P, et al. Coffee and amyotrophic lateral sclerosis: a possible preventive role. *Am J Epidemiol*. 2011;174(9):1002–1008.
127. Potenza RL, Armida M, Ferrante A, et al. Effects of chronic caffeine intake in a mouse model of amyotrophic lateral sclerosis. *J Neurosci Res*. 2013;91(4):585–592.

128. Nascimento F, Pousinha PA, Correia AM, Gomes R, Sebastião AM, Ribeiro JA. Adenosine A2A receptors activation facilitates neuromuscular transmission in the pre-symptomatic phase of the SOD1(G93A) ALS mice, but not in the symptomatic phase. *PLoS One*. 2014;9(8):e104081.
129. Nascimento F, Sebastião AM, Ribeiro JA. Presymptomatic and symptomatic ALS SOD1(G93A) mice differ in adenosine A1 and A2A receptor-mediated tonic modulation of neuromuscular transmission. *Purinergic Signal*. 2015;11(4):471–480.
130. Rocha MC, Pousinha PA, Correia AM, Sebastião AM, Ribeiro JA. Early changes of neuromuscular transmission in the SOD1(G93A) mice model of ALS start long before motor symptoms onset. *PLoS One*. 2013;8(9):e73846.

CHAPTER

5

Gene Regulation of Adenosine A_{2A} Receptors in the Central Nervous System

Marta Barrachina[1], *Chien-Yu Lin*[2] *and Yijuang Chern*[2]

[1]Bellvitge Biomedical Research Institute, Barcelona, Spain [2]Academia Sinica, Taipei, Taiwan

OUTLINE

Adenosine Receptors in Neurodegenerative Diseases.
DOI: http://dx.doi.org/10.1016/B978-0-12-803724-9.00005-3

GENERAL INTRODUCTION

The A_{2A} adenosine receptor ($A_{2A}R$) is a seven-transmembrane receptor that plays modulatory roles in many pathophysiological functions of the brain and various peripheral tissues, including those of the immune system. The $A_{2A}R$ exists in many brain areas (such as the cortex, hippocampus, amygdala) at low levels, but it is highly enriched in the striatum, which plays a major role in movement coordination. Expression of the $A_{2A}R$ in the brain is tightly regulated during neuronal development,[1] supporting the idea that $A_{2A}R$ functions are critical during neuronal development and may affect cognitive functions in adults.[2]

Accumulating evidence has revealed that the level of $A_{2A}R$ transcript is up-regulated in the brains of patients with a neurodegenerative diseases, such as Alzheimer's disease (AD),[3,4] or Parkinson's disease (PD),[5,6] (Table 5.1). Enhancement of the $A_{2A}R$ in AD occurs in astrocytes and appears to be detrimental. How astrocytic $A_{2A}Rs$ are altered during AD progression, however, is currently unknown.[4] Up-regulation of the $A_{2A}R$ in PD happens not only in the putamen but also in the peripheral blood cells, suggesting a possible common regulatory mechanism for the $A_{2A}R$ in central nervous system and immune cells.[6] DNA methylation has been implicated in the modulation of $A_{2A}R$ levels in PD, and it may be used to develop therapeutic treatments for PD.[22] Conversely, the amount of $A_{2A}Rs$ was markedly reduced in striatal neurons of Huntington's disease (HD). This suppression may result from the interference of CREB binding to the core promoter of $A_{2A}R$ by the disease-causing polyQ-expanded Huntingtin protein.[7] In addition, changes in DNA methylation patterns in the $A_{2A}R$ gene may also contribute to down-regulation of $A_{2A}Rs$ in HD.[12]

To date, a handful of transcriptional factors (including PPARγ, CREB, NFκB, NF-1, YY-1, and ZBP-89), pro-inflammatory cytokines (IL-1β and TNF-α), several miRNAs (including miRNA-214, miRNA-34b, miRNA-15, and miRNA-16), and DNA methylation have been implicated in the regulation of the $A_{2A}R$ gene (Table 5.1), all of which will be elaborated in the following sections. Among these regulators, the mechanism through which NFκB mediates alterations in $A_{2A}Rs$ during inflammation is of great interest, as it may contribute to many inflammation-related diseases.[6,19,23–26]

$A_{2A}R$ PROMOTER AND GENE STRUCTURE

The genes encoding the $A_{2A}R$ were identified more than two decades ago.[27–33] Briefly, the coding regions for the $A_{2A}R$ exist in two exons

TABLE 5.1 Regulation of A_{2A} Adenosine Receptor ($A_{2A}R$)

Regulatory factor/ machinery	$A_{2A}R$ expression	Experimental model/ phenotype	Tissue type/ cell line	Species	References
CENTRAL NERVOUS SYSTEM					
CREB	Increase	Transient transfection	PC12 cell line	Rat	7
NF-κB	Decrease	NF-κB/p50-deficient mice	Striatum	Mouse	8
NF-1	Decrease	Transient transfection	RBA-2 cell line	Rat	9
YY-1	Decrease	Transient transfection	SH-SY5Y cell line	Human	10
ZBP-89	Increase	Transient transfection	SH-SY5Y cell line	Human	10
miRNA-34b	Decrease	Transient transfection	SH-SY5Y cell line	Human	11
miRNA-34b	Decrease	PD patients	Putamen	Human	11
DNA methylation	Decrease	R6/1 and R6/2 HD mice	Striatum	Mouse	12
DNA methylation	Decrease	HD patients	Putamen	Human	12
DNA methylation	Increase	Subgroup of schizophrenia patients		Human	13
DNA methylation	Increase	5-Azacytidine treatment	SH-SY5Y cell line	Human	14
DNA methylation	Decrease	SAM treatment	U87-MG cell line	Human	14
DNA methylation	Decrease	SAM treatment	Striatum	Rat	15
ND	Decrease	HD patients	Basal ganglion	Human	16
ND	Decrease	HD mice	Striatum	Mouse	16
ND	Increase	AD patients	Cerebral cortex	Human	3
ND	Increase	AD patients	Astrocyte	Human	4
ND	Increase	PD patients	Striatopallidal neuron	Human	5
ND	Increase	PD patients	Putamen	Human	6

(*Continued*)

TABLE 5.1 (Continued)

Regulatory factor/ machinery	$A_{2A}R$ expression	Experimental model/ phenotype	Tissue type/ cell line	Species	References
IMMUNE SYSTEM					
PPARγ	Increase	LPS stimulation	Macrophage	Mouse	17
NF-κB	Increase	Hypoxia stimulation	Lymphocyte	Human	18
TNF-α	Increase	TNF-α stimulation	A549 cell line	Human	19
IL-1β	Increase	IL-1β stimulation	A549 cell line	Human	19
miRNA-15	Decrease	In vitro stimulation	PMNs	Human	20
miRNA-16	Decrease	In vitro stimulation	PMNs	Human	20
miRNA-214	Decrease	In vitro stimulation	PMNs	Human	20
miRNA-214	Decrease	LPS stimulation	Macrophage	Mouse	21
ND	Increase	PD patients	Lymphocyte and neutrophil	Human	6

CREB, cAMP response element-binding protein; NF-κB, nuclear factor-kappaB; NF-1, nuclear factor-1; YY-1, Yin Yang-1; ZBP-89, zinc finger DNA-binding protein 89; miRNA, microRNA; PPARγ, peroxisome proliferator activated receptor gamma; TNF-α, tumor necrosis factor-alpha; IL-1β, interleukin-1 Beta; SAM, *S*-adenosyl–L-methionine; LPS, lipopolysaccharides; HD, Huntington's disease; AD, Alzheimer's disease; PD, Parkinson's disease; PC12, rat adrenal gland pheochromocytoma; RBA-2, rat brain astrocyte-2; SH-SY5Y, human neuroblastoma; U87-MG, human glioblastoma astrocytoma; A549, human lung adenocarcinoma; PMNs, human polymorphonuclear neutrophil; ND, not determined.

(exons 2 and 3; Fig. 5.1) that are interrupted with one intron. The $A_{2A}R$ gene has multiple promoters that lead to the production of various $A_{2A}R$ transcripts.[27–29] Each $A_{2A}R$ transcript contains the same coding region plus an identical 3′ untranslated region (UTR) and a distinct 5′UTR. Such gene features of $A_{2A}Rs$ are conserved among species (Fig. 5.1) and are similar to those of several important genes (e.g., brain-derived neurotrophic factor, BDNF).[34] Regulation of BDNF is of particular interest because different stimuli have been shown to specifically regulate distinct BDNF promoters.[34,35] Such a strategy for alternative promoter usage might lead to the production of transcripts with various 5′UTRs that control the transport, the translation efficiency, and the selection of the translational start site of the targeted transcript.[36,37]

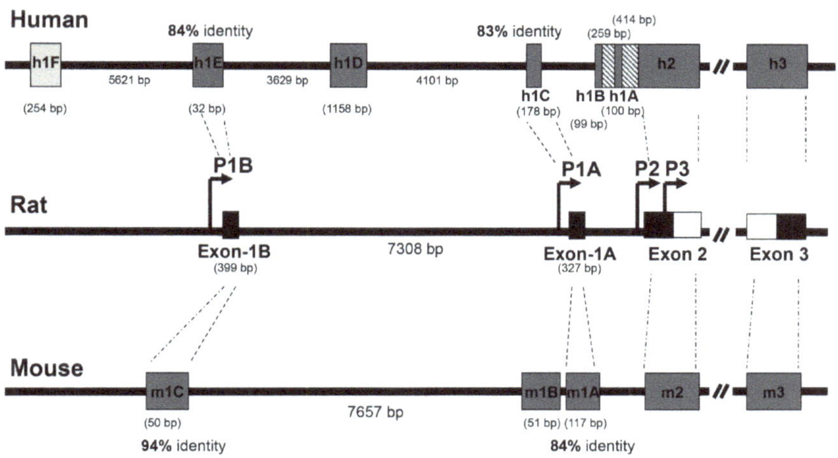

FIGURE 5.1 **The gene structures of human, rat, and mouse A_{2A} adenosine receptors ($A_{2A}R$).** The genomic feature of the $A_{2A}R$ gene is conserved among humans, rats, and mice. Boxes denote exons of the $A_{2A}R$ genes as previously reported.[27–33] The coding regions of $A_{2A}R$ are located in exons 2 and 3, which are interrupted with an intron of approximately 7 kb. The white and black boxes of the rat $A_{2A}R$ gene represent its coding and noncoding regions. At least four independent promoters, which produce multiple transcripts that contain the same coding and 3′ untranslated regions, but distinct 5′ untranslated regions (5′UTR), were identified in the rat $A_{2A}R$ gene. Likewise, multiple human $A_{2A}R$ transcripts with various lengths of 5′UTRs were also reported as detailed elsewhere. A functional GpG-rich region (namely CpG island) exists around h1E, suggesting that DNA methylation plays a critical role in the gene regulation of the $A_{2A}R$.[10,14]

Consistent with this hypothesis, Kreth and colleagues reported that human polymer-phonuclear neutrophils (PMNs) mainly express $A_{2A}R$ transcripts with a long 5′UTR at the resting level. Conversely, LPS-stimulated PMNs mostly express $A_{2A}R$ transcripts with short 5′UTRs, which can be translated more efficiently than those with long 5′UTRs.[29] This finding is consistent with an earlier study reporting that the 5′UTRs of the rat $A_{2A}R$ gene play a negative role in the translation of $A_{2A}R$ protein.[31] Multiple $A_{2A}R$ promoters and transcripts were also reported in mice. In a murine macrophage cell line (RAW264.7), treatment with LPS activated the P1B promoter via an NF-κB-dependent pathway and produced an $A_{2A}R$ transcript comprising m1C, m2, and m3.[38] Further investigation of the alternative promoter usage strategy of the $A_{2A}R$ gene may reveal novel insights into the transcriptional control of $A_{2A}R$ under various physiological or pathological conditions.

Given the importance of the $A_{2A}R$ in the striatum, the element that dictates the enriched expression of $A_{2A}R$ in the striatum has attracted much attention. Exogenous expression of a 4.8-kb DNA fragment that contains three promoters of the rat $A_{2A}R$ gene (P1A, P2, and P3;

Fig. 5.1) in vivo does not lead to a strong expression of the β-galactosidase reporter,[39] suggesting that the element that governs the high striatal expression of the A_{2A}R is missing. Conversely, this 4.8-kb DNA fragment drives the expression of the transgene in most A_{2A}R-expressing brain cells, but not in several major peripheral tissues. Thus, the expression pattern of A_{2A}R in the brain is likely controlled by elements located in this fragment. Further studies on the potential usage of P1B in the striatum are of great interest because a functional GpG-rich region (i.e., CpG island) exists around the exon h1E of the human A_{2A}R gene (Fig. 5.1).[10,14] In addition, DNA methylation appears to modulate the striatal level of A_{2A}Rs.[12,15] Ample evidence also suggests that DNA methylation plays a critical role in the regulation of A_{2A}R gene in various diseases, which will be further elaborated in the following sections.

EPIGENETICS OF *ADORA2A*

The use of agonists and antagonists for A_{2A}R has some limitations due to its ubiquity and multifunctionality in the organism, leading to the appearance of side effects. As an example, A_{2A}R has a vasodilating and an antiinflammatory effect (see Ref. 40; Cronstein, 1994). Therefore, the modulation of A_{2A}R expression levels could be a good strategy to modulate its cellular signaling. The role of epigenetics in *ADORA2A* was an unknown field, but the study of miRNAs, transcription factors, and DNA methylation involved in its gene regulation has evolved in recent years (Table 5.1). For instance, in polymorphonuclear leukocytes, the increase in A_{2A}R mRNA expression upon stimulation is inversely correlated with the expression levels of miRNA-214, miRNA-15, and miRNA-16.[20] However, these miRNAs where not found to be altered in the putamen of PD patients, where reduced miRNA34b levels were related to increased A_{2A}R levels.[11]

Similar data were shown for DNA methylation in different neurological scenarios. Increased 5-methylcytosine levels were found in the putamen of HD patients and in a subgroup of patients with schizophrenia suffering motor disturbances with respect to control cases.[12,13] By contrast, reduced 5-methylcytosine levels in specific CpG sites were identified in PD patients.[11]

USE OF DNA METHYLATING AGENTS AS A THERAPEUTIC TOOL

The DNA methylation and demethylation agents have a potential use as therapeutic drugs in gene modulation. As mentioned before, A_{2A}R is particularly enriched in the striatum, and it is related to the percentage of DNA in its 5′UTR. This percentage is inversely correlated with $A_{2A}R$ mRNA levels; it is highly hypomethylated in the putamen with respect to the cerebellum, where $A_{2A}R$ levels are very low.[10] Moreover, in vitro and in vivo studies have revealed that the use of a methylating agent (*S*-adenosylmethionine, SAM) increases the percentage of DNA methylation in *ADORA2A*, increasing the binding of a methyl-binding protein ($MeCP_2$) and, in turn, reducing the $A_{2A}R$ mRNA expression levels.[14] The same effect was observed in an in vivo study, which also demonstrated experimentally that SAM is able to cross the brain–blood barrier.[15]

As $A_{2A}R$ shows particular cerebral region enrichment, the use of DNA methylating agents can confer certain tissue specificity in its striatal expression. For instance it could be a good strategy in PD where striatal $A_{2A}R$ levels are increased.[22] By contrast, the use of DNA demethylating agents may be a tool to increase striatal $A_{2A}R$ levels in HD. Unfortunately, DNA methylating and demethylating agents lack gene specificity. However, azacytidine has been approved for some cancer treatments,[41] while SAM has been used to treat depression.[42–44]

THE DUAL CODON PROPERTIES OF THE $A_{2A}R$ GENE

Evidence from different laboratories using various technologies (including ribosome profiling and mass spectrometry-based proteomics) have clearly demonstrated that multiple proteins might be translated from different frames of a single transcript in mammalian cells.[36,42–44] Proteins resulting from such dual-coding genes may functionally interact either directly or indirectly.[45–47] In addition to the $A_{2A}R$ protein, a novel 134-amino acid protein (designated uORF5) can be translated from a different reading frame of the rat $A_{2A}R$ gene.[31,46,48] Briefly, uORF5 is translated from an upstream open reading frame (ORF) that overlaps with that for the $A_{2A}R$ protein in the rat *Adora2a* gene (Fig. 5.2). Expression of uORF5 is found in the rat striatum and a rat pheochromocytoma line (PC12), where the $A_{2A}R$ mRNA is abundant. Most interestingly, activation of the $A_{2A}R$ leads to an elevation of uORF5 protein at the posttranscriptional level in a protein kinase A-dependent pathway,[31] suggesting that uORF5 might function downstream of $A_{2A}R$ and serve as a tightly orchestrated set of signals that mediate the actions of $A_{2A}R$.

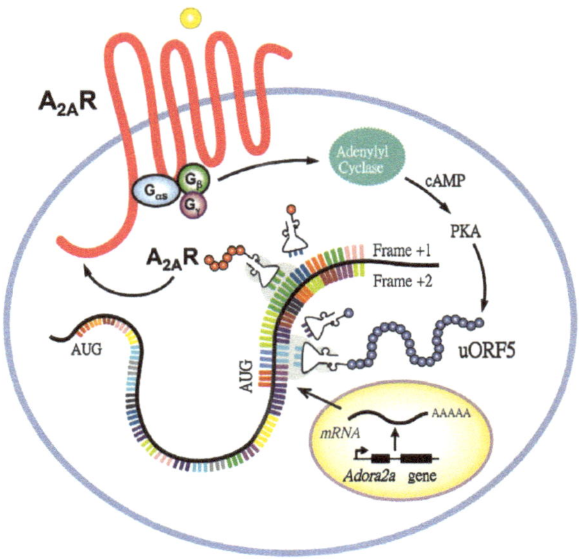

FIGURE 5.2 **The A_{2A} adenosine receptor (A_{2A}R) is a dual-coding gene.** A_{2A}R is a major target of caffeine. The transcript of the A_{2A}R contains two open reading frames and produces two structurally irrelevant proteins, A_{2A}R and uORF5, by using two different translational start sites. The uORF5 protein might function downstream of A_{2A}R because stimulation of the A_{2A}R enhances production of uORF5 through a protein kinase A (PKA)-dependent pathway.[46]

The dual-coding property of the *Adora2a* gene is conserved evolutionarily. The human and mouse *Adora2a* genes both contain coding regions for uORF5-like proteins. At present, the function of uORF5 in vivo is largely unknown. Exogenous expression of uORF5 protein in PC12 cells inhibited the nerve growth factor-evoked AP1-mediated transcript and altered the expression of several proteins that have been annotated in the mitogen-activated protein kinase pathway. In addition to A_{2A}R stimulation, expression of uORF5 can also be regulated by increasing the expression of the A_{2A}R transcript. For example, hypoxia up-regulates the transcript level of the A_{2A}R and the protein levels of both A_{2A}R and uORF5.[46] This finding is of great interest because a recent study reported that deprivation of oxygen and/or glucose during tissue damage affects the translation efficiency of many mRNAs and changes the recognition of translation initiation sites for some transcripts (including those of a few dual-coding genes).[37] Further studies using antibodies to detect the expression of uORF5 protein under various pathophysiological conditions and in genetic animal models to assess the function of uORF5 proteins in vivo would greatly enhance our current understanding of the role the A_{2A}R gene plays in human health and disease.

CONCLUDING REMARKS

Although *ADORA2A* gene is made up of three exons, exon 1 is non-translated, showing a great dynamism in its transcription, resulting in six tissue-specific isoforms regulated by several transcription factors and a CpG island located in the 5′UTR region. In this sense, several reports have revealed its gene modulation by DNA methylating and demethylating drugs. However, these drugs show a great spectrum of gene action, and therefore a great effort is still needed to identify epigenetic drugs with gene specificity.

Regarding its gene regulation by miRNAs and transcription factors, more studies are necessary to unravel its identification and its role in each pathological context. Most of the data obtained have been drawn in cell lines or in different pathological contexts, which have made it clear that gene regulation depends on the cellular scenario. Therefore, further studies are necessary to elucidate the molecular mechanisms that control the expression levels of $A_{2A}R$ in the entire organism in order to be able to identify specific targets to take advantage of its pharmacological control in a cell-specific manner.

Acknowledgments

This work was supported by grants from the Ministerio de Ciencia e Innovación, Instituto de Salud Carlos III (PI05/1631 and CP08/00095 to M.B.) (Spain), the Ministrant of Technology and Science of Taiwan (NSC96–2321-B-001–015, NSC97–2321-B-001–012, and 100–2320-B-001–0110MY3 to Y. Chern), and the Institute of Biomedical Sciences of Academia Sinica (103-Academia Sinica Investigation Award-06 to Y. Chern). We are grateful to Mr. Tsung-Hung Hung for his assistance in preparing the illustrations.

References

1. Weaver DR. A2a adenosine receptor gene expression in developing rat brain. *Brain Res Mol Brain Res*. 1993;20:313–327.
2. Soellner DE, Grandys T, Nunez JL. Chronic prenatal caffeine exposure impairs novel object recognition and radial arm maze behaviors in adult rats. *Behav Brain Res*. 2009;205:191–199.
3. Albasanz JL, Perez S, Barrachina M, Ferrer I, Martín M. Up-regulation of adenosine receptors in the frontal cortex in Alzheimer's disease. *Brain Pathol*. 2008;18:211–219.
4. Orr AG, Hsiao EC, Wang MM, Ho K, Kim DH, Wang X, et al. Astrocytic adenosine receptor A2A and Gs-coupled signaling regulate memory. *Nat Neurosci*. 2015;18:423–434.
5. Calon F, Dridi M, Hornykiewicz O, Bedard PJ, Rajput AH, Di Paolo T. Increased adenosine A2A receptors in the brain of Parkinson's disease patients with dyskinesias. *Brain*. 2004;127:1075–1084.
6. Varani K, Vincenzi F, Tosi A, Gessi S, Casetta I, Granieri G, et al. A2A adenosine receptor overexpression and functionality, as well as TNF-alpha levels, correlate with motor symptoms in Parkinson's disease. *FASEB J*. 2010;24:587–598.

7. Chiang MC, Lee YC, Huang CL, Chern Y. cAMP-response element-binding protein contributes to suppression of the A2A adenosine receptor promoter by mutant Huntingtin with expanded polyglutamine residues. *J Biol Chem*. 2005;280: 14331–14340.
8. Xie X, Jhaveri KA, Ding M, Hughes LF, Toth LA, Ramkumar V. Expression of striatal adenosine and dopamine receptors in mice deficient in the p50 subunit of NF-kappaB. *Life Sci*. 2007;81:1031–1041.
9. Lee YC, Lai HL, Sun CN, Chien CL, Chern Y. Identification of nuclear factor 1 (NF1) as a transcriptional modulator of rat A(2A) adenosine receptor. *Brain Res Mol Brain Res*. 2003;111:61–73.
10. Buira SP, Dentesano G, Albasanz JL, Moreno J, Martin M, Ferrer I, et al. DNA methylation and Yin Yang-1 repress adenosine A2A receptor levels in human brain. *J Neurochem*. 2010;115:283–295.
11. Villar-Menendez I, Porta S, Buira SP, Pereira-Veiga T, Diaz-Sanchez S, Albasanz JL, et al. Increased striatal adenosine A2A receptor levels is an early event in Parkinson's disease-related pathology and it is potentially regulated by miR-34b. *Neurobiol Dis*. 2014;69:206–214.
12. Villar-Menendez I, Blanch M, Tyebji S, Pereira-Veiga T, Albasanz JL, Martin M, et al. Increased 5-methylcytosine and decreased 5-hydroxymethylcytosine levels are associated with reduced striatal A2AR levels in Huntington's disease. *Neuromol Med*. 2013;15:295–309.
13. Villar-Menéndez I, Díaz-Sánchez S, Blanch M, Albasanz JL, Pereira-Veiga T, Monje A, et al. Reduced striatal adenosine A2A receptor levels define a molecular subgroup in schizophrenia. *J Psychiatr Res*. 2014;51:49–59.
14. Buira SP, Albasanz JL, Dentesano G, Moreno J, Martin M, Ferrer I, et al. DNA methylation regulates adenosine A(2A) receptor cell surface expression levels. *J Neurochem*. 2010;112:1273–1285.
15. Villar-Menendez I, Nunez F, Diaz-Sanchez S, Albasanz JL, Taura J, Fernandez-Duenas V, et al. Striatal adenosine A2A receptor expression is controlled by S-adenosyl-L-methionine-mediated methylation. *Purinergic Signal*. 2014;10:523–528.
16. Lee CF, Chern Y. Adenosine receptors and Huntington's disease. *Inter Rev Neurobiol*. 2014;119:195–232.
17. He X, Hu JL, Li J, Zhao L, Zhang Y, Zeng YJ, et al. A feedback loop in PPARgamma-adenosine A2A receptor signaling inhibits inflammation and attenuates lung damages in a mouse model of LPS-induced acute lung injury. *Cell Signal*. 2013;25:1913–1923.
18. Bruzzese L, Fromonot J, By Y, Durand-Gorde JM, Condo J, Kipson N, et al. NF-kappaB enhances hypoxia-driven T-cell immunosuppression via upregulation of adenosine A(2A) receptors. *Cell Signal*. 2014;26:1060–1067.
19. Morello S, Ito K, Yamamura S, Lee KY, Jazrawi E, Desouza P, et al. IL-1 beta and TNF-alpha regulation of the adenosine receptor (A2A) expression: differential requirement for NF-kappa B binding to the proximal promoter. *J Immunol*. 2006;177: 7173–7183.
20. Heyn J, Ledderose C, Hinske LC, Limbeck E, Mohnle P, Lindner HA, et al. Adenosine A2A receptor upregulation in human PMNs is controlled by miRNA-214, miRNA-15, and miRNA-16. *Shock*. 2012;37:156–163.
21. Zhao L, Liu YW, Yang T, Gan L, Yang N, Dai SS, et al. The mutual regulation between miR-214 and A2AR signaling plays an important role in inflammatory response. *Cell Signal*. 2015;27:2026–2034.
22. Barrachina M, Martín M, Ciruela F, Ferrer I. Towards new therapies for Parkinson's disease. ISBN: 978–953–307–463–4, InTech In: David Finkelstein, ed. *Epigenetic Modulation of Adenosine A2A Receptor: A Putative Therapeutical Tool for the Treatment of Parkinson's Disease*. 2011:295–312.

23. Fortin A, Harbour D, Fernandes M, Borgeat P, Bourgoin S. Differential expression of adenosine receptors in human neutrophils: up-regulation by specific Th1 cytokines and lipopolysaccharide. *J Leukoc Biol*. 2006;79:574–585.
24. Khoa ND, Montesinos MC, Reiss AB, Delano D, Awadallah N, Cronstein BN. Inflammatory cytokines regulate function and expression of adenosine A(2A) receptors in human monocytic THP-1 cells. *J Immunol*. 2001;167:4026–4032.
25. Trincavelli ML, Costa B, Tuscano D, Lucacchini A, Martini C. Up-regulation of A(2A) adenosine receptors by proinflammatory cytokines in rat PC12 cells. *Biochem Pharmacol*. 2002;64:625–631.
26. Bshesh K, Zhao B, Spight D, Biaggioni I, Feokistov I, Denenberg A, et al. The A2A receptor mediates an endogenous regulatory pathway of cytokine expression in THP-1 cells. *J Leukoc Biol*. 2002;72:1027–1036.
27. Chu YY, Tu KH, Lee YC, Kuo ZJ, Lai HL, Chern HY. Characterization of the rat A2a adenosine receptor gene. *DNA Cell Biol*. 1996;15:329–337.
28. Fredholm BB, Chern Y, Franco R, Sitkovsky M. Aspects of the general biology of adenosine A2A signaling. *Prog Neurobiol*. 2007;83:263–276.
29. Kreth S, Ledderose C, Kaufmann I, Groeger G, Thiel M. Differential expression of 5′-UTR splice variants of the adenosine A2A receptor gene in human granulocytes: identification, characterization, and functional impact on activation. *FASEB J*. 2008;22:3276–3286.
30. Le F, Townsend-Nicholson A, Baker E, Sutherland GR, Schofield PR. Characterization and chromosomal localization of the human A2a adenosine receptor gene: ADORA2A. *Biochem Biophys Res Commun*. 1996;223:461–467.
31. Lee YC, Chang CW, Su CW, Lin TN, Sun SH, Lai HL, et al. The 5′ untranslated regions of the rat A2A adenosine receptor gene function as negative translational regulators. *J Neurochem*. 1999;73:1790–1798.
32. Peterfreund RA, MacCollin M, Gusella J, Fink JS. Characterization and expression of the human A2a adenosine receptor gene. *J Neurochem*. 1996;66:362–368.
33. Yu L, Frith MC, Suzuki Y, Peterfreund RA, Gearan T, Sugano S, et al. Characterization of genomic organization of the adenosine A2A receptor gene by molecular and bioinformatics analyses. *Brain Res*. 2004;1000:156–173.
34. Kim DS, Lee SJ, Cho HJ. Differential usage of multiple brain-derived neurotrophic factor promoter in rat dorsal root ganglia following peripheral nerve injuries and inflammation. *Brain Res Mol Brain Res*. 2001;92:167–171.
35. Rattiner LM, Davis M, Ressler KJ. Differential regulation of brain-derived neurotrophic factor transcripts during the consolidation of fear learning. *Learn Mem*. 2004;11:727–731.
36. Mouilleron H, Delcourt V, Roucou X. Death of a dogma: eukaryotic mRNAs can code for more than one protein. *Nucleic Acids Res*. 2016;44:14–23.
37. Andreev DE, O'Connor PB, Zhdanov AV, Dmitriev RI, Shatsky IN, Papkovsky DB, et al. Oxygen and glucose deprivation induces widespread alterations in mRNA translation within 20 minutes. *Genome Biol*. 2015;16:90.
38. Elson G, Eisenberg M, Garg C, Outram S, Ferrante CJ, Hasko G, et al. Induction of murine adenosine A(2A) receptor expression by LPS: analysis of the 5′ upstream promoter. *Genes Immun*. 2013;14:147–153.
39. Lee YC, Chien CL, Sun CN, Huang CL, Huang NK, Chiang MC, et al. Characterization of the rat A2A adenosine receptor gene: a 4.8-kb promoter-proximal DNA fragment confers selective expression in the central nervous system. *Eur J Neurosci*. 2003;18:1786–1796.
40. Barraco M, Phillis JW. Subtypes of adenosine receptors in the brainstem mediate opposite blood pressure responses. *Neuropharmacology*. 1991;30:403–407.

41. Kaminskas E, Farrell AT, Wang YC, Sridhara R, Pazdur R. FDA drug approval summary: azacitidine (5-azacytidine, Vidaza) for injectable suspension. *Oncologist.* 2005;10:176–182.
42. Bottiglieri T, Godfrey P, Flynn T, et al. Cerebrospinal fluid S-adenosylmethionine in depression and dementia: effects of treatment with parenteral and oral S-adenosylmethionine. *J. Neurol. Neurosurg. Psychiatry.* 1990;53(12):1096–1098.
43. Di Rocco A, Rogers JD, Brown R, et al. S-adenosyl-methionine improves depression in patients with Parkinson's disease in an open-label clinical trial. *Mov. Disord.* 2000 Nov;15(6):1225–1229.
44. Papakostas GI. Evidence for S-adenosyl-L-methionine (SAM-e) for the treatment of major depressive disorder. *J. Clin. Psychiatry.* 2009;70(Suppl 5):18–22.
45. Chung WY, Wadhawan S, Szklarczyk R, Pond SK, Nekrutenko A. A first look at ARFome: dual-coding genes in mammalian genomes. *PLoS Comput Biol.* 2007;3:e91.
46. Liang H, Landweber LF. A genome-wide study of dual coding regions in human alternatively spliced genes. *Genome Res.* 2006;16:190–196.
47. Ribrioux S, Brungger A, Baumgarten B, Seuwen K, John MR. Bioinformatics prediction of overlapping frameshifted translation products in mammalian transcripts. *BMC Genomics.* 2008;9:122.
48. Klemke M, Kehlenbach RH, Huttner WB. Two overlapping reading frames in a single exon encode interacting proteins—a novel way of gene usage. *EMBO J.* 2001;20:3849–3860.
49. Lee CF, Lai HL, Lee YC, Chien CL, Chern Y. The A2A adenosine receptor is a dual coding gene: a novel mechanism of gene usage and signal transduction. *J Biol Chem.* 2014;289:1257–1270.
50. Bergeron D, Lapointe C, Bissonnette C, Tremblay G, Motard J, Roucou X. An out-of-frame overlapping reading frame in the Ataxin-1 coding sequence encodes a novel ataxin-1 interacting protein. *J Biol Chem.* 2013;288:21824–21835.
51. Chern Y, King K, Lai HL, Lai HT. Molecular cloning of a novel adenosine receptor gene from rat brain. *Biochem Biophys Res Commun.* 1992;185:304–309.
52. Cronstein BN. Adenosine, an endogenous anti-inflammatory agent. *J. Appl. Physiol. (1985).* 1994;76:5–13.

SECTION II

PATHOLOGY

CHAPTER

6

Sleep, Adenosine, and Neurodegeneration

Theresa E. Bjorness[1,2] *and Robert W. Greene*[1,2]

[1]University of Texas Southwestern Medical Center, Dallas, TX, United States [2]Veterans Administration Medical Center, Dallas, TX, United States

OUTLINE

Adenosine Receptors in Neurodegenerative Diseases.
DOI: http://dx.doi.org/10.1016/B978-0-12-803724-9.00006-5

SLEEP PHENOMENOLOGY AND ADENOSINE

Sleep is a behavioral state characterized by four main criteria, including: characteristic changes in the electroencephalogram (EEG), species-specific posture, increased response threshold to external stimuli, and quick reversibility.[1] Sleep consists of two separate states: Rapid eye movement sleep (REM) and nonrapid eye movement sleep (NREM). REM is also called paradoxical sleep or dream sleep. EEG and electromyogram (EMG) signals are used to differentiate sleep and waking states. REM is characterized by desynchronized cortical EEG activity similar to waking, loss of tone in most skeletal muscles, and phasic activities such as the rapid eye movements for which this state is named. NREM sleep is characterized by relatively synchronized cortical EEG with reduced muscle tone. In humans, NREM is divided into 3 stages that have differential prominent activity in the EEG. Stages 1 and 2 consist of low amplitude, mixed frequency EEG activity, with spindles and K complexes occurring during stage 2. Stage 3–4 (formerly stages 3 and 4) consists of high amplitude, low frequency (0.5–4.5 Hz) EEG activity and is often called slow wave sleep (SWS) due to the prominence of this slow wave activity (SWA).

Sleep is controlled by two processes[2,3]: (1) arousal driven either by internal stimuli, including a circadian process which provides control on the basis of the time of day (e.g., humans and other primates are diurnal and sleep primarily at night), or by external stimuli, and (2) a homeostatic process which provides control on the basis of time awake (e.g., there is a stronger sleep drive after 20 h of waking compared to 10 h of waking). There is accumulating evidence that these two processes are not entirely independent[4]; at the molecular level, circadian rhythm-related genes such as Per and Cry show homeostatic changes that may influence sleep need.[5,6] Homeostatic influence has not been linked to a single brain region; however, the neuromodulator adenosine (ado) is known to influence sleep homeostasis on the basis of several lines of evidence.[7,8] Ado levels progressively accumulate with prolonged waking in the cholinergic basal forebrain(BF),[9] a nucleus involved in wake maintenance. Acting through ado A_1 receptors (ADORA1), ado inhibits wake-active neurons in this region resulting in decreased arousal.[10–15] Additionally, ado can also promote sleeping behavior through stimulatory action via A_{2a} receptors (ADORA2) on sleep active neurons in the posterior hypothalamus.[16] Ado agonists and antagonists increase and decrease sleep, respectively.[17] People often self-medicate with caffeine in order to reduce sleepiness, making caffeine one of the most commonly used psychoactive substances. Although caffeine is a nonselective ado antagonist, ADORA2 blockade provides the strongest wake-promoting effects of caffeine as determined by the use of selective *Adora1* and *Adora2* knockout mice.[18] Recently, Lazarus and colleagues showed that caffeine promotes waking by blocking ADORA2s on GABAergic neurons in the shell of the nucleus accumbens thereby reducing inhibition of wake-promoting brain regions such the locus coeruleus (LC) and lateral hypothalamus (LH).[19]

Sleep homeostasis is often indicated by SWA power, also called delta power, a 0.5–4.5 Hz oscillation in membrane potential, i.e., generally accepted as a biomarker for sleep need; SWA power is increased with progressive waking and decreases during subsequent sleep.[20] Ado activity through ADORA1 is necessary for expression of rebound SWA power following sleep deprivation (SD).[21] Recently, it has been suggested that astrocytes are an important source of ado through the release of ATP that is degraded into ado by ectonucleotidases; blocking astrocytic release of ATP attenuates extracellular ado levels and sleep need in response to prolonged waking.[22,23] Mice lacking 5′ ectonucleotidase through genetic knockout show reduced sleep homeostasis.[24] However, this does not address the source of the ATP as either neuronal or glial or both.

Adenosine kinase (AdK) is the highest affinity enzyme metabolizing Ado by binding ATP and Ado and releasing AMP and ADP. It is predominantly expressed in astrocytes in adults,[25] and a glial specific knockout of this enzyme increases SWA power under baseline conditions and slows the decay of SWA during an average SWS episode.[26] Since SWA rebound in response to SD is lost in a neuronal specific *Adora1* knockout; a glial-neuronal circuit mediated by Ado is suggested. In sum, ado is a key mediator of homeostatic sleep need expression influencing sleep need through a glial-neuronal circuit.

Sleep and EEG parameters are under genetic control with trait-like stability within individuals, and variability between individuals.[27] The ado system is a source of variability in baseline sleep and susceptibility to sleep loss-induced deficits that occur across the general population. Polymorphisms for ado deaminase (ADA), which converts ado to inosine, and in *Adora2* confer vulnerability to sleep loss-induced performance deficits[28] as well as differences in EEG activity under baseline conditions,[29] and in the case of ADA polymorphisms, SWS amount.

SLEEP AND ADENOSINE CHANGES ACROSS NORMAL AGEING

Many neurodegenerative disorders are more common in middle aged and elderly populations, therefore it is important to consider the changes in sleep/waking that occur naturally across the lifecycle.

There have been disparate reports regarding the influence of ageing on sleep, however a meta-analysis of polysomnography (PSG) studies found increases in stage 1 and 2 NREM, decreases in SWS and REM, decreases in REM latency, decreases in total sleep time (TST) and sleep efficiency (SE), and increased wake after sleep onset (WASO). The increase in WASO across ageing showed the highest effect size with waking during the sleep period increasing by about 10 min/decade of life from 30 years onward.[30] These authors, and others, suggest that sleep disturbance in elderly adults is often the result of medical issues that impact sleep/waking behavior.[31,32] Alternatively, there is a profound reduction in SWA power with ageing,[33] which along with EEG changes during waking, makes sleep scoring more difficult.

Homeostatic sleep rebound, measured as the increase in SWA power or sleep time following SD, is decreased in aged animals,[34,35] although aged humans show relatively similar homeostatic sleep response compared to younger people, though the time course of recovery is altered.[32,36] These changes in the homeostatic sleep response are likely due, at least in part, to age-related changes in the ado system. Ado levels are higher in the BF under both baseline and SD conditions in aged animals; however, signaling through ADORA1s is reduced as determined by sleep response following ado and ADORA1 agonist administration into the BF.[37] There have also been reports of decreased ADORA1 expression across ageing in humans and other mammals.[38,39] Interestingly, Meerlo and colleagues measured ADORA1 expression via autoradiography in young (3 months), old (24 months) and very old (30 months) rats and found that ADORA1 expression showed variability across brain regions, but also that the pattern of expression varied with some brain regions showing a gradual decline across age, while others showed stability through 24 months followed by a steep decline by 30 months.[40] Furthermore, coupling of ADORA1 to inhibitory G protein is reduced in peripheral tissue of aged animals,[41,42] thereby decreasing ADORA1-mediated changes in adenylate cyclase activity.[42] In addition, ecto-5′-nucleotidase and AdK levels (protein and mRNA) increase in aged animals compared to young.[43] Although ecto-5′-nucleotidase and AdK have opposing outcomes for ado levels (increase and decrease, respectively), increased basal levels of ado seen in ageing would reduce AdK activity through allosteric inhibition,[44,45] thereby resulting in a net increase in basal ado levels. Finally, reductions in daily sleep homeostasis in aged humans can occur because of increased napping shortly before bedtime,[46] which would serve to reduce the drive for subsequent sleep at night.

SLEEP CHANGES IN ALZHEIMER'S DISEASE

Alzheimer's disease (AD) is a progressive from of dementia that occurs with increased prevalence across ageing and is characterized by progressive memory loss and neuropathological features such as senile plaques and neurofibrillary tangles.[47] AD is diagnosed on the basis of decreases in cognitive function and changes in molecular and neurodegenerative biomarkers such as Aβ42, phosphorylated tau, and fibrillary Aβ deposits. However, caution is necessary when considering biomarker evidence; cerebrospinal fluid (CSF) biomarkers show both intralaboratory and interlaboratory variability,[48] neuropathic signs of AD are found in about 30% of elderly people who did not have an AD diagnosis[49] and Aβ pathology increases can occur up to 20 years prior to the onset of dementia.[50]

Sleep disturbances are commonly reported in AD,[51,52] with severity of disruptions progressively increasing across AD stages.[53] PSG studies have found decreased SWS, decreased REM sleep, increased sleep fragmentation, and decreased SE, which is the TST relative to the total time in bed.[54,55] In addition to changes in sleep/waking amount, there are also alterations in underlying EEG activity during these states. AD is marked by an EEG slowing during waking and REM.[56,57] There is also a reduction in spindle and K complex activity during stage 2 NREM that can make it difficult to discriminate NREM stage 2 from NREM stage 1, leading to the use of "indeterminate NREM" when staging sleep in AD patients.[54,55,58] Overall, these EEG changes can make it more difficult to differentiate sleep/waking states in AD patients. Sleep changes, while common in AD, are not considered useful for diagnosing or staging AD.[59]

MECHANISMS OF ALZHEIMER'S DISEASE INFLUENCE ON SLEEP

Several neurotransmitter and neuromodulator systems implicated in sleep/waking induction/maintenance are altered in AD. Most notably, cholinergic and noradrenergic systems which promote arousal and REM sleep in the case of brainstem cholinergic neurons[60] are greatly reduced in AD.[61–63] While remaining LC neurons show compensatory responses,[64,65] there is disagreement in the literature as to whether there are alterations in overall noradrenergic tone.[66,67] Data gathered using mouse models of AD suggests that a reduction of LC neurons by a selective toxin may worsen the progression of AD by exacerbating cholinergic deficits[68] and increasing amyloid plaque formation.[69] Evidence from rodent studies suggests that cholinergic neurons may be particularly vulnerable to the neurotoxic effects of Aβ,[70] while noradrenergic neurons may protect against Aβ toxicity by acting through neurotrophic TrkB receptors.[71] It is clear that cholinergic and noradrenergic neuron loss is a major contributor to AD's onset and progression.

Additionally, the orexin (also called hypocretin) system, which is important for waking maintenance and sleep consolidation,[72] has been implicated in AD-induced sleep changes based on a reduction in orexin positive neurons in postmortem tissue via immunostaining, and reduction in orexin levels in the ventricular CSF from this same tissue.[73] However, orexin levels measured in the CSF of living AD subjects have been reported to either be within the normal range[74] or increased.[75] Alternatively, changes in sleep-promoting regions have also been reported in AD. The intermediate nucleus of the hypothalamus has been proposed as the human homolog of the VLPO,[76] which is a galaninergic nucleus in the hypothalamus that contains sleep active neurons and promotes sleep.[77,78] Decreases in galaninergic neurons in the intermediate nucleus have been reported in people with AD along with an association between the number of galanin positive neurons in the intermediate nucleus and sleep fragmentation as measured by actigraphy.[79]

CHANGES IN ADORA1 AND ADORA2 EXPRESSION

With respect to ado system changes in AD, ADORA1 and ADORA2 binding and expression have been investigated in postmortem frontal cortex and hippocampus. Increases in ADORA1 and ADORA2 protein as measured by autoradiography and western blots have been reported in the frontal cortex,[80] though there was a decrease in Kd for the ADORA2 receptor, suggesting a decrease in binding affinity. There were no changes in ADORA1 and ADORA2 mRNA along with no changes in adenylate cyclase, though there was an increase in response to forskolin stimulation suggesting a sensitization of the ado/adenylate cyclase system. Both increases[81] and decreases[82,83] in ADORA1 binding by immunohistochemistry (increases) and autoradiography (decreases) have been reported in the hippocampus of AD patients. Additionally, Angelo and colleagues found that ADORA1s were localized to plaques and tangles.[81] These same authors reported a slight increase in ADORA2s, these being localized to glia. Of note, there is evidence that caffeine intake (primarily studied as coffee and tea consumption) is inversely associated with AD incidence.[84] Evidence from animal studies suggest that caffeine reduces interstitial Aβ levels and Aβ deposits[85] and that caffeine-associated protective effects on cognitive function are dependent on ADORA2 blockade.[86,87]

ASSOCIATION OF ALZHEIMER'S DISEASE BIOMARKERS AND SLEEP

Several AD-related biomarkers have been associated with sleep alterations. In a population of middle-aged adults, some of which had a family history of sporadic AD, those with presumptive Aβ deposits as determined by CSF Aβ42 levels showed decreased SE, increased WASO, and more frequent napping.[88] Lower subjective sleep amount was associated with higher plaque burden as determined by PiB staining in a set of healthy adults.[89] In a mouse model of AD, sleep/waking and Aβ fluctuation across the light/dark cycle was normal prior to the development of plaques, but following plaque development there was an increase in waking and a decreased fluctuation of Aβ across the circadian period.[90] Furthermore, immunizing animals against Aβ42 reduced the plaque burden, normalized sleep/waking activity and Aβ fluctuation.[90]

Additionally, SD can influence AD-related biomarkers. In healthy humans, one night of SD altered the trajectory of CSF Aβ42 compared to controls such that the typical decrease in CSF Aβ42 after the sleep period was not present,[91] although the authors point out that the time course of CSF sampling between groups was slightly different. In a mouse model of AD, chronic, partial sleep restriction for 20 h/day for 3 weeks increased the relative abundance of Aβ plaques,[92] while sleep restriction for 6 h/day for 6 weeks resulted in a nonsignificant increase in Aβ and phosphorylated tau levels in a different AD mouse model.[93]

ROLE OF SLEEP LOSS IN ALZHEIMER'S DISEASE

There have been a few recent seminal papers investigating the role of sleep/waking activity on AD biomarker expression. First, Kang and colleagues showed that humans display a diurnal rhythm of CSF Aβ levels similar to the sleep/waking rhythm, while in mice Aβ levels correlate with waking time, increase following 6 h of SD, and decrease during subsequent sleep.[94] Additionally, these authors showed that (1) administering orexin, which increases waking, also increased CSF Aβ, (2) chronic sleep restriction allowing 4 h of sleep/day for 21 days increased plaque burden in multiple mouse models of AD, and (3) administering a dual orexin receptor antagonist increased sleep, reduced CSF Aβ, and blocked the increase formation of plaques.[94] Next, Xie and colleagues investigated the volume of interstitial space and clearance of Aβ across sleep/waking and anesthetic states. Relative to waking, interstitial space volume was 60% greater during sleep and anesthesia and the rate constant of Aβ clearance was tripled.[95] Administration of noradrenergic receptor antagonists, which increased SWA during waking, thereby mimicking SWS EEG activity, replicated the increase in interstitial space volume and faster Aβ clearance. Importantly, age-related decreases in volume transmission have been reported.[96] Thirdly, several groups have shown that synaptic activity, which is higher during waking compared to sleep,[97] modulates Aβ production and secretion with increased synaptic activity increasing Aβ and decreased synaptic activity decreasing Aβ in vivo.[98,99] Together, these results provide intriguing evidence on the basis of AD-related biomarkers that waking (and the synaptic activity that occurs therein) may promote AD development while sleep, particularly NREM sleep (and the relative decrease in synaptic activity and increase in interstitial space volume therein), may be protective.

Chronic or prolonged sleep loss may also influence AD development by challenging protective systems that counter waking-related cellular, metabolic, and oxidative stress that occur as a consequence of neuronal activity. Relatively short SD (3 h) in mice induces upregulation of SirT3, a protein deacetylase that regulates metabolism and oxidative stress responses, while relatively long SD (8 h) induces neuronal injury in the LC suggesting that the initial protective measures that occur with short SD cannot be maintained with longer deprivation.[100] Additionally, SD induces the unfolded protein response[101] (UPR) which is a protective mechanism to counter oxidative stress in the endoplasmic reticulum (ER) and prevent accumulation of protein aggregates; however, the UPR following SD is impaired in aged animals.[102] Interestingly, there is evidence of UPR activation in postmortem tissue from people with AD,[103] although it is unknown whether these subjects had any sleep disruption. The ER shows age-related changes, including increased susceptibility to oxidative damage[104,105] and ER stress, have been implicated in various neurodegenerative diseases, including AD.[106] Interestingly, wake-active orexinergic and noradrenergic neurons show ER dyshomeostasis with ageing.[107] Furthermore, sleep loss induces oxidative DNA damage in an organ-specific manner, while recovery sleep rescues this effect.[108] One theory of sleep function posits that sleep reduces the concentration of free radicals, thereby acting as an antioxidant,[109] however, this remains controversial due to conflicting evidence in the literature including reports of cell damage following chronic SD[110] or lack thereof.[111] Specific cell types/systems may be more susceptible to SD-related injury; in particular, Katsuki and colleagues found a significant decrease in orexin immunoreactivity in the hypothalamus following 7 days of SD, while melanin-concentrating hormone immunoreactivity was more modestly decreased.[112] This group also demonstrated that orexin neurons are vulnerable to ER stress in culture[113] and high fat diet-induced pathology in vivo.[114] Interestingly, oxidative stress may influence sleep need; infusion of a relatively low dose of an organic peroxide increased sleep and ado levels in rats,[115] while administration of oxidized glutathione promoted sleep in rabbits.[116] In sum, oxidative stress is a known cause of age-related physiological decline[117] and SD likely further taxes the already burdened antioxidant systems, thereby further contributing to neurodegeneration.

Briefly, other possible mechanisms by which sleep loss could influence the development of AD or other neurodegenerative diseases may include a stress response or impairments in the blood–brain barrier (BBB). Sleep loss, particularly chronic sleep loss, is a stressor[118] and induces exaggerated HPA axis responses to other stressors.[119] Acute and chronic stress increased interstitial Aβ levels in a mouse model of AD[120]; effects on Aβ levels were mimicked by microdialysis of corticotropin-releasing factor (CRF) and blocked by microdialysis of a CRF antagonist. In mice, chronic sleep restriction increased BBB permeability and decreased BBB function[121]; BBB permeability recovered following sleep. Increased BBB permeability has been found in people with a high plaque burden[122] and in a mouse model of AD, with BBB changes preceding the appearance of senile plaques.[123]

SLEEP CHANGES IN PARKINSON'S DISEASE

Parkinson's disease (PD),[124] is a degenerative α-synucleinopathy characterized by the presence of both trait motor symptoms such as tremor, rigidity, and akinesia, and nonmotor symptoms, including sleep disturbance.[125] Important pathological features of PD include the appearance of Lewy bodies and death of projection neurons of a variety of cell types, including dopaminergic, cholinergic, and noradrenergic, amongst others,[126] of which the loss of dopaminergic neurons of the substantia nigra (SN) cause the motor symptoms.[127] While most cases of PD disease are of idiopathic origin, PD can be inherited through either dominant or recessive patterns of genes involved in critical cell functions that, when mutated, can result in neurodegeneration through multiple pathways.[128]

Sleep disturbance is a relatively common nonmotor symptom of PD with a wide prevalence range (~20–98%), which may be a consequence of the specific population studied and the method used to determine sleep disturbance. Chaudhari and colleagues created a specific questionnaire to evaluate subjective sleep disturbance in PD, the Parkinson's

Disease Sleep Scale.[129] Using this scale, sleep disturbance was found in ~22% of patients from a UK population, with more severe PD correlating with lower sleep scale scores.[130] The PDSS also correlates with quality of life measures, indicating that sleep dysfunction contributes to overall decreases in wellbeing.[131] Using PSG, SWS, and REM is decreased compared to controls and sleep is highly fragmented,[132] particularly in advanced PD.[133] REM latency is not significantly shorter in PD patient; however, sleep onset REMs have been reported.[133,134] Sleep spindle density is decreased in PD,[132,135] particularly in those with dementia.[136]

Several sleep disorders are commonly reported in patients with PD including insomnia, REM sleep behavior disorder (RBD), and excessive daytime sleepiness (EDS), amongst others.[125] Sleep induction and sleep maintenance problems have been reported and use of sleeping pills is more common than in age matched controls.[137] One relatively large (124 subjects) study of sleep in PD patients and controls found a significant decrease in SE as measured by PSG and an increase in the Insomnia Severity Index for PD patients compared to controls.[138] RBD is a parasomnia in which muscle atonia, the loss of skeletal muscle tone during REM sleep due to active inhibition of motoneurons,[139] is absent, allowing people to act out their dreams, which can lead to severe self or bed partner injury.[140] Reported RBD prevalence in PD patients varies by means of identification; one study in which both PSG and clinical interview were performed found that about 30% of PD patients met the clinical standard for RBD, while only about half of these same PD patients reached the clinical standard for RBD by interview alone.[141] Although RBD can occur secondarily in a wide variety of neurological disorders, this parasomnia often precedes development of PD and other synucleinopathies, such that it is often considered an "early marker" of neurodegenerative disease.[142] Additionally, even PD patients who do not meet the clinical criteria for a diagnosis of RBD show more REM sleep without atonia compared to controls.[141] RBD is also associated with increased risk of psychotic symptoms such as hallucinations and delusions.[143] Like RDB, EDS can precede PD indicating increased risk of PD development,[144] and is associated with more severe PD.[133,145] Furthermore, daytime sleepiness in PD can occur in patients with relatively undisturbed nocturnal sleep, suggesting hypersomnia.[134]

PARKINSON'S DISEASE ROLE IN SLEEP PATHOLOGY

As with AD, PD features neurodegeneration of arousal-promoting brain regions, including those that provide noradrenergic, cholinergic, and dopaminergic signaling. Cell loss in the LC precedes that of DA cell loss[146] and occurs throughout the entire extent of the LC,[147] while cholinergic neurons in both the BF and pedunculopontine tegmentum (PPT) are variably reduced.[148] Loss of DA neurons is often associated with motor dysfunction[127]; however, DA also influences sleep/waking activity[149] and DA loss is hypothesized to contribute to the occurrence of EDS and RBD in PD patients.[150] DA neurons in the SN and ventral tegmental area (VTA) do not change overall firing rate across sleep/waking state[151]; however, within the VTA, dopaminergic neurons switch to a burst firing pattern in REM sleep[152] which is associated with greater DA release. Additionally, DA levels in the nucleus accumbens are higher during waking and REM compared to NREM.[153] Recently a wake-active population of DA neurons was identified in the ventral periaqueductal gray (vPAG); the vPAG has afferent and efferent connections to several different regions involved in sleep/waking activity.[154]

Both arousal-promoting orexin and sleep-promoting MCH neurons in the LH are progressively reduced in PD with increasing disease severity,[155] though there are conflicting reports as to possible changes in orexin levels.[156,157] Alternatively, histaminergic neuron number[158] and expression of L-histidine decarboxylase mRNA[159] are unchanged in PD, while histaminergic fiber density is increased in the SN of PD patients.[160] There is evidence that histamine may exacerbate degeneration of DA neurons[161] making local changes in histamine levels a potential source of DA neurodegeneration.

ROLE OF SLEEP LOSS IN PARKINSON'S DISEASE PATHOLOGY

There have been few reports of the effect of SD on people with PD with conflicting results as to whether SD improves motor symptoms and depression.[162–164] Overall, these findings suggest that there may be subpopulations of PD patients that respond differently to sleep loss.

TREATMENT RELATED ISSUES

Treatment of PD itself can influence sleep; DA modulators can increase or disrupt sleep generally in a dose-dependent manner,[165] while subthalamic stimulation can improve sleep.[166] Disease severity may also influence likelihood of sleep disruption with LDOPA and other DA modulators, disrupting sleep in mild and moderate stages[167] and improving sleep in late stage PD.[168]

Briefly, ADORA2 antagonists have been investigated as potential motor symptom therapeutic targets for several reasons, including: (1) their colocalization with D_2 receptors on GABAergic neurons of the indirect pathway which, when blocked, may correct the imbalance caused by DA depletion,[169] (2) their increased density in the putamen in PD,[170] particularly in patients with dyskinesia[171] that may contribute to further imbalances in the direct-indirect pathway output, and (3) their ability to provide benefits in nonhuman models of PD.[172] However, results in clinical trials have been mixed such that FDA approval has not been given,[173] though one ADORA2 antagonist, istradefylline, has received approval in Japan.[174] Importantly, evening administration of this drug resulted in EDS in several people such that morning administration has been recommended.[175] Additionally, there is evidence that ADORA2 antagonists may play a preventative role for PD based on the reduced risk of developing PD with higher caffeine consumption.[176] Finally, recent in vitro evidence provides mechanistic insight into how caffeine or other ADORA2 antagonists may protect against PD; ADORA2 antagonists decrease α-synuclein aggregation, block α-synuclein-induced synaptic plasticity deficits, and rescue cell death.[177]

SUMMARY

Sleep disturbance is a common symptom across neurodegenerative diseases, above and beyond that which occurs as part of natural ageing. Emerging evidence from AD research suggests a bidirectional influence of sleep and disease risk/progression with sleep loss potentially promoting AD-related pathology as determined by AD biomarkers and AD pathology impacting sleep through the targeted destruction of brain nuclei/transmitter systems involved in sleep/waking activity. Alternatively, parasomnias serve as an earlier warning system for PD and other synucleinopathies. In addition, the ado system continues to be an intriguing therapeutic target for diseases such as PD.

References

1. Zeppelin H. Mammalian sleep. In: Kryger MH, Roth T, Dement W, eds. *Principles and Practice of Sleep Medicine*. Philadelphia: W.B. Saunders Company; 2004.
2. Borbely AA. A two process model of sleep regulation. *Hum Neurobiol*. 1982;1 (3):195–204.
3. Borbely AA, Achermann P, Trachsel L, Tobler I. Sleep initiation and initial sleep intensity: interactions of homeostatic and circadian mechanisms. *J Biol Rhythms*. 1989;4 (2):149–160.
4. Franken P, Dijk DJ. Circadian clock genes and sleep homeostasis. *Eur J Neurosci*. 2009;29(9):1820–1829.

5. Wisor JP, O'Hara BF, Terao A, et al. A role for cryptochromes in sleep regulation. *BMC Neurosci*. 2002;3:20.
6. Franken P, Thomason R, Heller HC, O'Hara BF. A non-circadian role for clock-genes in sleep homeostasis: a strain comparison. *BMC Neurosci*. 2007;8:87.
7. Bjorness TE, Greene RW. Adenosine and sleep. *Curr Neuropharmacol*. 2009;7 (3):238–245.
8. Porkka-Heiskanen T, Kalinchuk AV. Adenosine, energy metabolism and sleep homeostasis. *Sleep Med Rev*. 2011;15(2):123–135.
9. Porkka-Heiskanen T, Strecker RE, McCarley RW. Brain site-specificity of extracellular adenosine concentration changes during sleep deprivation and spontaneous sleep: an in vivo microdialysis study. *Neuroscience*. 2000;99(3):507–517.
10. Alam MN, Szymusiak R, Gong H, King J, McGinty D. Adenosinergic modulation of rat basal forebrain neurons during sleep and waking: neuronal recording with microdialysis. *J Physiol*. 1999;521(Pt 3):679–690.
11. Thakkar MM, Winston S, McCarley RW. A1 receptor and adenosinergic homeostatic regulation of sleep-wakefulness: effects of antisense to the A1 receptor in the cholinergic basal forebrain. *J Neurosci*. 2003;23(10):4278–4287.
12. Rainnie DG, Grunze HC, McCarley RW, Greene RW. Adenosine inhibition of mesopontine cholinergic neurons: implications for EEG arousal. *Science*. 1994;263 (5147):689–692.
13. Chamberlin NL, Arrigoni E, Chou TC, Scammell TE, Greene RW, Saper CB. Effects of adenosine on gabaergic synaptic inputs to identified ventrolateral preoptic neurons. *Neuroscience*. 2003;119(4):913–918.
14. Morairty S, Rainnie D, McCarley R, Greene R. Disinhibition of ventrolateral preoptic area sleep-active neurons by adenosine: a new mechanism for sleep promotion. *Neuroscience*. 2004;123(2):451–457.
15. Liu ZW, Gao XB. Adenosine inhibits activity of hypocretin/orexin neurons by the A1 receptor in the lateral hypothalamus: a possible sleep-promoting effect. *J Neurophysiol*. 2007;97(1):837–848.
16. Gallopin T, Luppi PH, Cauli B, et al. The endogenous somnogen adenosine excites a subset of sleep-promoting neurons via A2A receptors in the ventrolateral preoptic nucleus. *Neuroscience*. 2005;134(4):1377–1390.
17. Radulovacki M. Role of adenosine in sleep in rats. *Rev Clin Basic Pharmacol*. 1985;5 (3–4):327–339.
18. Huang ZL, Qu WM, Eguchi N, et al. Adenosine A2A, but not A1, receptors mediate the arousal effect of caffeine. *Nat Neurosci*. 2005;8(7):858–859.
19. Lazarus M, Shen HY, Cherasse Y, et al. Arousal effect of caffeine depends on adenosine A2A receptors in the shell of the nucleus accumbens. *J Neurosci*. 2011;31 (27):10067–10075.
20. Achermann P, Borbely AA. Simulation of human sleep: ultradian dynamics of electroencephalographic slow-wave activity. *J Biol Rhythms*. 1990;5(2):141–157.
21. Bjorness TE, Kelly CL, Gao T, Poffenberger V, Greene RW. Control and function of the homeostatic sleep response by adenosine A1 receptors. *J Neurosci*. 2009;29 (5):1267–1276.
22. Halassa MM, Florian C, Fellin T, et al. Astrocytic modulation of sleep homeostasis and cognitive consequences of sleep loss. *Neuron*. 2009;61(2):213–219.
23. Schmitt LI, Sims RE, Dale N, Haydon PG. Wakefulness affects synaptic and network activity by increasing extracellular astrocyte-derived adenosine. *J Neurosci*. 2012;32 (13):4417–4425.
24. Zielinski MR, Taishi P, Clinton JM, Krueger JM. 5′-Ectonucleotidase-knockout mice lack non-REM sleep responses to sleep deprivation. *Eur J Neurosci*. 2012;35 (11):1789–1798.

25. Studer FE, Fedele DE, Marowsky A, et al. Shift of adenosine kinase expression from neurons to astrocytes during postnatal development suggests dual functionality of the enzyme. *Neuroscience*. 2006;142(1):125–137.
26. Bjorness TE, Mettlach N, Sonneborn A, et al. An adenosine mediated glial-neuronal circuit for homeostatic sleep. *J Neurosci*. 2016;36(13):3709–3721.
27. Landolt HP. Genotype-dependent differences in sleep, vigilance, and response to stimulants. *Curr Pharm Des*. 2008;14(32):3396–3407.
28. Rupp TL, Wesensten NJ, Newman R, Balkin TJ. PER3 and ADORA2A polymorphisms impact neurobehavioral performance during sleep restriction. *J Sleep Res*. 2013;22 (2):160–165.
29. Retey JV, Adam M, Honegger E, et al. A functional genetic variation of adenosine deaminase affects the duration and intensity of deep sleep in humans. *Proc Natl Acad Sci U S A*. 2005;102(43):15676–15681.
30. Ohayon MM, Carskadon MA, Guilleminault C, Vitiello MV. Meta-analysis of quantitative sleep parameters from childhood to old age in healthy individuals: developing normative sleep values across the human lifespan. *Sleep*. 2004;27(7):1255–1273.
31. Vitiello MV, Moe KE, Prinz PN. Sleep complaints cosegregate with illness in older adults: clinical research informed by and informing epidemiological studies of sleep. *J Psychosom Res*. 2002;53(1):555–559.
32. Prinz PN. Sleep and sleep disorders in older adults. *J Clin Neurophysiol*. 1995;12 (2):139–146.
33. Bliwise DL. Sleep in normal ageing and dementia. *Sleep*. 1993;16(1):40–81.
34. Shiromani PJ, Lu J, Wagner D, et al. Compensatory sleep response to 12 h wakefulness in young and old rats. *Am J Physiol Regul Integr Comp Physiol*. 2000;278(1):R125–R133.
35. Mendelson WB, Bergmann BM. Age-dependent changes in recovery sleep after 48 hours of sleep deprivation in rats. *Neurobiol Ageing*. 2000;21(5):689–693.
36. Carskadon MA, Dement WC. Sleep loss in elderly volunteers. *Sleep*. 1985;8(3):207–221.
37. Murillo-Rodriguez E, Blanco-Centurion C, Gerashchenko D, Salin-Pascual RJ, Shiromani PJ. The diurnal rhythm of adenosine levels in the basal forebrain of young and old rats. *Neuroscience*. 2004;123(2):361–370.
38. Meyer PT, Elmenhorst D, Boy C, et al. Effect of ageing on cerebral A1 adenosine receptors: a [18F]CPFPX PET study in humans. *Neurobiol Ageing*. 2007;28 (12):1914–1924.
39. Pagonopoulou O, Angelatou F. Reduction of A1 adenosine receptors in cortex, hippocampus and cerebellum in ageing mouse brain. *Neuroreport*. 1992;3(9):735–737.
40. Meerlo P, Roman V, Farkas E, Keijser JN, Nyakas C, Luiten PG. Ageing-related decline in adenosine A1 receptor binding in the rat brain: an autoradiographic study. *J Neurosci Res*. 2004;78(5):742–748.
41. Mackiewicz M, Nikonova EV, Zimmermann JE, et al. Age-related changes in adenosine metabolic enzymes in sleep/wake regulatory areas of the brain. *Neurobiol Ageing*. 2006;27(2):351–360.
42. Bhat SG, Wilson M, Ramkumar V. Age-dependent reductions in A1 adenosine receptor expression in rat testes. *Am J Physiol*. 1998;274(4 Pt 1):C1057–C1064.
43. Cai G, Wang HY, Gao E, et al. Reduced adenosine A1 receptor and G alpha protein coupling in rat ventricular myocardium during ageing. *Circ Res*. 1997;81(6):1065–1071.
44. Arch JR, Newsholme EA. Activities and some properties of 5′-nucleotidase, adenosine kinase and adenosine deaminase in tissues from vertebrates and invertebrates in relation to the control of the concentration and the physiological role of adenosine. *Biochem J*. 1978;174(3):965–977.
45. Mimouni M, Bontemps F, Van den Berghe G. Kinetic studies of rat liver adenosine kinase. Explanation of exchange reaction between adenosine and AMP. *J Biol Chem*. 1994;269(27):17820–17825.

46. Yoon IY, Kripke DF, Youngstedt SD, Elliott JA. Actigraphy suggests age-related differences in napping and nocturnal sleep. *J Sleep Res*. 2003;12(2):87–93.
47. Holtzman DM, Morris JC, Goate AM. Alzheimer's disease: the challenge of the second century. *Sci Transl Med*. 2011;3(77):77sr1.
48. Vos SJ, Visser PJ, Verhey F, et al. Variability of CSF Alzheimer's disease biomarkers: implications for clinical practice. *PLoS One*. 2014;9(6):e100784.
49. Price JL, McKeel Jr DW, Buckles VD, et al. Neuropathology of nondemented ageing: presumptive evidence for preclinical Alzheimer disease. *Neurobiol Ageing*. 2009;30 (7):1026–1036.
50. Jansen WJ, Ossenkoppele R, Knol DL, et al. Prevalence of cerebral amyloid pathology in persons without dementia: a meta-analysis. *JAMA*. 2015;313(19):1924–1938.
51. Lim MM, Gerstner JR, Holtzman DM. The sleep-wake cycle and Alzheimer's disease: what do we know? *Neurodegener Dis Manag*. 2014;4(5):351–362.
52. Peter-Derex L, Yammine P, Bastuji H, Croisile B. Sleep and Alzheimer's disease. *Sleep Med Rev*. 2015;19:29–38.
53. Prinz PN, Vitaliano PP, Vitiello MV, et al. Sleep, EEG and mental function changes in senile dementia of the Alzheimer's type. *Neurobiol Ageing*. 1982;3(4):361–370.
54. Prinz PN, Peskind ER, Vitaliano PP, et al. Changes in the sleep and waking EEGs of nondemented and demented elderly subjects. *J Am Geriatr Soc*. 1982;30(2):86–93.
55. Reynolds III CF, Kupfer DJ, Taska LS, et al. EEG sleep in elderly depressed, demented, and healthy subjects. *Biol Psych*. 1985;20(4):431–442.
56. Jeong J. EEG dynamics in patients with Alzheimer's disease. *Clin Neurophysiol*. 2004;115(7):1490–1505.
57. Hassainia F, Petit D, Nielsen T, Gauthier S, Montplaisir J. Quantitative EEG and statistical mapping of wakefulness and REM sleep in the evaluation of mild to moderate Alzheimer's disease. *Eur Neurol*. 1997;37(4):219–224.
58. Reynolds III CF, Kupfer DJ, Houck PR, et al. Reliable discrimination of elderly depressed and demented patients by electroencephalographic sleep data. *Arch Gen Psych*. 1988;45(3):258–264.
59. Bliwise DL, Tinklenberg J, Yesavage JA, et al. REM latency in Alzheimer's disease. *Biol Psych*. 1989;25(3):320–328.
60. Jones BE. From waking to sleeping: neuronal and chemical substrates. *Trends Pharmacol Sci*. 2005;26(11):578–586.
61. Bondareff W, Mountjoy CQ, Roth M. Selective loss of neurones of origin of adrenergic projection to cerebral cortex (nucleus locus coeruleus) in senile dementia. *Lancet*. 1981;1(8223):783–784.
62. Whitehouse PJ, Price DL, Struble RG, Clark AW, Coyle JT, Delon MR. Alzheimer's disease and senile dementia: loss of neurons in the basal forebrain. *Science*. 1982;215 (4537):1237–1239.
63. German DC, Manaye KF, White III CL, et al. Disease-specific patterns of locus coeruleus cell loss. *Ann Neurol*. 1992;32(5):667–676.
64. Hoogendijk WJ, Feenstra MG, Botterblom MH, et al. Increased activity of surviving locus ceruleus neurons in Alzheimer's disease. *Ann Neurol*. 1999;45(1):82–91.
65. Szot P, White SS, Greenup JL, Leverenz JB, Peskind ER, Raskind MA. Compensatory changes in the noradrenergic nervous system in the locus ceruleus and hippocampus of postmortem subjects with Alzheimer's disease and dementia with Lewy bodies. *J Neurosci*. 2006;26(2):467–478.
66. Adolfsson R, Gottfries CG, Roos BE, Winblad B. Changes in the brain catecholamines in patients with dementia of Alzheimer type. *Br J Psych*. 1979;135:216–223.
67. Raskind MA, Peskind ER, Holmes C, Goldstein DS. Patterns of cerebrospinal fluid catechols support increased central noradrenergic responsiveness in ageing and Alzheimer's disease. *Biol Psych*. 1999;46(6):756–765.

68. Heneka MT, Ramanathan M, Jacobs AH, et al. Locus ceruleus degeneration promotes Alzheimer pathogenesis in amyloid precursor protein 23 transgenic mice. *J Neurosci.* 2006;26(5):1343–1354.
69. Kalinin S, Gavrilyuk V, Polak PE, et al. Noradrenaline deficiency in brain increases beta-amyloid plaque burden in an animal model of Alzheimer's disease. *Neurobiol Ageing.* 2007;28(8):1206–1214.
70. Harkany T, Penke B, Luiten PG. Beta-Amyloid excitotoxicity in rat magnocellular nucleus basalis. Effect of cortical deafferentation on cerebral blood flow regulation and implications for Alzheimer's disease. *Ann N Y Acad Sci.* 2000;903:374–386.
71. Liu X, Ye K, Weinshenker D. Norepinephrine protects against amyloid-beta toxicity via TrkB. *J Alzheimers Dis.* 2015;44(1):251–260.
72. Siegel JM. The neurotransmitters of sleep. *J Clin Psych.* 2004;65(suppl 16):4–7.
73. Fronczek R, van Geest S, Frolich M, et al. Hypocretin (orexin) loss in Alzheimer's disease. *Neurobiol Ageing.* 2012;33(8):1642–1650.
74. Ripley B, Overeem S, Fujiki N, et al. CSF hypocretin/orexin levels in narcolepsy and other neurological conditions. *Neurology.* 2001;57(12):2253–2258.
75. Liguori C, Romigi A, Nuccetelli M, et al. Orexinergic system dysregulation, sleep impairment, and cognitive decline in Alzheimer disease. *JAMA Neurol.* 2014;71 (12):1498–1505.
76. Gaus SE, Strecker RE, Tate BA, Parker RA, Saper CB. Ventrolateral preoptic nucleus contains sleep-active, galaninergic neurons in multiple mammalian species. *Neuroscience.* 2002;115(1):285–294.
77. Sherin JE, Shiromani PJ, McCarley RW, Saper CB. Activation of ventrolateral preoptic neurons during sleep. *Science.* 1996;271(5246):216–219.
78. McGinty D, Szymusiak R. Brain structures and mechanisms involved in the generation of NREM sleep: focus on the preoptic hypothalamus. *Sleep Med Rev.* 2001;5 (4):323–342.
79. Lim AS, Ellison BA, Wang JL, et al. Sleep is related to neuron numbers in the ventrolateral preoptic/intermediate nucleus in older adults with and without Alzheimer's disease. *Brain.* 2014;137(Pt 10):2847–2861.
80. Albasanz JL, Perez S, Barrachina M, Ferrer I, Martin M. Up-regulation of adenosine receptors in the frontal cortex in Alzheimer's disease. *Brain Pathol.* 2008;18(2):211–219.
81. Angulo E, Casado V, Mallol J, et al. A(1) adenosine receptors accumulate in neurodegenerative structures in Alzheimer disease and mediate both amyloid precursor protein processing and tau phosphorylation and translocation. *Brain Pathol.* 2003;13 (4):440–451.
82. Kalaria RN, Sromek S, Wilcox BJ, Unnerstall JR. Hippocampal adenosine-A1-receptors are decreased in Alzheimers-disease. *Neurosci Lett.* 1990;118(2):257–260.
83. Ulas J, Brunner LC, Nguyen L, Cotman CW. Reduced density of adenosine-A1-receptors and preserved coupling of Adenosine-A1-receptors to G-proteins in Alzheimer hippocampus - a quantitative autoradiographic study. *Neuroscience.* 1993;52 (4):843–854.
84. Maia L, de Mendonca A. Does caffeine intake protect from Alzheimer's disease? *Eur J Neurol.* 2002;9(4):377–382.
85. Cao CH, Cirrito JR, Lin X, et al. Caffeine suppresses amyloid-beta levels in plasma and brain of Alzheimer's disease transgenic mice. *J Alzheimers Dis.* 2009;17(3):681–697.
86. Dall'Igna OP, Fett P, Gomes MW, Souza DO, Cunha RA, Lara DR. Caffeine and adenosine A(2a) receptor antagonists prevent beta-amyloid (25-35)-induced cognitive deficits in mice. *Exp Neurol.* 2007;203(1):241–245.
87. Canas PM, Porciuncula LO, Cunha GMA, et al. Adenosine A(2A) receptor blockade prevents synaptotoxicity and memory dysfunction caused by beta-amyloid peptides via p38 mitogen-activated protein kinase pathway. *J Neurosci.* 2009;29(47):14741–14751.

88. Ju YE, McLeland JS, Toedebusch CD, et al. Sleep quality and preclinical Alzheimer disease. *JAMA Neurol*. 2013;70(5):587–593.
89. Spira AP, Gamaldo AA, An Y, et al. Self-reported sleep and beta-amyloid deposition in community-dwelling older adults. *JAMA Neurol*. 2013;70(12):1537–1543.
90. Roh JH, Huang Y, Bero AW, et al. Disruption of the sleep-wake cycle and diurnal fluctuation of beta-amyloid in mice with Alzheimer's disease pathology. *Sci Transl Med*. 2012;4(150):150ra122.
91. Ooms S, Overeem S, Besse K, Rikkert MO, Verbeek M, Claassen JA. Effect of 1 night of total sleep deprivation on cerebrospinal fluid beta-amyloid 42 in healthy middle-aged men: a randomized clinical trial. *JAMA Neurol*. 2014;71(8):971–977.
92. Roh JH, Jiang H, Finn MB, et al. Potential role of orexin and sleep modulation in the pathogenesis of Alzheimer's disease. *J Exp Med*. 2014;211(13):2487–2496.
93. Rothman SM, Herdener N, Frankola KA, Mughal MR, Mattson MP. Chronic mild sleep restriction accentuates contextual memory impairments, and accumulations of cortical Abeta and pTau in a mouse model of Alzheimer's disease. *Brain Res*. 2013;1529:200–208.
94. Kang JE, Lim MM, Bateman RJ, et al. Amyloid-beta dynamics are regulated by orexin and the sleep-wake cycle. *Science*. 2009;326(5955):1005–1007.
95. Xie L, Kang H, Xu Q, et al. Sleep drives metabolite clearance from the adult brain. *Science*. 2013;342(6156):373–377.
96. Sykova E. Glia and volume transmission during physiological and pathological states. *J Neural Transm (Vienna)*. 2005;112(1):137–147.
97. Steriade M, Timofeev I, Grenier F. Natural waking and sleep states: a view from inside neocortical neurons. *J Neurophysiol*. 2001;85(5):1969–1985.
98. Cirrito JR, Yamada KA, Finn MB, et al. Synaptic activity regulates interstitial fluid amyloid-beta levels in vivo. *Neuron*. 2005;48(6):913–922.
99. Bero AW, Yan P, Roh JH, et al. Neuronal activity regulates the regional vulnerability to amyloid-beta deposition. *Nat Neurosci*. 2011;14(6):750–756.
100. Zhang J, Zhu Y, Zhan G, et al. Extended wakefulness: compromised metabolics in and degeneration of locus ceruleus neurons. *J Neurosci*. 2014;34(12):4418–4431.
101. Naidoo N, Giang W, Galante RJ, Pack AI. Sleep deprivation induces the unfolded protein response in mouse cerebral cortex. *J Neurochem*. 2005;92(5):1150–1157.
102. Naidoo N, Ferber M, Master M, Zhu Y, Pack AI. Ageing impairs the unfolded protein response to sleep deprivation and leads to proapoptotic signaling. *J Neurosci*. 2008;28(26):6539–6548.
103. Hoozemans JJ, Veerhuis R, Van Haastert ES, et al. The unfolded protein response is activated in Alzheimer's disease. *Acta Neuropathol*. 2005;110(2):165–172.
104. Rabek JP, Boylston III WH, Papaconstantinou J. Carbonylation of ER chaperone proteins in aged mouse liver. *Biochem Biophys Res Commun*. 2003;305(3):566–572.
105. van der Vlies D, Woudenberg J, Post JA. Protein oxidation in ageing: endoplasmic reticulum as a target. *Amino Acids*. 2003;25(3-4):397–407.
106. Stefani IC, Wright D, Polizzi KM, Kontoravdi C. The role of ER stress-induced apoptosis in neurodegeneration. *Curr Alzheimer Res*. 2012;9(3):373–387.
107. Naidoo N, Zhu J, Zhu Y, et al. Endoplasmic reticulum stress in wake-active neurons progresses with ageing. *Ageing Cell*. 2011;10(4):640–649.
108. Everson CA, Henchen CJ, Szabo A, Hogg N. Cell injury and repair resulting from sleep loss and sleep recovery in laboratory rats. *Sleep*. 2014;37(12):1929–1940.
109. Reimund E. The free radical flux theory of sleep. *Med Hypotheses*. 1994;43(4):231–233.
110. Eiland MM, Ramanathan L, Gulyani S, et al. Increases in amino-cupric-silver staining of the supraoptic nucleus after sleep deprivation. *Brain Res*. 2002;945(1):1–8.
111. Cirelli C, Shaw PJ, Rechtschaffen A, Tononi G. No evidence of brain cell degeneration after long-term sleep deprivation in rats. *Brain Res*. 1999;840(1–2):184–193.

112. Obukuro K, Nobunaga M, Takigawa M, et al. Nitric oxide mediates selective degeneration of hypothalamic orexin neurons through dysfunction of protein disulfide isomerase. *J Neurosci*. 2013;33(31):12557–12568.
113. Michinaga S, Hisatsune A, Isohama Y, Katsuki H. Orexin neurons in hypothalamic slice cultures are vulnerable to endoplasmic reticulum stress. *Neuroscience*. 2011;190:289–300.
114. Nobunaga M, Obukuro K, Kurauchi Y, et al. High fat diet induces specific pathological changes in hypothalamic orexin neurons in mice. *Neurochem Int*. 2014;78:61–66.
115. Ikeda M, Ikeda-Sagara M, Okada T, et al. Brain oxidation is an initial process in sleep induction. *Neuroscience*. 2005;130(4):1029–1040.
116. Kimura M, Kapas L, Krueger JM. Oxidized glutathione promotes sleep in rabbits. *Brain Res Bull*. 1998;45(6):545–548.
117. Sohal RS, Weindruch R. Oxidative stress, caloric restriction, and ageing. *Science*. 1996;273(5271):59–63.
118. McEwen BS, Karatsoreos IN. Sleep deprivation and circadian disruption: stress, allostasis, and allostatic load. *Sleep Med Clin*. 2015;10(1):1–10.
119. Minkel J, Moreta M, Muto J, et al. Sleep deprivation potentiates HPA axis stress reactivity in healthy adults. *Health Psychol*. 2014;33(11):1430–1434.
120. Kang JE, Cirrito JR, Dong H, Csernansky JG, Holtzman DM. Acute stress increases interstitial fluid amyloid-beta via corticotropin-releasing factor and neuronal activity. *Proc Natl Acad Sci U S A*. 2007;104(25):10673–10678.
121. He J, Hsuchou H, He Y, Kastin AJ, Wang Y, Pan W. Sleep restriction impairs blood-brain barrier function. *J Neurosci*. 2014;34(44):14697–14706.
122. Wisniewski HM, Kozlowski PB. Evidence for blood-brain barrier changes in senile dementia of the Alzheimer type (SDAT). *Ann N Y Acad Sci*. 1982;396:119–129.
123. Ujiie M, Dickstein DL, Carlow DA, Jefferies WA. Blood-brain barrier permeability precedes senile plaque formation in an Alzheimer disease model. *Microcirculation*. 2003;10(6):463–470.
124. Parkinson J. An essay on the shaking palsy. 1817. *J Neuropsych Clin Neurosci*. 2002;14 (2):223–236:discussion 222.
125. Raggi A, Bella R, Pennisi G, Neri W, Ferri R. Sleep disorders in Parkinson's disease: a narrative review of the literature. *Rev Neurosci*. 2013;24(3):279–291.
126. Braak H, Braak E. Pathoanatomy of Parkinson's disease. *J Neurol*. 2000;247(suppl 2): II3–10.
127. Xia R, Mao ZH. Progression of motor symptoms in Parkinson's disease. *Neurosci Bull*. 2012;28(1):39–48.
128. De Rosa P, Marini ES, Gelmetti V, Valente EM. Candidate genes for Parkinson disease: lessons from pathogenesis. *Clin Chim Acta*. 2015;449:68–76.
129. Chaudhuri KR, Pal S, DiMarco A, et al. The Parkinson's disease sleep scale: a new instrument for assessing sleep and nocturnal disability in Parkinson's disease. *J Neurol Neurosurg Psych*. 2002;73(6):629–635.
130. Porter B, Macfarlane R, Walker R. The frequency and nature of sleep disorders in a community-based population of patients with Parkinson's disease. *Eur J Neurol*. 2008;15(1):50–54.
131. Scaravilli T, Gasparoli E, Rinaldi F, Polesello G, Bracco F. Health-related quality of life and sleep disorders in Parkinson's disease. *Neurol Sci*. 2003;24(3):209–210.
132. Emser W, Brenner M, Stober T, Schimrigk K. Changes in nocturnal sleep in Huntington's and Parkinson's disease. *J Neurol*. 1988;235(3):177–179.
133. Wienecke M, Werth E, Poryazova R, et al. Progressive dopamine and hypocretin deficiencies in Parkinson's disease: is there an impact on sleep and wakefulness? *J Sleep Res*. 2012;21(6):710–717.
134. Arnulf I, Konofal E, Merino-Andreu M, et al. Parkinson's disease and sleepiness: an integral part of PD. *Neurology*. 2002;58(7):1019–1024.

135. Christensen JA, Nikolic M, Warby SC, et al. Sleep spindle alterations in patients with Parkinson's disease. *Front Hum Neurosci*. 2015;9:233.
136. Latreille V, Carrier J, Lafortune M, et al. Sleep spindles in Parkinson's disease may predict the development of dementia. *Neurobiol Ageing*. 2015;36(2):1083–1090.
137. Tandberg E, Larsen JP, Karlsen K. A community-based study of sleep disorders in patients with Parkinson's disease. *Mov Disord*. 1998;13(6):895–899.
138. Yong MH, Fook-Chong S, Pavanni R, Lim LL, Tan EK. Case control polysomnographic studies of sleep disorders in Parkinson's disease. *PLoS One*. 2011;6(7):e22511.
139. Chase MH, Morales FR. The atonia and myoclonia of active (REM) sleep. *Annu Rev Psychol*. 1990;41:557–584.
140. Schenck CH, Bundlie SR, Ettinger MG, Mahowald MW. Chronic behavioral disorders of human REM sleep: a new category of parasomnia. *Sleep*. 1986;9(2):293–308.
141. Gagnon JF, Bedard MA, Fantini ML, et al. REM sleep behavior disorder and REM sleep without atonia in Parkinson's disease. *Neurology*. 2002;59(4):585–589.
142. Iranzo A, Molinuevo JL, Santamaria J, et al. Rapid-eye-movement sleep behaviour disorder as an early marker for a neurodegenerative disorder: a descriptive study. *Lancet Neurol*. 2006;5(7):572–577.
143. Pacchetti C, Manni R, Zangaglia R, et al. Relationship between hallucinations, delusions, and rapid eye movement sleep behavior disorder in Parkinson's disease. *Mov Disord*. 2005;20(11):1439–1448.
144. Abbott RD, Ross GW, White LR, et al. Excessive daytime sleepiness and subsequent development of Parkinson disease. *Neurology*. 2005;65(9):1442–1446.
145. Gjerstad MD, Aarsland D, Larsen JP. Development of daytime somnolence over time in Parkinson's disease. *Neurology*. 2002;58(10):1544–1546.
146. Braak H, Del Tredici K, Rub U, de Vos RA, Jansen Steur EN, Braak E. Staging of brain pathology related to sporadic Parkinson's disease. *Neurobiol Ageing*. 2003;24(2):197–211.
147. German DC, Manaye KF, White III CL, et al. Disease-specific patterns of locus coeruleus cell loss. *Ann Neurol*. 1992;32(5):667–676.
148. Hirsch EC, Graybiel AM, Duyckaerts C, Javoy-Agid F. Neuronal loss in the pedunculopontine tegmental nucleus in Parkinson disease and in progressive supranuclear palsy. *Proc Natl Acad Sci U S A*. 1987;84(16):5976–5980.
149. Monti JM, Monti D. The involvement of dopamine in the modulation of sleep and waking. *Sleep Med Rev*. 2007;11(2):113–133.
150. Lima MM. Sleep disturbances in Parkinson's disease: the contribution of dopamine in REM sleep regulation. *Sleep Med Rev*. 2013;17(5):367–375.
151. Miller JD, Farber J, Gatz P, Roffwarg H, German DC. Activity of mesencephalic dopamine and non-dopamine neurons across stages of sleep and walking in the rat. *Brain Res*. 1983;273(1):133–141.
152. Dahan L, Astier B, Vautrelle N, Urbain N, Kocsis B, Chouvet G. Prominent burst firing of dopaminergic neurons in the ventral tegmental area during paradoxical sleep. *Neuropsychopharmacology*. 2007;32(6):1232–1241.
153. Lena I, Parrot S, Deschaux O, et al. Variations in extracellular levels of dopamine, noradrenaline, glutamate, and aspartate across the sleep–wake cycle in the medial prefrontal cortex and nucleus accumbens of freely moving rats. *J Neurosci Res*. 2005;81(6):891–899.
154. Lu J, Jhou TC, Saper CB. Identification of wake-active dopaminergic neurons in the ventral periaqueductal gray matter. *J Neurosci*. 2006;26(1):193–202.
155. Thannickal TC, Lai YY, Siegel JM. Hypocretin (orexin) cell loss in Parkinson's disease. *Brain*. 2007;130(Pt 6):1586–1595.
156. Drouot X, Moutereau S, Nguyen JP, et al. Low levels of ventricular CSF orexin/hypocretin in advanced PD. *Neurology*. 2003;61(4):540–543.

157. Compta Y, Santamaria J, Ratti L, et al. Cerebrospinal hypocretin, daytime sleepiness and sleep architecture in Parkinson's disease dementia. *Brain*. 2009;132(Pt 12):3308–3317.
158. Nakamura S, Ohnishi K, Nishimura M, et al. Large neurons in the tuberomammillary nucleus in patients with Parkinson's disease and multiple system atrophy. *Neurology*. 1996;46(6):1693–1696.
159. Shan L, Liu CQ, Balesar R, Hofman MA, Bao AM, Swaab DF. Neuronal histamine production remains unaltered in Parkinson's disease despite the accumulation of Lewy bodies and Lewy neurites in the tuberomamillary nucleus. *Neurobiol Ageing*. 2012;33(7):1343–1344.
160. Anichtchik OV, Rinne JO, Kalimo H, Panula P. An altered histaminergic innervation of the substantia nigra in Parkinson's disease. *Exp Neurol*. 2000;163(1):20–30.
161. Liu CQ, Chen Z, Liu FX, Hu DN, Luo JH. Involvement of brain endogenous histamine in the degeneration of dopaminergic neurons in 6-hydroxydopamine-lesioned rats. *Neuropharmacology*. 2007;53(7):832–841.
162. Bertolucci PH, Andrade LA, Lima JG, Carlini EA. Total sleep deprivation and Parkinson disease. *Arq Neuropsiquiatr*. 1987;45(3):224–230.
163. Hogl B, Peralta C, Wetter TC, Gershanik O, Trenkwalder C. Effect of sleep deprivation on motor performance in patients with Parkinson's disease. *Mov Disord*. 2001;16 (4):616–621.
164. Lauterbach EC. Sleep benefit and sleep deprivation in subgroups of depressed patients with Parkinson's disease. *Am J Psych*. 1994;151(5):782–783.
165. Schafer D, Greulich W. Effects of parkinsonian medication on sleep. *J Neurol*. 2000;247(suppl 4):IV/24–IV/27.
166. Arnulf I, Bejjani BP, Garma L, et al. Improvement of sleep architecture in PD with subthalamic nucleus stimulation. *Neurology*. 2000;55(11):1732–1734.
167. van Hilten B, Hoff JI, Middelkoop HA, et al. Sleep disruption in Parkinson's disease. Assessment by continuous activity monitoring. *Arch Neurol*. 1994;51(9):922–928.
168. Askenasy JJ, Yahr MD. Reversal of sleep disturbance in Parkinson's disease by antiparkinsonian therapy: a preliminary study. *Neurology*. 1985;35(4):527–532.
169. Hickey P, Stacy M. Adenosine A2A antagonists in Parkinson's disease: what's next? *Curr Neurol Neurosci Rep*. 2012;12(4):376–385.
170. Villar-Menendez I, Porta S, Buira SP, et al. Increased striatal adenosine A2A receptor levels is an early event in Parkinson's disease-related pathology and it is potentially regulated by miR-34b. *Neurobiol Dis*. 2014;69:206–214.
171. Calon F, Dridi M, Hornykiewicz O, Bedard PJ, Rajput AH, Di Paolo T. Increased adenosine A2A receptors in the brain of Parkinson's disease patients with dyskinesias. *Brain*. 2004;127(Pt 5):1075–1084.
172. Jenner P, Mori A, Hauser R, Morelli M, Fredholm BB, Chen JF. Adenosine, adenosine A 2A antagonists, and Parkinson's disease. *Parkinsonism Relat Disord*. 2009;15 (6):406–413.
173. Navarro G, Borroto-Escuela DO, Fuxe K, Franco R. Purinergic signaling in Parkinson's disease. Relevance for treatment. *Neuropharmacology*. 2015;104:161–168.
174. Dungo R, Deeks ED. Istradefylline: first global approval. *Drugs*. 2013;73(8):875–882.
175. Matsuura K, Tomimoto H. Istradefylline is recommended for morning use: a report of 4 cases. *Intern Med*. 2015;54(5):509–511.
176. Schwarzschild MA, Chen JF, Ascherio A. Caffeinated clues and the promise of adenosine A(2A) antagonists in PD. *Neurology*. 2002;58(8):1154–1160.
177. Ferreira DG, Batalha VL, Vicente Miranda H, et al. Adenosine A2A receptors modulate alpha-synuclein aggregation and toxicity. *Cereb Cortex*. 2015;pii: bhv268.

CHAPTER

7

Neuronal vs Glial Cell Contribution to Adenosine A_{2A} Receptor-Induced Neurodegeneration

Antonella Ferrante, Maria T. Tebano, Alberto Martire, Maria R. Domenici and Patrizia Popoli

Istituto Superiore di Sanità, Rome, Italy

OUTLINE

Adenosine Receptors in Neurodegenerative Diseases.
DOI: http://dx.doi.org/10.1016/B978-0-12-803724-9.00007-7

NEURODEGENERATION AND NEURODEGENERATIVE DISEASES

Neurodegeneration is an evolving process which involves different cell types in specific time windows after the noxious stimulus occurred; even if the death of neurons is the final result of this process, at earlier stages different cell types (neurons, astrocytes, microglia) cooperate in an attempt to counteract the toxic stimulus and to repair tissue damage. When the results of damage are too widespread to be controlled, a selective loss of anatomically or physiologically related specific neuronal systems with consequent progressive dysfunction occurs. Cognitive disturbances (dementia) and movement disorders are the more representative impairments, and prototypical examples of such dysfunctions are found in disorders like Alzheimer's disease (AD), Parkinson's disease (PD), amyotrophic lateral sclerosis (ALS), and Huntington's disease (HD). Ageing is a major risk factor, and the extended life expectancy is driving a dramatic increase in incidence and prevalence of many neurodegenerative diseases (NDs) in all industrialized countries. The demographics of dementia published by the World Health Organization (WHO) predict that NDs will reach over 70 million in 2030 and 106 million in 2050. Therefore, it is clear that NDs represent a major challenge for today's society in terms of the large and rapidly increasing number of patients and the general lack of effective treatments.

Even though it is still unknown whether the NDs follow a unifying mechanism for disease initiation and propagation,[1] identifying the precise steps in the pathological cascade that lead to neuronal death in each ND, and determining which of these steps represent possible therapeutic targets, could facilitate the development of new drugs.[1–3]

In recent years, adenosine receptors are attracting much attention as possible targets for NDs. Indeed, adenosine modulates many different physiological processes and in the central nervous system (CNS) it fulfills a dual role[4]: on the one hand, it acts as a neuromodulator able to fine-tune the neuronal excitability and neurotransmitter release; and on the other hand, it exerts a homeostatic control of cell metabolism. Indeed, during stressful or injuring conditions (e.g., inflammation, hypoxia, ischemia), adenosine levels rapidly increase[5]; this rise has long been recognized as a neuroprotectant "signal of life" necessary to prepare cells to better handle the noxious stimulus.[4] However, a detrimental effect of adenosine on neuronal injury has also been described.[6] In fact, under particular circumstances, when nervous damage has no possibility to be repaired, adenosine can act as a "retaliatory metabolite."[7] This apparently inconsistent action of adenosine in the CNS can be explained by the fact that its overall effect on tissue damage depends on many factors such as the particular cell type expressing it (neurons, glia, peripheral inflammatory cells), the mechanisms responsible for the damage and, last but not least, the receptor subtype engaged. Adenosine A_{2A} receptors (A_{2A}Rs), in particular, have been extensively studied as potential targets for treating NDs, revealing in most cases very complex, and even opposite, effects.[8] Recently, the evidence that glia actively signals with neurons and influences synaptic activity in healthy brains suggested that alterations of neuron-glia signaling pathways could contribute to both the early stages of the disease and the symptomatic phase.[9]

Several mechanisms are involved in the pathogenesis of NDs: excitotoxicity,[10] neuroinflammation,[11] accumulation of abnormal proteins,[12] oxidative stress, and mitochondrial dysfunction.[13] In the present chapter, we will focus on excitotoxicity and neuroinflammation, since in these mechanisms the intercellular interaction between neurons and glia is particularly significant, and since adenosine—mainly by activating A_{2A} receptors—is able to modulate both of them. Before considering the complex role of adenosine in these neurodegenerative processes, it could be useful to briefly explain their molecular basis.

Excitotoxicity

The term excitotoxicity was introduced by Olney[14] in the 1960s (1969) to indicate the neurotoxicity due to excessive and sustained neuronal exposure to L-glutamate, which is the major excitatory neurotransmitter and is functionally involved in virtually all activities of the nervous system, including cognition and memory.[15] Under normal conditions, the extracellular glutamate concentration is kept in the low micromolar range by the action of excitatory aminoacid transporters (EAATs) that import glutamate into astrocytes and neurons. When uncontrolled glutamate release surpasses the capacity of astrocyte clearance mechanisms, or when the function or expression of EAATs is decreased, excessive activation of glutamate receptors can occur.[16]

Among the ionotropic glutamate receptors, the NMDA receptors, essential mediators of synaptic plasticity and transmission in physiological conditions, are highly involved in the process of excitotoxicity.[17] For many years, it was thought that the degree of Ca^{2+} entry through NMDARs was solely responsible for these differences in cellular outcome: Moderate levels of NMDAR activity were considered to be beneficial for neurons, whereas excessive activation of NMDARs, with subsequent "Ca^{2+} overload" of the neurons, was deleterious. More recently, however, different studies have shown rather that the location of the NMDAR dictates whether it is coupled to prodeath or prosurvival signals. In fact, NMDA receptors can be found synaptically, perisynaptically, or extrasynaptically.[18] Synaptic NMDA receptors (S-NMDARs) are activated by presynaptic glutamate release,[19] whereas perisynaptic NMDAs and extrasynaptic NMDA receptors (E-NMDARs) require high glutamate concentrations.[18] It has been demonstrated that NMDAR-induced responses depend on the receptor location. Activation of S-NMDARs mediates neuronal survival via antiapoptotic and antioxidant effects. In contrast, activation of E-NMDARs is associated with neurotoxicity.[18] These differences between E-NMDAs and S-NMDAs result from the opposing actions on intracellular signaling pathways.[17] In addition, since many E-NMDARs are adjacent to glia it is possible that astrocytic release of glutamate may result in activation of E-NMDARs.[20]

Inappropriate levels of NMDAR activity can contribute to neuronal death or dysfunction both in acute excitotoxic insults (e.g., stroke) and certain NDs, such as HD and AD.[21]

The "paradox of NMDA" and the implications of synaptic versus extrasynaptic NMDA receptor signaling for NDs have been emphasized in different papers (for review see 17). Cell death associated with NDs (including AD and HD) may be partly due to an imbalance of synaptic and extrasynaptic NMDAR signaling caused by synapse loss, failure to transduce Ca^{2+} signals from the synapse to the nucleus, or by redistributions of NMDARs from synaptic to extrasynaptic site. The antagonistic signaling of synaptic and extrasynaptic NMDARs provides a novel conceptual basis for future developments of neuroprotective therapies.

Neuroinflammation

Inflammation is associated with many NDs, and even though it does not typically represent a *primummovens* of the disease, it plays a crucial role in the progression of neuronal degeneration and death.

In neuroinflammation, the microglial activation involves a crosstalk with neurons, astrocytes, and peripheral T-cells which infiltrate the CNS parenchyma trough the blood–brain barrier (BBB). Such a cross talk is characterized by activation of microglial cells (which interact with astrocytes and neurons), secretion of inflammatory mediators, and changes in the permeability of the BBB, which is followed by the infiltration of peripheral immune cells into the CNS parenchyma.

In this chapter we will provide a short synthesis of the main findings; for a detailed analysis see refs. 22 and 23. Microglial cells, the resident tissue macrophages in the CNS, control brain homeostasis by monitoring the surrounding environment, by secreting trophic factors which influence astrocyte and neuron functions, thus contributing to the protection and remodeling of synapses.[24] Activated microglia acquire the features of a toxic phenotype (M1-like) with ameboid shape, production of high amounts of proinflammatory cytokines, high mobility, and strong phagocytic capacity.[25] This early activation plays a beneficial role, and it is necessary for the clearance of cellular debris, which is required for subsequent repair of lesions. Then, M1-like microglia can acquire two different fates in the recovery phase. The beneficial fate is when, during late stages, the neuroprotective M2-like microglia phenotypes participate in attenuating inflammation and, concomitantly, produce neurotrophic factors, thus promoting tissue repair.[24] The detrimental fate is when an uncontrolled activation or overactivation of M1-like microglia occurs, thus triggering chronic inflammation, which results in the production of neurotoxic factors and eventually leads to neuronal loss.[26] Furthermore, activated microglia can release large amounts of glutamate which may lead to excitotoxicity trough the activation of neuronal ionotropic glutamate receptors.[27]

Although the role of glia in neuroinflammation has been considered prominent, in recent years evidence of an active bidirectional crosstalk between microglial cells and another local population of cells has been shown. Both astrocytes and neurons are actively involved in the regulation of damage/repair of the nervous system and act synergistically with microglia, promoting chronic neuroinflammation or favoring neuroprotection. In fact, both beneficial and detrimental roles have been demonstrated for each of them.[28,29] On the one hand, neurons actively produce antiinflammatory signals or modulate antiinflammatory mechanisms in microglia; on the other hand, cellular mediators released by dying/injured neurons can continue to stimulate M1-like microglial activation with subsequent release of mediators (ROS and RNS, glutamate and TNF-α) and instauration of a vicious cycle mediating chronic and progressive neurodegeneration.[30]

Importantly, during the past decade, several studies have shown that the CNS infiltration of lymphocytes and CD4+ T-cells has a pivotal role in the regulation of neuroinflammation and consequent neurodegeneration.[31] CD4+ T-cells can acquire different functional phenotypes, endowed with inflammatory (Th1, Th17) or antiinflammatory (Th2, Tregs) properties and able to produce different groups of mediators and to recruit different sets of immune cells (for review see 32), thus favoring the persistence of a M1-like microglia phenotype or the switch to a M2-like phenotype. Interestingly, whereas some NDs such as MS, PD, and AD involve the participation of inflammatory CD4+ T-cells, the physiopathology of other NDs, such as ALS, is associated with the participation of antiinflammatory CD4+ T-cells. Therefore, the precise phenotype acquired by CD4+ T-cells participating in neuroinflammation seems critical for the outcome of NDs or CNS recovery after injury.[32]

ROLE OF A_{2A} RECEPTORS IN NEURODEGENERATION

As mentioned above, the role of A_{2A}Rs in neurodegeneration is still a matter of hot debate; in fact, the overall effect of their engagement on tissue injury depends on the specific cell type (neuronal vs nonneuronal) prominently involved in that particular stage of the pathological process, and on the type and extent of damage.[33] A_{2A}Rs are expressed on the membrane of all cell types involved in the neurodegenerative process, i.e., on neurons, astrocytes, and microglia.[34] The observation that they are localized mainly on neurons at synapses[35] and that they facilitate presynaptic glutamate release in cerebral cortex,[36] hippocampus,[37] and striatum,[38] formerly prompted the idea that they could control neuronal death induced by excitotoxic insults. Indeed, it is well accepted that A_{2A}R blockade is beneficial in different conditions in which increased glutamate outflow (and the subsequent excitotoxicity) plays a pathogenic role like in brain ischemia,[39] epilepsy,[40] and in quinolinic acid, 3-nitropropionic acid (3-NP), or kainic acid (KA)-induced brain lesions.[33,41] However, also nonneuronal A_{2A}Rs, especially those expressed on astrocytes, play an important role in controlling neuronal damage outcome after the excitotoxic insult. In fact, astrocytic A_{2A}Rs regulate glutamate uptake not only in normal conditions[42] but also during a pathological state[43]; in particular, the activation of the receptor by its selective agonist CGS21680 increases glutamate release from hippocampal astrocytes,[42] while its blockade by antagonists completely prevents the glutamate outflow induced by inhibitors of glial glutamate uptake.[44] From these pieces of evidence, it could be deduced that the activation of both neuronal and astrocytic A_{2A}Rs is detrimental, at least against the excitotoxic insult. On the "postsynaptic side", however, neuronal A_{2A}Rs are known to control the activation and expression of NMDARs[45,46] and they thus seem to exert a protective role. In fact, their stimulation, rather than their blockade, is found to be neuroprotective against 3-NP-induced striatal damage[33] and against NMDA-induced toxicity in striatal cells.[47] It is thus evident that A_{2A}Rs exert a multifaceted action on the excitotoxic mechanism of neurodegeneration, since their activation could be detrimental on glutamate release from presynaptic neurons and on its uptake by astrocytes, but also protective at the postsynaptic level by inhibiting the toxic effects of NMDA.

This intricate action of A_{2A}Rs is evident also in neuroinflammation. Activated resident microglia and infiltrating neutrophils and macrophages from blood vessels are the major players of the immune response following injury in CNS[48]; glial cells, namely oligodendrocytes, astrocytes, and microglia, form a glial scar necessary to limit the lesion site and to protect surrounding neurons from the toxic environment;

this induces an antiinflammatory switch in the phenotype of microglial cells which starts the removal of cell debris and, if possible, the regeneration of damaged neurons through the release of growth factors.[48] A_{2A}Rs are able to control both inflammation and growth factors in a complex way. As concerning inflammation, A_{2A}Rs are unequivocally the main OFF signal of peripheral inflammation[49] but their effect in the CNS is currently debatable. During inflammation, cytokines released in the site of insult induce an upregulation of A_{2A}Rs in microglial cells and their activation is necessary to induce microglial proliferation, to control levels of inflammatory enzymes and to increase the release of inflammatory mediators.[50,51] These pieces of evidence suggested that A_{2A}Rs on microglia could contribute to the detrimental consequences of inflammation on neuronal cells. Indeed, the blockade of A_{2A}Rs is able to prevent the recruitment of microglia to the hippocampal CA3 region of rats injected with KA,[52] to blunt the toxic effect of IL-1β (a master regulator of neuroinflammation) on hippocampal neurons[53] and to prevent LPS-induced neuroinflammation and synaptic transmission impairment in rat hippocampus in vivo.[54] However, it has been demonstrated that also the activation of A_{2A}Rs could suppress inflammation in the brain and that the overall effect of A_{2A}Rs on neuroinflammation depends on the environmental conditions and on the interplay with other neurotransmitters such as glutamate.[55] In fact, in their elegant paper, Dai and coworkers demonstrated that glutamate levels are able to control the switch of A_{2A}Rs from a proinflammatory action to an antiinflammatory one both in vitro and in vivo. In particular, in the presence of low levels of glutamate, their activation reduces the LPS-induced inflammation in primary microglial cells while at high concentrations of glutamate, A_{2A}Rs activation increases the inflammatory response. The same influence of glutamate on A_{2A}R inflammatory functions was found also in an in vivo model of traumatic brain injury. To further complicate the story, peripheral A_{2A}Rs could also contribute to neuroinflammation and, more generally, to neurodegeneration. In fact, during chronic neuroinflammation, released cytokines increase the permeability of BBB and the infiltration into the CNS of systemic immune effector cells[56] which have a strong impact on in situ inflammation. As mentioned above, A_{2A}Rs have a strong antiinflammatory effect[49] but also a promigratory action on peripheral lymphocytes.[57] The importance of peripheral A_{2A}Rs in controlling neurodegeneration was demonstrated in an elegant study by the group of Jiang-Fan Chen.[58] By using an animal model of focal ischemia, they demonstrated that the ischemic area in the cerebral cortex of γ-irradiated wild-type mice receiving a bone marrow transplant from A_{2A}R knock-out mice ($A_{2A}-/-$) was increased with respect to that observed in γ-irradiated $A_{2A}-/-$ mice receiving bone marrow from wild-type mice. In other words, neuroprotection was superior in mice expressing A_{2A}Rs in myeloid cells but not in the brain. More recently, similar results were obtained also in mouse models of multiple sclerosis (MS),[59] spinal cord injury,[60] and cerebral ischemia,[61] indicating that A_{2A}Rs on peripheral immune cells play a neuroprotective role by downregulating inflammatory cascade inside the CNS.

As mentioned above, A_{2A}Rs control neurotrophin actions and, in particular, their interplay with Brain-Derived Neurotrophic Factor (BDNF) is well known. In fact, the presence and the tonic activation of A_{2A}Rs are necessary to maintain normal BDNF levels in the brain,[62] they transactivate the BDNF receptor TrkB and promote its translocation to lipid rafts.[63] Indeed, in vivo blockade of A_{2A}Rs induces a significant decrease in the levels of TrkB and impairment of BDNF signaling at synapses.[64] Since a decrease in the level or in the action of BDNF has been implicated in many NDs,[65] restoring a normal function of these factors could have a therapeutic impact. Moreover, since neurotrophins are not able to cross the BBB, activating their receptors trough the activation of A_{2A}Rs could be a valuable option. Indeed, an increase in neurotrophic factor expression through A_{2A}R stimulation is hypothesized to contribute, at least in part, to the restoration of neurological functions and to the reduction of cerebral damage in a mouse model of ischemia.[66] However, BDNF has also been shown to support microglial activation in vivo,[67] which could result in a potential amplification of neuroinflammation.[68] Therefore, BDNF could mediate the switch between acute neuroinflammation (which is beneficial in resolving brain-noxious stimuli) and chronic neuroinflammation, which could contribute to the progression of NDs.[69] As hypothesized by Sebastião and Ribeiro (2009),[70] a precise time window for the use of A_{2A}Rs agonists should be considered if neuroprotection exerted by neurotrophic factor action is desired. In particular, they proposed that in early stages of neurodegeneration, when neurons are struggling for life, an enhancement of neurotrophic factors could be beneficial and, thus, A_{2A}R agonist could be considered of therapeutic impact; on the contrary, in late stages of the process, antagonists could be more advantageous.

The multifaceted action of A_{2A}Rs is evident also in many neurodegenerative disorders such as PD, HD, AD, ischemia, and in neuroinflammation-related disorders like MS and ALS. As we will briefly described below, distinct neuronal/glial mechanisms are often involved in the modulatory role exerted by of A_{2A}Rs in these diseases.

Adenosine A_{2A} Receptors and Parkinson's Disease

The role played by A_{2A}Rs in the pathophysiology of PD will be extensively discussed elsewhere in this book, and we will thus briefly mention it. The neuropathological hallmarks of PD are the degeneration of the nigrostriatal dopaminergic neurons.[71] This causes dopamine depletion and a reduction in the inhibitory tonus of the medium-sized spiny GABA neurons (MSNs) of the caudate putamen,[72] which is the major CNS area for the control of movements. Since MSNs express A_{2A}Rs at high levels, and since an antagonistic functional interaction exists between A_{2A} and dopamine D2 receptors,[73] the depletion of dopamine in the striatum causes an A_{2A}Rs oversignaling that contributes to the hypokinetic symptoms of PD. For all these reasons, A_{2A}Rs antagonists have been considered an attractive alternative (or in adjunctive) to the dopamine-based therapies for PD. Furthermore, epidemiological studies demonstrated an inverse relationship between the consumption of caffeine (a nonselective A_{2A}R antagonist) and the risk of developing PD.[74] These data suggested that A_{2A}R antagonists could slow down disease progression and, in support of this hypothesis, the specific blockade of A_{2A}Rs was shown to be protective in several PD models.[75] In particular, A_{2A}R antagonists improved motor function in different animal models[76] but also attenuated cell loss and counteracted striatal dopamine depletion.[77] Yu et al. (2008),[78] by using an MPTP model of PD, demonstrated that A_{2A}Rs antagonists stimulated motor activity and protected against loss of dopaminergic neurons by distinct cellular mechanism. In fact, by using mice lacking A_{2A}Rs selectively in forebrain neurons (fb-A_{2A}R KO, still expressing receptors in glial cells), or mice with a global deletion of the receptor in all cell types of CNS (gb-A_{2A}R KO), they demonstrated that fb-A_{2A}R KO mice did not show the locomotor effect of the antagonist KW6002 which, on the contrary, was still able to exert its protective effect against MPTP-induced neural toxicity. The authors concluded that the effect of A_{2A}R inactivation on neurons could be responsible for its motor effects while the blockade of receptors present on microglial and astrocytic cells was able to attenuate the inflammatory response induced by MPTP-toxicity and to rescue neuronal cells. However, more recent findings showed that also A_{2A}Rs expressed on neurons play a role in the protective effect of A_{2A}R antagonists in PD. The pharmacological blockade of A_{2A}Rs improved synaptic deficits and reduced cell death in hippocampal models of PD by decreasing the number of α-synuclein aggregates.[79] The ability of A_{2A}Rs to control synuclein-induced toxicity was already demonstrated by using gb-A_{2A}R KO mice expressing the human form of α-synuclein, gene which displayed a significantly lower loss of striatal dopamine neurons compared to WT mice expressing the receptor.[80]

In conclusion, A_{2A}Rs blockade is considered a promising strategy to manage PD symptoms and many clinical trials for the use of antagonists as monotherapy or in association with the conventional L-DOPA treatments are ongoing (see 81 for a review).

Adenosine A_{2A} Receptor in Huntington's disease

The same striatal GABAergic MSNs that, in the indirect pathway, play such a relevant role in PD, are particularly vulnerable in HD. HD is a dominantly inherited ND caused by the expansion of a cytosine-adenine-guanine trinucleotide repeat encoding a polyglutamine tract within the huntingtin protein. The selective vulnerability of GABAergic enkephalin/A_{2A} receptors-expressing neurons of the striatum, which occurs despite ubiquitous expression of mutant and normal huntingtin, has suggested that A_{2A}Rs might play a pathogenetic role in HD (see 8, for review). In agreement, it has been demonstrated that a genetic variation in the ADORA$_{2A}$ gene modifies age at onset in HD patients.[82] Furthermore, dysfunctions of transcriptional factors or DNA methylation induced by mutant huntingtin contribute to the downregulation of A_{2A}Rs in HD; for instance, both reduced A_{2A}R protein levels and increased 5-Methylcytosine levels in the 5′ UTR of the ADORA$_{2A}$ gene have been found in HD patients and animal models.[83] Moreover, changes in A_{2A}R expression and signaling have been reported in various experimental models of HD, as extensively reviewed in 8. Such an aberrant A_{2A}R phenotype was accompanied by profound functional changes since A_{2A}R activation oppositely modulated NMDA-induced toxicity in the striatum of R6/2 vs wild-type mice.[84] Similarly, in the same experimental paradigm, an opposite, and A_{2A}R-dependent, influence of BDNF toward NMDA-induced synaptic effects has been found.[85]

What is the functional significance of the altered A_{2A}R phenotype in HD? This is a very difficult question to answer, because a highly complex profile has emerged in experimental models of HD, with both A_{2A}R agonists and antagonists showing beneficial effects depending on the model used, the type of pharmacological treatment, and the time window of drug administration (see 8, for review). Indeed, a protective influence on emotional/anxious responses and NMDA-induced toxicity was observed in R6/2 mice after subchronic treatment, in early phases, with the A_{2A}R antagonist SCH 58261.[86] The same antagonist, however, was no longer effective when administered at later stages of the disease.[87] The different effect observed when blocking A_{2A}Rs in different periods could depend on changes on the functional states of the receptor accordingly to the stage of the disease.

Puzzling results have been obtained also in HD animal models, where pharmacological blockade or genetic inactivation of A_{2A}Rs resulted as neuroprotective in R6/2,[88] but detrimental in N171-82Q HD mice.[89] Even more intriguingly, a recent retrospective analysis in HD patients suggested that a consumption of caffeine, a nonselective A_{2A}R antagonist, is associated with an earlier disease onset.[90]

It has been reported that when the treatment is prolonged up to the last weeks of the lifespan of R6/2 mice, the use of an A_{2A}R agonist, rather than an antagonist, was frankly beneficial.[91] More recently, it has been shown that T1-11, an A_{2A}R agonist of herbal origin, improved motor performance in adult R6/2 mice, where also reduced the striatal level of expanded polyglutamine tract-containing exon 1 fragment of huntingtin.[92]

Recent advances in human-induced pluripotent stem cell (iPSC) technology have enabled the generation of HD patient-specific iPSC lines and their differentiation into disease-relevant cell types for studying disease pathogenesis; also in this innovative in vitro model it has been recently demonstrated that activation of the A_{2A}R reduced DNA damage and oxidative stress-induced apoptosis neurons through a cAMP/PKA-dependent pathway.[93]

Finally, given the recent finding regarding a dysfunctional astrocyte glutamate and Ca^{2+} signaling in R6/2 mice,[94] which could in turn contribute to altered MSNs in the striatum, it will be of paramount importance in the near future to elucidate the role played by A_{2A}Rs expressed on glial cells in HD.

Adenosine A_{2A} Receptor in Alzheimer Disease

AD is a neurodegenerative disorder characterized by a selective synaptic and neuronal loss in the cerebral cortex and hippocampus, by senile plaques composed of amyloid-β (Aβ) peptide and by neurofibrillary tangles.[95] The resulting synaptic dysfunction is the cause of the progressive impairment of memory and other cognitive skills distinctive of the disease.[96] Given that A_{2A}Rs are able to control and integrate cognition and memory,[97] and that they are overexpressed in the frontal cortex of AD patients in either early and later stages of the disease,[98] it was hypothesized that they could represent a new therapeutic target for AD. Indeed, the first evidence for a possible protective role exerted by A_{2A}R blockade in AD came from an epidemiological study demonstrating that the incidence of AD was inversely associated with the consumption of coffee in the previous two decades of life.[99,100] Another recent study demonstrated that coffee drinking of 3–5 cups per day in midlife is associated with a decreased risk of dementia/AD.[101] The epidemiological evidence that A_{2A}R blockade could be beneficial against AD was also supported by results obtained in in vitro and in vivo models of AD. In particular, A_{2A}Rs antagonism prevented synaptic loss and neuronal death triggered by amyloid-β (Aβ) peptide both in cerebellar granule[102] and hippocampal cultures.[103] In AD transgenic mice, a six-month period of caffeine intake reduced their cognitive deficits and restored the levels of soluble Aβ.[104] Moreover, memory deficits induced by intracerebral administration of Aβ were prevented by both caffeine and selective $A_{2A}R$ antagonists[105] and by genetic $A_{2A}R$ ablation.[103] A possible mechanism for such a neuroprotection exerted by $A_{2A}R$ blockade could be the stimulation of prosurvival cascades (i.e., phospho-CREB) and the inhibition of the proapoptotic one (i.e., phospho-p38 and phospho-JNK)[103] in neurons. However, it was recently demonstrated that also the astrocytic $A_{2A}R$ plays a major role in regulating memory in AD.[106] In this interesting paper, the authors described how the conditional ablation of $A_{2A}R$ in astrocytes was enough to enhance memory in old hAPP mice (i.e., a mouse model of AD expressing the human APP minigene) that had many amyloid plaques and increased levels of astrocytic $A_{2A}Rs$; in contrast, the receptor deletion was devoid of any effects in young hAPP mice, with minimal plaques and unaltered $A_{2A}R$ expression. The authors concluded that alterations in astrocytic $A_{2A}Rs$ could contribute to memory deficit in ageing hAPP mice and that different pathogenic mechanisms may control different stages of AD. For these reasons, they suggested that targeting astrocytic $A_{2A}Rs$ could represent a therapeutic strategy for memory enhancement. In spite of all the evidence suggesting a beneficial role of $A_{2A}R$ antagonists in AD, it was recently pointed out that caution is needed in chronically blocking this receptor.[64] In fact, in vivo chronic $A_{2A}R$ blockade impairs the facilitatory action of BDNF on Long-Term Potentiation in the hippocampus, which is considered a prototypical experimental model for learning and memory.[107] Given that the levels of BDNF and of its receptor TrkB are reduced in AD,[108] that TrkB activation in AD mice improves cognitive functions,[109] and that the facilitatory effect of BDNF on synaptic transmission are dependent on the activation of $A_{2A}R$,[62] it can be hypothesized that its stimulation, instead of its blockade, could be beneficial in AD. Indeed, it has been demonstrated that $A_{2A}R$ activation promotes acetylcholine release and cholinergic signaling which results in decreased AD.[110] In summary, even if a protective role exerted by $A_{2A}R$ antagonists in AD is evident, some caution is needed in chronic blocking since this can interfere with other neuroprotective mechanisms exerted by $A_{2A}R$, such as the protective neurotrophin action.

Adenosine A_{2A} Receptor in Ischemia

Many studies demonstrated a neuroprotective role of $A_{2A}R$ blockade in ischemia (see 66 for a review). However, also the $A_{2A}R$ agonist CGS21680 induced protection against neurological deficit, weight loss, cortical infarct volume, demyelination, and glial activation.[61] These apparently discrepant results can be explained by different factors.[66]

First of all, the time-window in which agonists or antagonists are administered is determinant for the improvement of the cerebral damage; in fact, immediately after ischemia $A_{2A}R$ antagonist could be protective; on the contrary, when secondary damage due to neuroinflammation and blood cell infiltration is occurring, $A_{2A}R$ agonist could be beneficial. Secondly, different cell types could be responsible for the dual $A_{2A}R$-mediated response. In fact, immediately after ischemia, the blockade of $A_{2A}Rs$ could be beneficial by reducing glutamate release from neuronal and glial cells. On the contrary, the later stimulation of $A_{2A}Rs$ expressed on peripheral vascular cells could provide protection by reducing brain infiltration of peripheral immune cells and thus by attenuating neuroinflammation.

Adenosine A_{2A} Receptor in Multiple Sclerosis

MS is an autoimmune inflammatory, degenerative disorder of the CNS characterized by demyelination, axonal degeneration, and secondary loss of neurons.[111] During MS, autoreactive immune cells from periphery infiltrate CNS and attack and destroy myelin both in the brain and in the spinal cord.[112] Contrasting data about the role played by $A_{2A}R$ in MS are present in the literature. In fact, in experimental autoimmune encephalomyelitis (EAE), in an animal model of MS, the $A_{2A}R$ antagonist SCH58261 was shown to protect from the disease, and genetic ablation of CD73, a molecule that generates adenosine, resulted in milder EAE course.[57] On the contrary, other results demonstrated that also the loss of $A_{2A}R$ exacerbates EAE pathology.[113] A more recent paper helps to clarify the role played by $A_{2A}Rs$ in MS[114] underlying the different effect of those receptors present in bone-marrow cells from those expressed in CNS. In fact, during the early stage of the disease, when the peripheral immune response is activated, the stimulation of $A_{2A}Rs$ is beneficial (effect mediated by a less proliferation and migratory activity of T-cells); on the other hand, in the later stage characterized by disruption of BBB, blockade of the receptor is found to be neuroprotective (probably acting on activation of microglia and macrophages). Thus, as for the above-described NDs, $A_{2A}Rs$ appears as Janus-faced molecules in MS since their effect on the neurodegenerative process depends on the specific cell type and body compartment more involved in a particular stage of this evolving process.

Adenosine A_{2A} Receptor in Amyotrophic Lateral Sclerosis

ALS is a progressive neurodegenerative and neuroinflammatory disease characterized by the loss of both upper and lower motor neurons, leading to muscular atrophy, dysarthria, dysphagia, and paralysis.[115] Recently a significant increase of A_{2A} receptors in lymphocytes derived from patients affected by ALS was demonstrated.[116] This is interesting since changes in the expression and signaling of A_{2A}Rs in blood cells has been found correlated with clinical phenotypes in other NDs, such as PD, in which A_{2A}R ligands have shown a therapeutic effect. A_{2A}R expression was also found increased in end-stage human ALS spinal cords.[117] In a study performed in SOD1-G93A mice, the most-used rodent ALS model, we found that chronic administration of caffeine, whose chronic effects are mainly ascribed to A_{2A}R blockade, dramatically reduced the survival of mice. In contrast with our data, a recent study showed that the selective A_{2A}R antagonist KW6002 significantly delayed disease progression of SOD1-G93A mice.[117] Such a discrepancy could be explained in part by the use of a nonselective versus a selective antagonist and by the fact that the treatments with caffeine and KW6002 were started at different time points. On the other hand, besides the beneficial influence exerted by the antagonist KW6002, the stimulation of A_{2A}Rs increased motor neuron survival in mice,[118] facilitated respiratory motoneuron functions in rats,[119] and up-regulated the expression of vascular endothelial growth factor, whose reduction leads to ALS-like phenotype in mice,[120] and whose delivery is neuroprotective in G93A-SOD1 models.[121] At the same time, A_{2A}R-mediated adenosine signaling has been reported to induce motor neuron cell death.[117] All these contrasting findings reported in different studies reveal that the role exerted by A_{2A}Rs in ALS is very complex and that—once more—changes in receptor expression/functioning in the different experimental models and/or in different stages of the disease may be crucial in determining the final outcome.

CONCLUSIONS

On the whole, it may be proposed that participation of A_{2A}Rs to neurodegeneration depends on the different subpopulations of cells involved, on the specific mechanism responsible for the degeneration, and on the duration and/or intensity of the noxious insult. In particular, in mild toxic insults there might be the main role of synaptic A_{2A}Rs, while in prolonged toxic insults, microglial receptors could play a predominant role.[122] Eventually, when the damage becomes so severe as to increase the permeability of BBB, those receptors expressed on inflammatory cells invading the brain parenchyma could be important. Thus, from a therapeutic point of view, the possibility to delineate precise time windows for a beneficial or a detrimental effect of A_{2A}Rs stimulation during the neurodegenerative process has to be taken into account.

References

1. Haass C. Initiation and propagation of neurodegeneration. *Nat Med*. 2010;16:1201–1204.
2. Golde TE. The therapeutic importance of understanding mechanisms of neuronal cell death in neurodegenerative disease. *Mol Neurodegener*. 2009;4:8.
3. Palop JJ, Chin J, Mucke L. A network dysfunction perspective on neurodegenerative diseases. *Nature*. 2006;443:768–773.
4. Cunha RA. Adenosine as neuromodulator and as a homeostatic regulator in the central nervous system: different roles, different sources and different receptors. *Neurochem Int*. 2001;38:107–125.
5. Pedata F, Corsi C, Melani A, Bordoni F, Latini S. Adenosine extracellular brain concentrations and role of A_{2A} receptors in ischemia. *Ann N Y Acad Sci*. 2001;939:74–84.
6. de Mendonça A, Sebastião AM, Ribeiro JA. Adenosine: does it have a neuroprotective role after all? *Brain Res Rev*. 2000;33:258–274.
7. Newby AC. Adenosine and the concept of "retaliatory metabolites." *Trends Biochem Sci*. 1984;9:42–44.
8. Popoli P, Blum D, Martire A, Ledent C, Ceruti S, Abbracchio MP. Functions, dysfunctions and possible therapeutic relevance of adenosine A_{2A} receptors in Huntington's disease. *Prog Neurobiol*. 2007;81:331–348.
9. Chung WS, Welsh CA, Barres BA, Stevens B. Do glia drive synaptic and cognitive impairment in disease? *Nat Neurosci*. 2015;18:1539–1545.
10. Lewerenz J, Maher P. Chronic glutamate toxicity in neurodegenerative diseases. What is the evidence? *Front Neurosci*. 2015;9:469–478.
11. Wyss-Coray T, Mucke L. Inflammation in neurodegenerative disease – a double edged sword. *Neuron*. 2002;35:419–432.
12. Nakamura T, Lipton SA. Cell death: protein misfolding and neurodegenerative diseases. *Apoptosis*. 2009;14:455–468.
13. Lin MT, Beal MF. Mitochondrial dysfunction and oxidative stress in neurodegenerative diseases. *Nature*. 2006;443:787–795.
14. Olney JW. Brain lesions, obesity, and other disturbances in mice treated with monosodium glutamate. *Science*. 1969;164:719–721.
15. Zhou Y, Danbolt NC. Glutamate as a neurotransmitter in the healthy brain. *J Neural Transm*. 2014;121:799–817.
16. Danbolt NC. Glutamate uptake. *Prog Neurobiol*. 2001;65:1–105.
17. Hardingham GE, Bading H. Synaptic versus extrasynaptic NMDA receptor signalling: implications for neurodegenerative disorders. *Nat Rev Neurosci*. 2010;11:682–696.
18. Groc L, Bard L, Choquet D. Surface trafficking of N-methyl-d-aspartate receptors: physiological and pathological perspectives. *Neuroscience*. 2009;158:4–18.
19. Newpher TM, Ehlers MD. Spine microdomains for postsynaptic signaling and plasticity. *Trends Cell Biol*. 2009;19:218–227.
20. Petralia RS, Wang YX, Hua F, et al. Organization of NMDA receptors at extrasynaptic locations. *Neuroscience*. 2010;167:68–87.
21. Kalia LV, Kalia SK, Salter MW. NMDA receptors in clinical neurology: excitatory times ahead. *Lancet Neurol*. 2008;7:742–755.
22. Glass CK, Saijo K, Winner B, Marchetto MC, Gage FH. Mechanisms underlying inflammation in neurodegeneration. *Cell*. 2010;140:918–934.
23. González H, Elguetaa D, Montoya A, Pacheco R. Neuroimmune regulation of microglial activity involved in neuroinflammation and neurodegenerative diseases. *J Neuroimmunol*. 2014;274:1–13.
24. Kettenmann H, Hanisch UK, Noda M, Verkhratsky A. Physiology of microglia. *Physiol Rev*. 2011;91:461–553.

25. Ransohoff RM, Perry VH. Microglial physiology: unique stimuli, specialized responses. *Annu Rev Immunol*. 2009;27:119–145.
26. Block ML, Zecca L, Hong JS. Microglia-mediated neurotoxicity: uncovering the molecular mechanisms. *Nat Rev Neurosci*. 2007;8:57–69.
27. Takeuchi H, Jin S, Wang J, et al. Tumor necrosis factor-alpha induces neurotoxicity via glutamate release from hemichannels of activated microglia in an autocrine manner. *J Biol Chem*. 2006;281:21362–21368.
28. Biber K, Neumann H, Inoue K, Boddeke HW. Neuronal 'on' and 'off' signals control microglia. *Trends Neurosci*. 2007;30:596–602.
29. Pekny M, Wilhelmsson U, Pekna M. The dual role of astrocyte activation and reactive gliosis. *Neurosci Lett*. 2014;565:30–38.
30. Lehnardt S, Schott E, Trimbuch T, et al. A vicious cycle involving release of heat shock protein 60 from injured cells and activation of toll-like receptor 4 mediates neurodegeneration in the CNS. *J Neurosci*. 2008;28:2320–2331.
31. Appel SH. CD4+ T cells mediate cytotoxicity in neurodegenerative diseases. *J Clin Invest*. 2009;119:13–15.
32. González H, Pacheco R. T-cell-mediated regulation of neuroinflammation involved in neurodegenerative diseases. *J Neuroinflammation*. 2014;11:201.
33. Blum D, Galas MC, Pintor A, et al. A dual role of adenosine A_{2A} receptors in 3-nitropropionic acid-induced striatal lesions: implications for the neuroprotective potential of A_{2A} antagonists. *J Neurosci*. 2003;23:5361–5369.
34. Boison D. Adenosine dysfunction in epilepsy. *Glia*. 2012;60:1234–1243.
35. Rebola N, Canas PM, Oliveira CR, Cunha RA. Different synaptic and subsynaptic localization of adenosine A_{2A} receptors in the hippocampus and striatum of the rat. *Neuroscience*. 2005;132:893–903.
36. Marcoli M, Raiteri L, Bonfanti A, et al. Sensitivity to selective adenosine A_1 and A_{2A} receptor antagonists of the release of glutamate induced by ischemia in rat cerebrocortical slices. *Neuropharmacology*. 2003;45:201–210.
37. Rebola N, Sebastião AM, de Mendonça A, Oliveira CR, Ribeiro JA, Cunha RA. Enhanced adenosine A_{2A} receptor facilitation of synaptic transmission in the hippocampus of aged rats. *J Neurophysiol*. 2003;90:1295–1303.
38. Popoli P, Pintor A, Domenici MR, et al. Blockade of striatal adenosine A_{2A} receptor reduces, through a presynaptic mechanism, quinolinic acid-induced excitotoxicity: possible relevance to neuroprotective interventions in neurodegenerative diseases of the striatum. *J Neurosci*. 2002;22:1967–1975.
39. Melani A, Dettori I, Corti F, Cellai L, Pedata F. Time-course of protection by the selective A_{2A} receptor antagonist SCH58261 after transient focal cerebral ischemia. *Neurol Sci*. 2015;36:1441–1448.
40. Cognato GP, Agostinho PM, Hockemeyer J, Müller CE, Souza DO, Cunha RA. Caffeine and an adenosine A_{2A} receptor antagonist prevent memory impairment and synaptotoxicity in adult rats triggered by a convulsive episode in early life. *J Neurochem*. 2010;112:453–462.
41. Reggio R, Pèzzola A, Popoli P. The intrastratial injection of an adenosine A(2) receptor antagonist prevents frontal cortex EEG abnormalities in a rat model of Huntington's disease. *Brain Res*. 1999;831:315–318.
42. Matos M, Augusto E, Santos-Rodrigues AD, et al. Adenosine A2A receptors modulate glutamate uptake in cultured astrocytes and gliosomes. *Glia*. 2012;60:702–716.
43. Matos M, Augusto E, Machado NJ, dos Santos-Rodrigues A, Cunha RA, Agostinho P. Astrocytic adenosine A_{2A} receptors control the amyloid-β peptide-induced decrease of glutamate uptake. *J Alzheimers Dis*. 2012;31:555–567.
44. Pintor A, Galluzzo M, Grieco R, Pèzzola A, Reggio R, Popoli P. Adenosine A_{2A} receptor antagonists prevent the increase in striatal glutamate levels induced by glutamate uptake inhibitors. *J Neurochem*. 2004;89:152–156.

45. Tebano MT, Martire A, Rebola N, et al. Adenosine A_{2A} receptors and metabotropic glutamate 5 receptors are co-localized and functionally interact in the hippocampus: a possible key mechanism in the modulation of N-methyl-D-aspartate effects. *J Neurochem*. 2005;95:1188–1200.
46. Higley MJ, Sabatini BL. Competitive regulation of synaptic Ca2+ influx by D2 dopamine and A_{2A} adenosine receptors. *Nat Neurosci*. 2010;13:958–966.
47. Tebano MT, Pintor A, Frank C, et al. Adenosine A_{2A} receptor blockade differentially influences excitotoxic mechanisms at pre- and postsynaptic sites in the rat striatum. *J Neurosci Res*. 2004;77:100–107.
48. Raposo C, Schwartz M. Glial scar and immune cell involvement in tissue remodeling and repair following acute CNS injuries. *Glia*. 2014;62:1895–1904.
49. Blackburn MR, Vance CO, Morschl E, Wilson CN. Adenosine receptors and inflammation. *Handb Exp Pharmacol*. 2009;(193):215–269.
50. Gomes C, Ferreira R, George J, et al. Activation of microglial cells triggers a release of brain-derived neurotrophic factor (BDNF) inducing their proliferation in an adenosine A_{2A} receptor-dependent manner: A_{2A} receptor blockade prevents BDNF release and proliferation of microglia. *J Neuroinflammation*. 2013;10:16.
51. Orr AG, Orr AL, Li X, Gross RE, Traynelis SF. Adenosine A_{2A} receptor mediates microglial process retraction. *Nat Neurosci*. 2009;12:872–878.
52. Lee HK, Choi SS, Han KJ, Han EJ, Suh HW. Roles of adenosine receptors in the regulation of kainic acid-induced neurotoxic responses in mice. *Brain Res Mol Brain Res*. 2004;125:76–85.
53. Stone TW, Behan WMH. Interleukin-1β but not tumor necrosis factor-α potentiates neuronal damage by quinolinic acid: protection by an adenosine A2A receptor antagonist. *J Neurosci Res*. 2007;85:1077–1085.
54. Rebola N, Simões AP, Canas PM, et al. Adenosine A_{2A} receptors control neuroinflammation and consequent hippocampal neuronal dysfunction. *J Neurochem*. 2011;117: 100–111.
55. Dai SS, Zhou YG, Li W, et al. Local glutamate level dictates adenosine A_{2A} receptor regulation of neuroinflammation and traumatic brain injury. *J Neurosci*. 2010;30: 5802–5810.
56. Lyons SA, Pastor A, Ohlemeyer C, et al. Distinct physiologic properties of microglia and blood-borne cells in rat brain slices after permanent middle cerebral artery occlusion. *J Cereb Blood Flow Metab*. 2000;20:1537–1549.
57. Mills JH, Thompson LF, Mueller C, et al. CD73 is required for efficient entry of lymphocytes into the central nervous system during experimental autoimmune encephalomyelitis. *Proc Natl Acad Sci U S A*. 2008;105:9325–9330.
58. Yu L, Huang Z, Mariani J, Wang Y, Moskowitz M, Chen JF. Selective inactivation or reconstitution of adenosine A_{2A} receptors in bone marrow cells reveals their significant contribution to the development of ischemic brain injury. *Nat Med*. 2004;10: 1081–1087.
59. Mills JH, Kim DG, Krenz A, Chen JF, Bynoe MS. A_{2A} adenosine receptor signaling in lymphocytes and the central nervous system regulates inflammation during experimental autoimmune encephalomyelitis. *J Immunol*. 2012;188:5713–5722.
60. Genovese T, Melani A, Esposito E, et al. The selective adenosine A_{2A} receptor agonist CGS21680 reduces JNK MAPK activation in oligodendrocytes in injured spinal cord. *Shock*. 2009;32(Suppl6):S578–S585.
61. Melani A, Corti F, Cellai L, Vannucchi MG, Pedata F. Low doses of the selective adenosine A_{2A} receptor agonist CGS21680 are protective in a rat model of transient cerebral ischemia. *Brain Res*. 2014;1551:59–72.
62. Tebano MT, Martire A, Potenza RL, et al. Adenosine A_{2A} receptors are required for normal BDNF levels and BDNF-induced potentiation of synaptic transmission in the mouse hippocampus. *J Neurochem*. 2008;104:279–286.

63. Assaife-Lopes N, Sousa VC, Pereira DB, Ribeiro JA, Sebastiao AM. Regulation of TrkB receptor translocation to lipid rafts by adenosine A_{2A} receptors and its functional implications for BDNF-induced regulation of synaptic plasticity. *Purinergic Signal.* 2014;10:251–267.
64. Jerónimo-Santos A, Batalha VL, Müller CE, et al. Impact of in vivo chronic blockade of adenosine A_{2A} receptors on the BDNF-mediated facilitation of LTP. *Neuropharmacology.* 2014;83:99–106.
65. Schulte-Herbrüggen O, Braun A, Rochlitzer S, Jockers-Scherübl MC, Hellweg R. Neurotrophic factors – a tool for therapeutic strategies in neurological, neuropsychiatric and neuroimmunological diseases? *Curr Med Chem.* 2007;14:2318–2329.
66. Pedata F, Dettori I, Coppi E, et al. Purinergic signalling in brain ischemia. *Neuropharmacology.* 2016;104:105–130.
67. Jiang Y, Wei N, Lu T, Zhu J, Xu G, Liu X. Intranasal brain-derived neurotrophic factor protects brain from ischemic insult via modulating local inflammation in rats. *Neuroscience.* 2011;172:398.
68. Zhou LJ, Yang T, Wei X, et al. Brain-derived neurotrophic factor contributes to spinal long-term potentiation and mechanical hypersensitivity by activation of spinal microglia in rat. *Brain Behav Immun.* 2011;25:322.
69. Tansey MG, Goldberg MS. Neuroinflammation in Parkinson's disease: its role in neuronal death and implications for therapeutic intervention. *Neurobiol Dis.* 2010;37:510–518.
70. Sebastião AM, Ribeiro JA. Tuning and fine-tuning of synapses with adenosine. *Curr Neuropharmacol.* 2009;7:180–194.
71. Fuxe K, Manger P, Genedani S, Agnati L. The nigrostriatal DA pathway and Parkinson's disease. *J Neural Transm.* 2006;(Suppl 70):71–83.
72. Gerfen CR. Indirect-pathway neurons lose their spines in Parkinson disease. *Nat Neurosci.* 2006;9:157–158.
73. Fuxe K, Marcellino D, Leo G, Agnati L. Molecular integration via allosteric interactions in receptor heteromers. *Curr Opin Pharmacol.* 2010;10:14–22.
74. Ascherio A, Zhang SM, Hernán MA, et al. Prospective study of caffeine consumption and risk of Parkinson's disease in men and women. *Ann Neurol.* 2001;50:56–63.
75. Xu K, Xu Y-H, Chen J-F, Schwarzschild MA. Neuroprotection by caffeine: time course and role of its metabolites in the MPTP model of Parkinson's disease. *Neuroscience.* 2010;167:475–481.
76. Morelli M, Carta AR, Jenner P. Adenosine A_{2A} receptors and Parkinson's disease. *Handb Exp Pharmacol.* 2009;(193):589–615.
77. Pierri M, Vaudano E, Sager T, Englund U. KW-6002 protects from MPTP induced dopaminergic toxicity in the mouse. *Neuropharmacology.* 2005;48:517–524.
78. Yu L, Shen HY, Coelho JE, et al. Adenosine A_{2A} receptor antagonists exert motor and neuroprotective effects by distinct cellular mechanisms. *Ann Neurol.* 2008;63:338–346.
79. Ferreira DG, Batalha VL, Vicente Miranda H, et al. Adenosine A_{2A} receptors modulate α-synuclein aggregation and toxicity. *Cereb Cortex.* 2015;pii: bhv268.
80. Kachroo A, Schwarzschild MA. Adenosine A_{2A} receptor gene disruption protects in an α-synuclein model of Parkinson's disease. *Ann Neurol.* 2012;71:278–282.
81. Pinna A. Adenosine A_{2A} receptor antagonists in Parkinson's disease: progress in clinical trials from the newly approved istradefylline to drugs in early development and those already discontinued. *CNS Drugs.* 2014;28:455–474.
82. Dhaenens CM, Burnouf S, Simonin C, et al. Huntington French Speaking Network. A genetic variation in the $ADORA_{2A}$ gene modifies age at onset in Huntington's disease. *Neurobiol Dis.* 2009;35:474–476.
83. Villar-Menéndez I, Blanch M, Tyebji S, et al. Increased 5-methylcytosine and decreased 5-hydroxymethylcytosine levels are associated with reduced striatal $A_{2A}R$ levels in Huntington's disease. *Neuromolecular Med.* 2013;15:295–309.

84. Martire A, Calamandrei G, Felici F, et al. Opposite effects of the A_{2A} receptor agonist CGS21680 in the striatum of Huntington's disease versus wild-type mice. *Neurosci Lett*. 2007;417:78–83.
85. Martire A, Pepponi R, Domenici MR, Ferrante A, Chiodi V, Popoli P. BDNF prevents NMDA-induced toxicity in models of Huntington's disease: the effects are genotype specific and adenosine A_{2A} receptor is involved. *J Neurochem*. 2013;125:225–235.
86. Domenici MR, Scattoni ML, Martire A, et al. Behavioral and electrophysiological effects of the adenosine A_{2A} receptor antagonist SCH 58261 in R6/2 Huntington's disease mice. *Neurobiol Dis*. 2007;28:197–205.
87. Martire A, Ferrante A, Potenza RL, et al. Remodeling of striatal NMDA receptors by chronic A_{2A} receptor blockade in Huntington's disease mice. *Neurobiol Dis*. 2010;37:99–105.
88. Li W, Silva HB, Real J, et al. Inactivation of adenosine A_{2A} receptors reverses working memory deficits at early stages of Huntington's disease models. *Neurobiol Dis*. 2015;79:70–80.
89. Mievis S, Blum D, Ledent C. A_{2A} receptor knockout worsens survival and motor behaviour in a transgenic mouse model of Huntington's disease. *Neurobiol Dis*. 2011;41:570–576.
90. Simonin C, Duru C, Salleron J, et al. Huntington French Speaking Network. Association between caffeine intake and age at onset in Huntington's disease. *Neurobiol Dis*. 2013;58:179–182.
91. Chou SY, Lee YC, Chen HM, et al. CGS21680 attenuates symptoms of Huntington's disease in a transgenic mouse model. *J Neurochem*. 2005;93:310–320.
92. Huang NK, Lin JH, Lin JT, et al. A new drug design targeting the adenosinergic system for Huntington's disease. *PLoS One*. 2011;6:e20934.
93. Chiu FL, Lin JT, Chuang CY, et al. Elucidating the role of the A_{2A} adenosine receptor in neurodegeneration using neurons derived from Huntington's disease iPSCs. *Hum Mol Genet*. 2015;24:6066–6079.
94. Jiang R, Diaz-Castro B, Looger LL, Khakh BS. Dysfunctional Calcium and Glutamate Signaling in Striatal Astrocytes from Huntington's Disease Model Mice. *J Neurosci*. 2016;36:3453–3470.
95. Haass C, Selkoe DJ. Soluble protein oligomers in neurodegeneration: lessons from the Alzheimer's amyloid beta-peptide. *Nat Rev Mol Cell Biol*. 2007;8:101–112.
96. Kalaria RN, Maestre GE, Arizaga R, et al. World Federation of Neurology Dementia Research Group. Alzheimer's disease and vascular dementia in developing countries: prevalence, management, and risk factors. *Lancet Neurol*. 2008;7: 812–826.
97. Cunha RA, Agostinho PM. Chronic caffeine consumption prevents memory disturbance in different animal models of memory decline. *J Alzheimers Dis*. 2010;20(Suppl 1): S95–S116.
98. Albasanz JL, Perez S, Barrachina M, Ferrer I, Martín M. Up-regulation of adenosine receptors in the frontal cortex in Alzheimer's disease. *Brain Pathol*. 2008;18:211–219.
99. Maia L, de Mendonça A. Does caffeine intake protect from Alzheimer's disease? *Eur J Neurol*. 2002;9:377–382.
100. Santos C, Lunet N, Azevedo A, de Mendonça A, Ritchie K, Barros H. Caffeine intake is associated with a lower risk of cognitive decline: a cohort study from Portugal. *J Alzheimers Dis*. 2010;20(Suppl 1):S175–S185.
101. Eskelinen MH, Kivipelto M. Caffeine as a protective factor in dementia and Alzheimer's disease. *J Alzheimers Dis*. 2010;20(Suppl 1):S167–S174.
102. Dall'Igna OP, Porciúncula LO, Souza DO, Cunha RA, Lara DR. Neuroprotection by caffeine and adenosine A_{2A} receptor blockade of beta-amyloid neurotoxicity. *Br J Pharmacol*. 2003;138:1207–1209.

103. Canas PM, Porciúncula LO, Cunha GM, et al. Adenosine A_{2A} receptor blockade prevents synaptotoxicity and memory dysfunction caused by b-amyloid peptides via p38 mitogen-activated protein kinase pathway. *J Neurosci.* 2009;29:14741–14751.
104. Arendash GW, Schleif W, Rezai-Zadeh K, et al. Caffeine protects Alzheimer's mice against cognitive impairment and reduces brain beta-amyloid production. *Neuroscience.* 2006;142:941–952.
105. Dall'Igna OP, Fett P, Gomes MW, Souza DO, Cunha RA, Lara DR. Caffeine and adenosine A_{2A} receptor antagonists prevent beta-amyloid (25-35)-induced cognitive deficits in mice. *Exp Neurol.* 2007;203:241–245.
106. Orr AG, Hsiao EC, Wang MM, et al. Astrocytic adenosine receptor A_{2A} and Gs-coupled signaling regulate memory. *Nat Neurosci.* 2015;18:423–434.
107. Martin SJ, Grimwood PD, Morris R. Synaptic plasticity and memory: an evaluation of the hypothesis. *Annu Rev Neurosci.* 2000;23:649–711.
108. Phillips HS, Hains JM, Armanini M, Laramee GR, Johnson SA, Winslow JW. BDNF mRNA is decreased in the hippocampus of individuals with Alzheimer's disease. *Neuron.* 1991;7:695–702.
109. Blurton-Jones M, Kitazawa M, Martinez-Coria H, et al. Neural stem cells improve cognition via BDNF in a transgenic model of Alzheimer disease. *Proc Natl Acad Sci U S A.* 2009;106:13594–13599.
110. Jin S, Fredholm BB. Adenosine A_{2A} receptor stimulation increases release of acetylcholine from rat hippocampus but not striatum, and does not affect catecholamine release. *Naunyn Schmiedebergs Arch Pharmacol.* 1997;355:48–56.
111. Zamvil SS, Steinman L. Diverse targets for intervention during inflammatory and neurodegenerative phases of multiple sclerosis. *Neuron.* 2003;38:685–688.
112. Boppana S, Huang H, Ito K, Dhib-Jalbut S. Immunologic aspects of multiple sclerosis. *Mt Sinai J Med.* 2011;78:207–220.
113. Yao S, Li Z, Huang Q, et al. Genetic inactivation of the adenosine A_{2A} receptor exacerbates brain damage in mice with experimental autoimmune encephalomyelitis. *J Neurochem.* 2012;123:100–112.
114. Ingwersen J, Wingerath B, Graf J, et al. Dual roles of the adenosine A_{2A} receptor in autoimmune neuroinflammation. *J Neuroinflammation.* 2016;13:48.
115. Orsini M, Oliveira AB, Nascimento OJ, et al. Amyotrophic lateral sclerosis: new perpectives and update. *Neurol Int.* 2015;7:5885.
116. Vincenzi F, Corciulo C, Targa M, et al. A_{2A} adenosine receptors are up-regulated in lymphocytes from amyotrophic lateral sclerosis patients. *Amyotroph Lateral Scler Frontotemporal Degener.* 2013;14:406–413.
117. Ng SK, Higashimori H, Tolman M, Yang Y. Suppression of adenosine 2a receptor (A2aR)-mediated adenosine signaling improves disease phenotypes in a mouse model of amyotrophic lateral sclerosis. *Exp Neurol.* 2015;267:115–122.
118. Wiese S, Jablonka S, Holtmann B, et al. Adenosine receptor A_{2A}-R contributes to motoneuron survival by transactivating the tyrosine kinase receptor TrkB. *Proc Natl Acad Sci U S A.* 2007;104:17210–17215.
119. Golder FJ, Ranganathan L, Satriotomo I, et al. Spinal adenosine A2a receptor activation elicits long-lasting phrenic motor facilitation. *J Neurosci.* 2008;28:2033–2042.
120. Oosthuyse B, Moons L, Storkebaum E, et al. Deletion of the hypoxia-response element in the vascular endothelial growth factor promoter causes motor neuron degeneration. *Nat Genet.* 2001;28:131–138.
121. Azzouz M, Ralph GS, Storkebaum E, et al. VEGF delivery with retrogradely transported lentivector prolongs survival in a mouse ALS model. *Nature.* 2004;429:413–417.
122. Cunha RA, Ferrè S, Vaugeois JM, Chen JF. Potential therapeutic interest of adenosine A_{2A} receptors in psychiatric disorders. *Curr Pharm Des.* 2008;14:1512–1524.

CHAPTER

8

Adenosine and Oxygen/Glucose Deprivation in the Brain

Felicita Pedata, Ilaria Dettori, Irene Fusco, Elisabetta Coppi, Anna M. Pugliese and Alessia Melani

University of Florence, Florence, Italy

OUTLINE

Adenosine Receptors in Neurodegenerative Diseases.
DOI: http://dx.doi.org/10.1016/B978-0-12-803724-9.00008-9

INTRODUCTION

Hypoxic–ischemic insult to the brain generally causes necrosis, although in most cases there exists also a process of delayed and apoptotic type injury in the region (penumbra) surrounding the area of most severe damage.[1] Stroke is today evaluated as the second most common cause of death and a major cause of long-term disability worldwide. Ischemic stroke commonly accounts for approximately 80% of all stroke cases, and is caused from occlusion of a major cerebral artery by a thrombus or an embolism, which leads to loss of cerebral blood flow, a condition of hypoxia and glucose deprivation (oxygen and glucose deprivation: OGD) and subsequently tissue damage in the affected region. Brain injury results from a complex sequence of pathophysiological events consequent to hypoxia/ischemia that evolve over time.[2] A primary acute mechanism of excitotoxicity and periinfarct depolarizations is due to increased extracellular concentration of glutamate (see Fig. 8.1). Excitotoxicity brings to activation of resident immune cells, i.e., microglia, and production or activation of inflammation mediators. Protracted neuroinflammation is now recognized as the

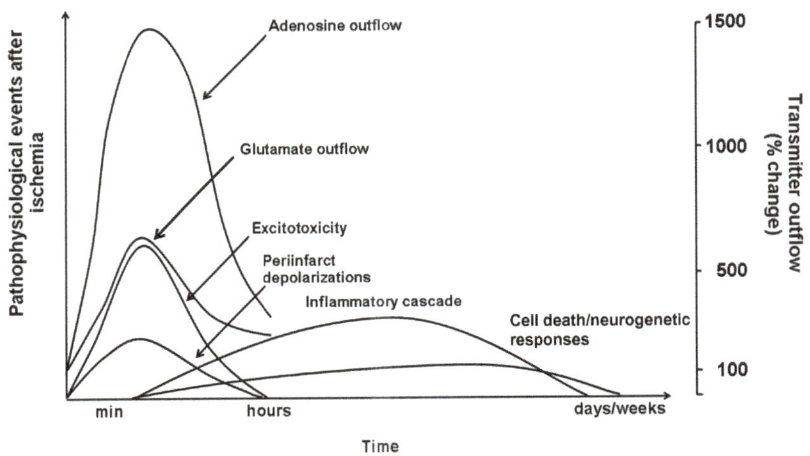

FIGURE 8.1 Cascade of pathogenetic mechanisms after ischemia. Primary mechanisms of excitotoxicity lead to acute cell death in the ischemic core. Depolarization spreads in the periinfarct areas. Glutamate and adenosine extracellular concentrations increase in the first four hours after ischemia.[9–11] The curves of increases of glutamate and adenosine evoked by ischemia induced by MCAo were drawn on the basis of values obtained by striatal microdialysis.[9] In the following several hours, activation of resident immune cells, i.e., microglia, and production of a cascade of inflammation mediators occur. Cell death/neurogenetic responses progress along the days/weeks after ischemia. *Figure modified from Dirnagl U, Iadecola C, Moskowitz MA. Pathobiology of ischaemic stroke: an integrated view.* Trends Neurosci. *1999;22:391–397.*

predominant mechanism of secondary brain injury progression.[3] Activated microglial cells proliferate, migrate, and by production of inflammatory substances and chemokines, trigger an inflammatory response.[4] Proinflammatory mediators and oxidative stress give rise to the endothelial expression of cellular adhesion molecules and to an altered permeability of the blood–brain barrier (BBB) that allows infiltration of leukocytes that in their turn exacerbate neuroinflammation and ischemic damage.[5,6]

The only successful pharmacological treatment approved to date is tissue plasminogen activator that aims to decrease ischemia-associated thrombosis risk. Yet, because of the narrow therapeutic time-window involved, thrombolytic application is very restricted in clinical settings.[7] Aspirin, other antiplatelets, and anticoagulants are used as preventive therapy of stroke.[8]

ADENOSINE IS RELEASED BY ISCHEMIA AND STIMULATE ALL P1 RECEPTOR SUBTYPES

Numerous authors have proposed adenosine and adenosine receptors as important targets for therapeutic implementation in the treatment of stroke. An increase of extracellular adenosine concentration during in vitro conditions that mimic ischemia, i.e., OGD, has been demonstrated in hippocampal slices.[12–15] An increase of extracellular adenosine concentration in in vivo ischemia was demonstrated first by the cortical cup technique and later on by the microdialysis technique.[9,16–19] The increase in adenosine extracellular levels after ischemia is accounted for by different mechanisms. As estimated in vivo by the microdialysis technique in the striatum, in the first 20 min after ischemia the increase of adenosine extracellular concentration is due for the major part to extracellularly released ATP that is hydrolyzed by ectonucleotidases.[11] Degradation of ATP by ectonucleotidases (NTPDases 1, 2, and 3 that convert ATP to ADP and AMP) and of AMP by ecto-5′-nucleotidase occurs in the first 20 min after onset of in vivo ischemia.[20] Then, in the hours after ischemia, adenosine per se is mainly released from cells.[11] Adenosine is transported out of cells likely by the equilibrative nucleoside transporter ENT2 and returns to basal values after about 4 h.[11] A down-regulation of the concentrative nucleoside transporters CNT2 might also contribute to the extracellular adenosine increase after stroke.[21]

After in vivo ischemia the extracellular concentrations of adenosine are high enough to stimulate all adenosine receptor subtypes (A_1, A_{2A}, A_{2B}, and A_3 receptors).[11] All receptor subtypes are expressed at significant levels in neurons and glial cells and in peripheral inflammatory cells.[22] The wide distribution is consistent with the multifaceted neurochemical and molecular effects of adenosine and suggests that the adenosine role in ischemia is the consequence of an interplay between different receptor activations in neuronal, glial, and inflammatory cells, which may vary depending on the time-related development of the pathological condition.

During ischemia, adenosine has long been known to act predominantly as a neuroprotectant endogenous agent.[23] In fact, adenosine infusion into the ischemic striatum has been shown to significantly ameliorate neurological outcome and reduce infarct volume after transient focal cerebral ischemia.[24]

ROLE OF ADENOSINE RECEPTORS IN ISCHEMIA

The increase in extracellular adenosine early after ischemia is considered an endogenous neuroprotective response due to stimulation of the A_1 receptor subtype, however important roles of the other three receptor subtypes have been outlined in the last 20 years.

Adenosine A_1 Receptors are Protective

One of the prime adaptive mechanisms in response to hypoxia-ischemia is the cellular activation of A_1 receptors, which inhibits excitatory synaptic transmission as demonstrated in vitro and in vivo.[25] Protective effects are greatly attributed to A_1 receptor activation due to reduced Ca^{2+} influx, thus lowering presynaptic release of excitatory neurotransmitters[26,27] and in particular glutamate, which exerts an excitotoxic effect during ischemia mainly by overstimulation of NMDA[28] and AMPA receptors.[29] In addition, by directly increasing the K^+ and Cl^- ion conductances, adenosine stabilizes the neuronal membrane potentials, thus reducing neuronal excitability.[28] Consequent reductions in cellular metabolism and energy consumption[30] and moderate lowering of the body/brain temperature[31] protect against ischemia. A continuous infusion of the A_1 adenosine receptor agonist (6) N-cyclohexyladenosine (CHA) that maintains the body temperature between 29 and 31°C for 24 h induces better survival and decreases the extent of brain damage in rats subjected to asphyxial cardiac arrest for 8 min.[32]

Consistent data demonstrate that adenosine acting on the A_1 receptor reduces the ischemia-evoked increase of excitatory transmission. In brain slices, the OGD-induced depression of synaptic transmission is reversed by administration of selective A_1 receptor antagonists[15] that also increase OGD-evoked aspartate and glutamate efflux,[33] impair the recovery of synaptic potentials,[34] and shorten the onset of anoxic depolarization induced by hypoxia.[35] Depression of excitatory synaptic transmission brought about by adenosine A_1 receptors during hypoxia/ischemia involves AMPA receptor downregulation[29] and is crucial for the functional recovery of hippocampal circuits upon reoxygenation. The A_1-mediated depression of excitatory synaptic transmission may also be secondary to the enhancement of inhibitory synaptic transmission in CA1 neurons.[36]

In in vitro studies, both adenosine and selective A_1 receptor agonists reduce neuronal damage following hypoxia and/or OGD in primary cortical or hippocampal cell cultures[37] and brain slices.[38] A_1 receptor agonists increase survival in anoxia and anoxia/reoxygenation and decrease reactive oxygen species (ROS) production, while A_1 receptor blockade increases ROS release and cell death in primary neuronal cultures.[39] Studies in support of the neuroprotective role of A_1 adenosine receptor stimulation demonstrate that hippocampal slices from A_1 receptor knock-out (KO) mice showed a markedly reduced and delayed protective response to hypoxia compared to slices from wild-type mice.[40] In astrocytes prepared from A_1 receptor KO mice, more pronounced hypoxic cytotoxicity was observed.[41]

In in vivo animal models of global cerebral ischemia, it has been demonstrated that systemic or intracerebroventricular injection of A_1 agonists is protective in ischemia.[42] Several different intimate molecular mechanisms might account for adenosine A_1 receptor-mediated neuroprotection in hypoxia/ischemia. Postischemic i.p. administration of adenosine amine congener (ADAC) resulted in preservation of microtubule associated protein 2 (MAP-2).[43] 2-chloro-N(6)-Cyclopentyladenosine (CCPA) administered i.c.v. before focal ischemia reduces lipid peroxidation.[44] Chronic co-administration of CCPA and Vitamin C i.p. after global ischemia, induced by common carotid arteries ligation, minimized ischemia-reperfusion damage by increasing the expression of antiapoptotic protein Bcl-2 and decreasing the expression of proapoptotic protein Bax in mice.[45] A_1 adenosine receptor stimulation inhibits lipopolysaccharide (LPS)-induced hypoxia-inducible factor-1 accumulation in murine astrocytes exposed to hypoxic injury.[46]

In accordance with a protective role of adenosine A_1 receptors in ischemia, acute administration of adenosine A_1 antagonists exacerbates the damage.[47]

Interestingly, chronic administration of adenosine receptor antagonists administered before an ischemic insult reduced the neuronal injury[48] and chronic administration of A_1 agonists worsened survival and increased neuronal loss.[49] It has been suggested these phenomena depend on A_1 receptor upregulation and desensitization respectively.

Plastic changes in A_1 receptors are critical to understanding the effects of adenosine A_1 agonists/antagonists but also whether adenosine maintains its neuroprotective efficiency after ischemia. Several studies have shown that short periods of focal or global ischemia produced a long lasting decrease in the density of A_1 receptors.[50] In rat hippocampal slices, hypoxia leads to a rapid (<90 min) desensitization of A_1 receptor that is likely due to an internalization of A_1 receptors in nerve terminals,[51] a process that may result in hyperexcitability and increased brain damage. On the contrary, it was reported that A_1 receptor KO mice, when exposed to global ischemia, do not show increased neuronal damage in the CA1 region of the hippocampus, in the cortex or in the striatum.[52] These discrepancies may reflect development of compensatory mechanisms after genetic deletion.

Plastic changes of A_1 receptors might account for the conflicting data on the role of A_1 receptors obtained in neonatal period. In models of hypoxia-ischemia in neonatal rats it was in fact reported that A_1 receptors contribute to protection of hypoxic brain[53] or exacerbates ischemic injury.[54] It should be remembered that in the formation of the central nervous system, A_1 receptor activation potently inhibits the development of axons and can lead to leukomalacia.[55] In this light, the decrease of adenosine A_1 receptors that has been described after neonatal hypoxia/ischemia[56,57] might be regarded as an endogenous protective feature. Notably, caffeine, that is a competitive antagonist of adenosine A_1, A_{2A}, and A_{2B} receptors, is commonly used in neonates against apnea of prematurity and has become a candidate for neuroprotection.[58]

Adenosine, by stimulating A_1 receptors, plays a crucial role in the "precondition phenomenon" consisting in protection by sublethal anoxic/ischemic insults from subsequent ischemic insults. An A_1 receptor agonist, 2-chloroadenosine (CADO) markedly enhanced, and A_1 receptor antagonists completely prevented, the protective effect of ischemic preconditioning in rat hippocampal slices.[59] In accordance, in in vivo models of ischemia, the selective A_1 antagonist, 8-cyclopentyl-1,3-dipropylxanthine (DPCPX), attenuated the neuroprotective effect of ischemic preconditioning[60] and CCPA pretreatment induced ischemic tolerance against cerebral ischemia/reperfusion injury induced by middle cerebral artery occlusion (MCAo) in the rat.[61]

Although the data demonstrate a neuroprotective effect of adenosine through A_1 receptors during ischemia, the use of selective A_1 agonists is hampered by undesirable peripheral effects such as sedation, bradycardia, and hypotension.

Interestingly, nowadays it is proposed that partial agonists at A_1 receptor may be devoid of hemodynamic effects, being therefore valuable drugs in ischemia.[62] The possibility that new adenosine A_1 receptor partial agonists are protective in ex vivo and in vitro experimental models of ischemia was recently discussed by Martire and coworkers (personal communication 2016).[63]

Adenosine A_{2A} Receptor Antagonists are Protective Against Ischemic Damage

Several studies have demonstrated that antagonists of adenosine A_{2A} receptors were protective in models of global ischemia.[64] In all these studies, adenosine A_{2A} receptor antagonists were administered pre-ischemia. However, postischemia administration is more relevant to a possible clinical use of drugs in stroke. The selective A_{2A} receptor antagonist, 7-(2-phenylethyl)-5-amino-2-(2-furyl)-pyrazolo-[4,3-e]-1,2,4,triazolo[1,5-c] pyrimidine (SCH58261), acutely administered after hypoxia/ischemia in neonatal rats, reduced brain damage[53] and acutely administered i.p. 5 min after focal ischemia in adult rats was protective from brain damage 24 h thereafter.[10,65] The same antagonist, administered subchronically (i.p., 5 min, 6 and 15 h) after focal ischemia, was protective against brain damage, neurological deficit,[66-68] and disorganization of myelin[67] 24 h after focal cerebral ischemia in the adult rat. In the model of global ischemia (i.e., 7 min Asphyxic Cardiac Arrest) in newborn piglets, posttreatment infusion with SCH58261 improved neurologic recovery and protected striatopallidal neurons 4 days after ischemia.[69] The protective effect of the A_{2A} antagonists has been attributed to a central effect, as recently proved by the observation that SCH58261 administered acutely i.p. or directly in the hippocampus[70] has ameliorated infarct size and behavioral parameter of damage 24 h after occlusion of both carotids in the rat.

SCH58261 behaves as a significant protective agent at a dose (0.01 mg/kg) that does not have cardiovascular effects. This low dose does not affect motor activity in naive animals but decreases controlateral turning behavior after MCAo induced by the monofilament technique.[10,66] At a higher dose, in the range that is effective in different models of Parkinson's Disease (PD), the same drug significantly increases motility and rearing in the rat.[71] A noxious role of A_{2A} receptors during ischemia is also supported by the observation that A_{2A} receptor KO mice show significantly decreased infarct volumes up to 48 h after transient MCAo.[72]

A_{2A} Receptor Antagonists Protect From the Increase of Glutamatergic Excitatory Transmission

The ability of adenosine A_{2A} receptor antagonists in protecting against ischemic damage is largely attributed to the control of excessive glutamatergic transmission and of the ensuing acute excitotoxicity after ischemia. A_{2A} receptors play in fact an important modulation of synaptic transmission counteracting depression brought about by A_1 receptor.[73] In the CA1 area of the rat hippocampus, the selective A_{2A} receptor agonist, 2-p-(2-carboxyethyl)phenethylamino-5′-n-ethylcarboxamidoadenosine hydrochloride (CGS21680), clearly reduces the OGD-induced depression of synaptic activity.[74] In agreement, the selective A_{2A} receptor antagonists, 4-(2-[7-amino-2-(2-furyl)[1,2,4]triazolo[2,3-a][1,3,5]triazin-5-yl-amino]ethyl)phenol (ZM241385) and SCH58261, delay the appearance of anoxic depolarization, a phenomenon strictly related to cell damage and death,[75] protect from the synaptic activity depression brought about by a severe (7 min) OGD period and protect CA1 neuron and astrocyte from injury.[76] The same effects of ZM241385 were observed after a severe 9 min OGD period in the gyrus dentatus of the hippocampus.[77]

Several mechanisms contribute to the A_{2A} receptor regulation of glutamatergic excitatory transmission. Adenosine, by stimulating A_{2A} receptors located presynaptically on glutamatergic terminals, can directly regulate glutamate outflow under normoxic[78] and ischemic conditions.[33] Moreover, A_{2A} receptors modulate glutamate uptake transporters. In particular, A_{2A} receptors located on astrocytes mediate inhibition of glutamate uptake by glutamate transporter-1 (GLT-1).[79] An imbalance of A_1/A_{2A} receptor expression might also contribute to inhibition of excitatory synaptic transmission under ischemia. Short periods of global ischemia decrease A_1 adenosine receptor density in the brain, likely due to an internalization of A_1 adenosine receptors in nerve terminals,[51] thus switching the balance toward A_{2A} receptor-mediated effects. Moreover adenosine acting on A_{2A} receptor increases AMPA[80] and NMDA receptor function.[81]

All of the above-described modulatory effects of the glutamatergic excitatory transmission by adenosine A_{2A} receptors might be relevant in in vivo ischemia. A definite overexpression of A_{2A} receptors was found in vivo in neurons of the striatum and cortex 24 h after focal ischemia.[82] The A_{2A} agonist CGS21680 increases excitatory aminoacid outflow from the ischemic cortex.[83] The low dose of SCH58261 that 24 h after ischemia has protected against tissue damage induced by MCAo[10] or quinolinic acid (QA) excitotoxicity,[84] has also reduced, in the first 4 h after ischemia, the increase of extracellular glutamate estimated by microdialysis in the striatum[10] and has reduced glutamate content in the hippocampus after occlusion of both carotids in the rat.[70] In agreement, adenosine A_{2A} receptor KO mice are protected from an excess of striatal glutamate outflow and damage induced by transient MCAo.[72]

A further mechanism by which A_{2A} receptor antagonism is protective may be due to the capability of increasing GABA extracellular concentration during ischemia.[85]

The observation that the A_{2A} receptor selective antagonist, ZM241385, injected peripherally or directly intra hippocampus is protective against excitotoxicity induced by kainate[86] confirms that the neuroprotective properties of A_{2A} receptor antagonists largely reside in a control of excitotoxicity mediated by A_{2A} receptors located on brain cells.

A_{2A} Receptor Antagonists Protect from Ischemia-Induced Activation of Intracellular Pathways

24 h after focal ischemia the A_{2A} receptor antagonist SCH58261 might have exerted protection also because it has decreased the ischemia-induced activation of p38 mitogen-activated protein kinase (MAPK)in activated microglia[66] and of JNK MAPK that is mainly expressed in mature oligodendrocytes and in oligodendrocyte progenitors (OPCs).[67] p38 is considered a death factor in ischemia[87] and phospho-JNK is a factor involved in oligodendrocyte death.[88] JNK MAPK KO mice are in fact protected from damage following cerebral ischemia.[89] Reduced activation of JNK might be directly due to A_{2A} receptors located on OPCs.[90] In fact in primary OPC culture, selective stimulation of A_{2A} receptors by CGS21680 inhibits maturation of OPCs[91] and inhibits "delayed rectifier" K^+ currents (K_{DR})[91] that are known to promote proliferation and differentiation of OPC to mature oligodendrocytes, thus preventing myelin deposition.

Direct intra hippocampus administration of SCH58261 after global ischemia, 24 h thereafter, has reduced also phospho-ERK 1/2 bringing to the reduction of different inflammation products and to the increase of the antiinflammatory cytokine interleukin-10 (IL-10).[70]

The reduced MAPK activation by SCH58261 might be due to a direct effect of the A_{2A} receptor antagonists on A_{2A} receptors located on oligodendrocytes or microglia, but also to the overall reduction of the excitotoxic cascade that in the initial hours after in vivo ischemia primes microglial activation and MAPK activation. In fact, oligodendroglia are extremely sensitive to glutamate receptor overactivation and ensuing oxidative stress, and p38 and ERK1/2 MAPK activation is definitely induced by glutamate receptor stimulation.[92]

The recent observation that the A_{2A} receptor antagonist SCH58261, chronically administered, has not maintained protection 7 days after transient focal ischemia,[93] however, supports the idea that the early protection offered by A_{2A} antagonism is overwhelmed on time by the secondary damage due to blood cell infiltration and neuroinflammation.

Adenosine A_{2A} Receptor Agonists are Protective Against Ischemic Damage

Considering that A_{2A} receptor antagonists are protective after ischemia, in an apparent paradoxical manner, also adenosine A_{2A} agonists were found protective under hypoxia/ischemia. An early study demonstrated that the adenosine A_{2A} receptor agonist 2-[(2-aminoethylamino)-carbonylethylphenylethylamino]-5′-N-ethylcarboxoamidoadenosine (APEC), administered systemically and chronically for 13 days, before a global 10 min ischemia in the adult gerbil, ameliorated animal and neuron survival.[94] Also the selective A_{2A} receptor agonist CGS21680, administered immediately after 5 min of global ischemia in the gerbil, at the high dose of 10 mg/kg i.p., exhibited highly significant protection against neuronal loss.[95] In agreement, A_{2A} receptor KO mice, subjected to chronic cerebral hypoperfusion by permanent stenosis of bilateral common carotid artery, showed impairment in working memory, increased demyelination, proliferation of glia, and increased levels of proinflammatory cytokines[96] and A_{2A} receptor KO neonatal mice show aggravated hypoxic/ischemic injury in comparison to wild-type mice.[97] Most recently, Melani et al. (2014)[98] have demonstrated that the A_{2A} receptor agonist, CGS21680, administered twice/day for 7 days i.p. (chronic protocol), starting from 4 h after transient (1 h) MCAo, induced protection from neurological deficit, weight loss, cortical infarct volume, myelin disorganization, and glial activation evaluated 7 days after ischemia.

In considering translation to clinic, a main problem of A_{2A} receptor agonists consists in their cardiovascular effect, because adenosine A_{2A} receptors located on vascular smooth muscle and endothelial cells exert a vasodilatory effect. Relevantly, Melani and coworkers (2014)[98] have demonstrated that the protective dose (0.01 mg/kg) of CGS21680 does not modify either mean blood pressure or heart frequency.

CGS21680 crosses the BBB, and protection after ischemia can be attributed to central effects. As a vasodilator agent, adenosine is implicated in cerebral blood flow regulation and might favor brain reperfusion after ischemia. CGS21680 administered directly into the rat striatum immediately prior to the induction of intracerebral hemorrhage reduces parenchymal neutrophil infiltration and tissue damage: an effect that was related to the inhibition of TNF-α expression.[99] Moreover, activation of central A_{2A} receptors is known to increase expression and release of neurotrophic factors such as nerve growth factor (NGF), brain-derived neurotrophic factor (BDNF), and glial cell line-derived neurotrophic factor (GDNF).[100] The increase in neurotrophic factor expression by adenosine A_{2A} receptor stimulation may contribute to the restoration of neurological functions and cerebral damage after brain ischemia.

A bulk of evidences indicate that A_{2A} receptors located also on blood cells greatly account for the protective effects of adenosine A_{2A} agonists after ischemia. The A_{2A} adenosine receptors are expressed in fact both on cells of innate (microglia, macrophages, mast cells, monocytes, dendritic cells, neutrophils) and on adaptive (lymphocytes) immunity. After ischemia, altered permeability of BBB allows infiltration of leukocytes (neutrophils, lymphocytes, and monocytes) that in turn exacerbate ischemic damage.[5] In the transient MCAo model in the rat, selective immunostaining for granulocytes, by anti-HIS-48 antibody, shows numerous infiltrated cells in ischemic striatal and cortical core, two days after tMCAo.[98] This is in agreement with the observation that after transient MCAo, a peak of neutrophil infiltration occurs at 6 and 48 h thereafter.[101] Seven days thereafter, infiltrated blood cells were anymore observed.[98] Chronic treatment with the A_{2A} adenosine receptor agonist, CGS21680, two days after transient MCAo, has definitely reduced the number of infiltrated blood cells in the ischemic areas.[98] The importance of a protracted treatment with the A_{2A} agonist in order to achieve protection is proved by the observation that the A_{2A} agonist administered subchronically (4 and 20 h after induction of MCAo) did not prove protective 24 h after permanent MCAo nor 7 days after transient MCAo.[102]

Many studies have reported that selective activation of A_{2A} receptors directly on blood cells, including platelets, monocytes, some mast cells, neutrophils, and T cells, inhibits proinflammatory responses, reduces production of adhesion cell factors, and reduces neutrophil activation, thereby exerting antioxidant and antiinflammatory effects. A_{2A} receptor activation is known to reduce ischemia-induced rolling, adhesion, and transmigration of various peripheral inflammatory cells (such as lymphocytes and neutrophils).[5] In support of A_{2A} receptors on blood cells being greatly responsible for the protective effects of A_{2A} agonists, protection of motor deficits by A_{2A} receptor agonists systemically administered after spinal trauma is lost in mice lacking A_{2A} receptors on bone marrow derived cells (BMDCs), but is restored in A_{2A} receptor KO mice reconstituted with A_{2A} receptors on BMDCs.[103] Moreover, in the spinal cord trauma model in the mouse, CGS21680 protected from damage when injected systemically, but not when centrally injected into the injured spinal cord.[104] Consistent with its anti-inflammatory and immunosuppressive role, the protective effect of adenosine A_{2A} receptor stimulation has been observed in different pathologies where the inflammatory process has an important role in tissue damage, such as ischemia/reperfusion liver injury,[105] spinal cord trauma,[104,106] rheumatoid arthritis,[107] acute lung inflammation,[108] intestine ischemia/reperfusion injury,[109,110] and experimental autoimmune encephalomyelitis.[111]

Adenosine A_{2B} Receptors in Brain Ischemia

A_{2B} receptor is expressed at low levels in the rat brain and has a low-affinity for adenosine, however it might be activated during conditions of hypoxia or ischemia when the extracellular adenosine levels rise.[112] Because of the paucity of A_{2B} selective agonists and antagonists, few studies are present on the role of A_{2B} receptors in brain ischemia. In primary murine astrocytes, the expression of A_{2B} receptor is strongly stimulated by LPS in concert with hypoxia.[46] In human astroglial cells, a selective A_{2B} antagonist, *N*-(4-acetylphenyl)-2-[4-(2,3,6,7-tetrahydro-2,6-dioxo-1,3-dipropyl-1*H*-purin-8-yl) henoxy]acetamide*N*-(4-Cyanophenyl)-2-[4-(2,3,6,7-tetrahydro-2,6-dioxo-1,3-dipropyl-1*H*-purin-8-yl)phenoxy]-acetamide (MRS1706), completely prevents elongation of astrocytic processes (a morphological hallmark of in vivo reactive astrogliosis) induced by selective stimulation of A_{2B} receptors.[113] The selective A_{2B} receptor antagonist, (MRS1754), reduced the ceramide production in astrocytes and attenuated inflammatory responses and neuronal damage after global cerebral ischemia.[114] Protective effects of A_{2B} antagonists might be due to an early reduction of p38 MAPK activation[115] or to reduced expression of the "regulators of g-protein signaling" (RGS) in particular RGS-3 as demonstrated in astrocytoma cells.[116] Thus a A_{2B} receptor desensitization described on astroglia might represent a cell defense mechanism in ischemia[82] and antagonism of A_{2B} receptor located on brain cells might be considered protective from ischemic brain damage.

Besides brain cells, A_{2B} receptors are present on blood immune cells and platelets,[117] where in most cases they are coexpressed with A_{2A} receptors. The recent introduction of new pharmacological and genetic tools led to the understanding of a role of A_{2B} receptors also in the regulation of inflammation, immunity, and tissue repair.[118] Attenuation of hypoxia-associated increases in tissue neutrophil number in different tissues including the brain largely depend on hematopoietic cell A_{2B} signaling.[119] Moreover, A_{2B} receptors are expressed on the surface of endothelial cells[120] where they are upregulated by hypoxia via the oxygen-regulated transcription factor, HIF-1α (hypoxia-inducible factor).[121] Studies in mice deleted of A_{2B} receptors on bone marrow cells indicate an important contribution of vascular A_{2B} receptors in attenuating vascular leakage during hypoxia.[119] The A_{2B} receptor antagonist MRS1754 increases adhesion in human microvascular endothelial cells (HMEC-1s) exposed to hypoxia[121] and adenosine A_{2B} receptor KO mice show increased basal levels of TNF-α and expression of adhesion molecules in lymphoid cells, resulting in increased leukocyte rolling and adhesion.[122] In agreement, the selective A_{2B} receptor agonist Bay 60-658380 systemically administered before normobaric hypoxia decreases vascular leak in lung, liver, and colon.[119]

Evidences indicate that A_{2B} receptors are also a valuable target to protect the heart[123] and kidney from ischemia.[124] However, no evidence up to now demonstrates a protective effect of A_{2B} agonists in animal models of stroke. All together these studies indicate that, similarly to the A_{2A} receptor, the A_{2B} receptor located on endothelial and blood immune cells is implicated in dampening vascular adhesion signals and hypoxia-induced inflammation.[125]

A further possible role of A_{2B} receptors in hypoxia/ischemia might be secondary to promotion of an angiogenic response because activation of A_{2B} receptors by adenosine increases endothelial cell proliferation, chemotaxis, capillary tube formation, and release of vascular endothelial growth factor (VEGF).[120]

Adenosine A_3 Receptors in Brain Ischemia

Adenosine A_3 receptor is widespread in the rat and mouse brain but compared to A_1 and A_{2A} receptors has less affinity for adenosine (10–30 nM vs 1 μM) and was detected at relatively low levels.[126] However, since extracellular adenosine concentrations in the first hours after ischemia reach a μM range,[9,25] the adenosine A_3 receptor is also involved in the tonic adenosine effects in ischemia.

The studies currently in the literature concerning the role of the A_3 adenosine receptor in the pathophysiology of cerebral ischemia are rather contradictory.[127,128] The use of mice with genetic deletion of the A_3 receptors has pointed out a neuroprotective function of A_3 adenosine receptors. Mice lacking A_3 receptors showed in fact increased neurodegeneration in response to repeated episodes of moderate hypoxia[129] and an increase in cerebral infarction after transient ligation of MCA.[130] Accordingly, a chronic administration (10-day preischemic) of the A_3 agonist N(6)-(3-iodobenzyl)-adenosine-5′-N-methylcarboxamide (IB-MECA) reduced ischemic damage after global forebrain ischemia in the gerbil[131] and pretreatment with a selective A_3 agonist, 1-[2-Chloro-6-[[(3-iodophenyl) methyl]amino]-9*H*-purin-9-yl]-1-deoxy-*N*-methyl-β-D-ribofuranuronamide (Cl-IB-MECA), decreased the size of infarction induced by transient MCAo.[130]

Under in vitro OGD (5 min), selective activation of A_3 adenosine receptors by a brief (5 min) application of IB-MECA brings about an inhibition of excitatory neurotransmission on cortical neurons[132] and application of the selective A_3 receptor antagonist, 3-propyl-6-ethyl-5-[(ethylthio) carbonyl]-2-phenyl-4-propyl-3-pyridine carboxylate (MRS1523), before a brief (2 min) OGD reduces the OGD-induced depression of fEPSP in the CA1 hippocampal area.[133] These findings indicated an inhibitory role of A_3 receptors on synaptic transmission during brief OGD periods and have suggested that A_3 receptors have a synergistic role with A_1 receptors in decreasing synaptic transmission, thus sustaining the neuroprotective effect of A_1 receptors.

On the other hand, when hippocampal slices are submitted to a severe (7 min) OGD, the selective antagonists of A_3 adenosine receptors abolish or delay the occurrence of anoxic depolarization (AD) and significantly protect from the irreversible disruption of neurotransmission caused by the severe ischemic episode in the CA1 region of rat hippocampal slices.[133-137] Depression of synaptic transmission following 15 min OGD was prevented by A_3 receptor antagonists also in the CA3 hippocampal area.[138]

To explain results above reported we should consider that rat cortical neurons exposed to hypoxia in vitro show an increase in activation of protein kinase C (PKC) after selective A_3 adenosine receptor stimulation.[139] If OGD is applied long enough to be considered severe, PKC activation induced by A_3 adenosine receptors could account for an increase in intracellular calcium, which may participate in increasing tissue excitability and thus lead to irreversible synaptic failure. Thus, while initially after OGD massive excitotoxicity may be controlled by A_3 adenosine receptors, later the ensuing cascade of cytotoxic events could be potentiated by prolonged A_3 adenosine receptor stimulation.

Moreover, ischemia-induced plasticity of A_3 receptors might be relevant in explaining the A_3 agonist effects in ischemia. A desensitization of A_3 receptors might account for the effect of a long application (before and during OGD) of Cl-IB-MECA and of new selective A_3 agonists[140,141] that, like A_3 antagonists, protect from the depression of synaptic activity brought about by prolonged OGD and delay the appearance of anoxic depolarization in the CA1 region of rat hippocampal slices.[133]

A_3 receptor mRNA has been identified in mouse astrocytes, in microglia and in oligodendrocytes. In human D384 astrocytoma cells, Cl-IB-MECA, at relatively low concentration (0.8 μM), reduced ATP depletion and apoptosis caused by hypoxic conditions.[41] Primary astrocytes prepared from A_3 adenosine receptor KO mice were more affected by hypoxia than those prepared from WT mice.[41] In cultured murine astrocytes, stimulation of A_3 receptors decreases HIF-1 expression induced by LPS under hypoxic conditions,[46] leading to inhibition of genes involved in inflammation injury.[46] In the in vivo model of transient MCAo, IB-MECA administered after ischemia proved to decrease the intensity of reactive gliosis involving microglia and astrocytes as evaluated 7 days after ischemia.[43]

Besides being localized on central cells, adenosine A_3 receptors are also localized on blood cells.[46] The state of the art about the role of adenosine A_3 receptors in inflammatory responses appears to be conflicting also because exposure of blood peripheral cell lines to selective A_3 adenosine receptor agonists results in both anti- and proinflammatory effects.[127] Choi and colleagues (2011)[142] have demonstrated that treatment with 2-chloro-N(6)-(3-iodobenzyl)-5′-N-methylcarbamoyl-4′-thioadenosine (LJ529), a selective A_3 agonist administered after transient MCAo, markedly reduced cerebral ischemic injury 24 h thereafter. LJ529 also prevented the infiltration/migration of microglia and monocytes occurring after MCAo. A_3 receptor agonists can mediate their protective effects via antiinflammatory signaling (inhibition of proinflammatory cytokines) and/or concomitant inhibition of innate immune cell trafficking because of A_3 receptor desensitisation.[143] A_3 agonists, under clinical evaluation for the treatment of inflammatory diseases and cancer, demonstrated excellent safety and efficacy.[144]

Overall, results raise the question of the time-related utility of A_3 receptor antagonists/agonists for treatment of ischemia. It may be speculated that after ischemia, a prolonged treatment with A_3 adenosine receptor agonists protects first by reducing glutamate-mediated excitotoxicity, and later on after ischemia, by desensitizing central A_3 receptors and via antiinflammatory effects mediated by A_3 receptors on blood cells.

CONCLUSIONS

Information acquired up to now indicates that adenosine receptors located on any cell type of the brain and on vascular and blood cells partake in either salvage or demise of the tissue after a stroke. They thus represent important targets for drugs having different therapeutic time-windows after stroke.

One of the prime adaptive mechanisms in response to hypoxia-ischemia is the cellular activation of adenosine A_1 receptors, which inhibits excessive excitatory synaptic transmission. At the same time, adenosine A_{2A} receptors, on the contrary, contribute to excessive excitotoxicity. Unfortunately, the use of selective A_1 agonists is hampered by undesirable peripheral effects such as sedation, bradycardia, and hypotension. Neuroprotective strategies with antagonists of adenosine A_{2A} receptors are aimed at targeting the brain parenchima to antagonize exitotoxicity and the ensuing production of harmful molecular events responsible for acute brain damage.

In the hours and days after ischemia, adenosine A_{2A} (possibly also A_{2B}) and A_3 receptors peripherally located on vascular and blood cells may be the targets of drugs aimed at dampening vascular adhesion signals and neuroinflammation.

On the all, a therapeutic strategy with adenosine receptor antagonists/agonists should be carefully evaluated in terms of time after ischemia due to the balance of central versus peripheral adenosine receptor-mediated effects over time after ischemia. Strategies aimed at targeting events in a longer time-window of days/weeks after ischemia appear to be promising in antagonizing inflammation and promoting neuroplasticity and neurogenesis.

Considering that compounds active at adenosine receptors are drugs under development and already existing in therapy or in clinical experimentation for other indications, some of them could enter in a reasonable time in clinical trials for stroke. Still, there is urgent need of novel compounds to be developed with higher selectivity, oral bioavailability, stability in vivo, longer half-life, and better capability to cross the BBB.

Acknowledgments

This work was supported by grants from the National Institute of Health Grant NS041083-10 and NS073947 (USA) and from University of Florence.

References

1. Lo EH. A new penumbra: transitioning from injury into repair after stroke. *Nat Med.* 2008;14:497–500.
2. Dirnagl U. Pathobiology of injury after stroke: The neurovascular unit and beyond. *Ann N Y Acad Sci.* 2012;285:39–45.
3. Tuttolomondo A, Di Sciacca R, Di Raimondo D, Renda C, Pinto A, Licata G. Inflammation as a therapeutic target in acute ischemic stroke treatment. *Curr Top Med Chem.* 2009;9:1240–1260.
4. Dirnagl U, Iadecola C, Moskowitz MA. Pathobiology of ischaemic stroke: an integrated view. *Trends Neurosci.* 1999;22:391–397.
5. Haskó G, Linden J, Cronstein B, Pacher P. Adenosine receptors: therapeutic aspects for inflammatory and immune diseases. *Nat Rev Drug Discov.* 2008;7:759–770.
6. Iadecola C, Anrather J. Stroke research at a crossroad: asking the brain for directions. *Nat Neurosci.* 2011;14:1363–1368.
7. Chen F, Qi Z, Luo Y, et al. Non-pharmaceutical therapies for stroke: Mechanisms and clinical implications. *Prog Neurobiol.* 2014;115:246–269.
8. Macrez R, Ali C, Toutirais O, et al. Stroke and the immune system: from pathophysiology to new therapeutic strategies. *Lancet Neurol.* 2011;10:471–480.
9. Melani A, Pantoni L, Corsi C, et al. Striatal outflow of adenosine, excitatory amino acids, gamma-aminobutyric acid, and taurine in awake freely moving rats after middle cerebral artery occlusion: correlations with neurological deficit and histopathological damage. *Stroke.* 1999;30:2448–2455.
10. Melani A, Pantoni L, Bordoni F, et al. The selective A_{2A} receptor antagonist SCH58261 reduces striatal transmitter outflow, turning behavior and ischemic brain damage induced by permanent focal ischemia in the rat. *Brain Res.* 2003;959:243–250.
11. Melani A, Corti F, Stephan H, et al. Ecto-ATPase inhibition: ATP and adenosine release under physiological and ischemic *in vivo* conditions in the rat striatum. *Exp Neurol.* 2012;233:193–204.
12. Dale N, Pearson T, Fringuelli BG. Direct measurement of adenosine release during hypoxia in the CA1 region of the rat hippocampal slice. *J Physiol.* 2000;526:143–155.
13. Frenguelli BG, Wigmore G, Llaudet E, Dale N. Temporal and mechanistic dissociation of ATP and adenosine release during ischaemia in the mammalian hippocampus. *J Neurochem.* 2007;101:1400–1413.
14. Latini S, Bordoni F, Corradetti R, Pepeu G, Pedata F. Temporal correlation between adenosine outflow and synaptic potential inhibition in rat hippocampal slices during ischemia-like conditions. *Brain Res.* 1998;794:325–328.

15. Pedata F, Latini S, Pugliese AM, Pepeu G. Investigations into the adenosine outflow from hippocampal slices evoked by ischemia-like conditions. *J Neurochem.* 1993;61:284–289.
16. Dux E, Fastbom J, Ungerstedt U, Rudolphi K, Fredholm BB. Protective effect of adenosine and a novel xanthine derivative propentofylline on the cell damage after bilateral carotid occlusion in the gerbil hippocampus. *Brain Res.* 1990;516:248–256.
17. Hagberg H, Andersson P, Lacarewicz J, Jacobson I, Butcher S, Sandberg M. Extracellular adenosine, inosine, hypoxanthine, and xanthine in relation to tissue nucleotides and purines in rat striatum during transient ischemia. *J Neurochem.* 1987;49:227–231.
18. Matsumoto K, Graf R, Rosner G, Shimada N, Heiss WD. Flow thresholds for extracellular purine catabolite elevation in cat focal ischemia. *Brain Res.* 1992;579:309–314.
19. Sciotti VM, Roche FM, Grabb MC, Van Wylen DG. Adenosine receptor blockade augments interstitial fluid levels of excitatory amino acids during cerebral ischemia. *J Cereb Blood Flow Metab.* 1992;12:646–655.
20. Zimmermann H. Extracellular metabolism of ATP and other nucleotides. *Naunyn Schmiedebergs Arch Pharmacol.* 2000;362:299–309.
21. Medina-Pulido L, Molina-Arcas M, Justicia C, et al. Hypoxia and P1 receptor activation regulate the high-affinity concentrative adenosine transporter CNT2 in differentiated neuronal PC12 cells. *Biochem J.* 2013;454:437–445.
22. Burnstock G, Boeynaems JM. Purinergic signalling and immune cells. *Purinergic Signal.* 2014;10:529–564.
23. Pedata F, Pugliese AM, Coppi E, et al. Adenosine in the central nervous system: effects on neurotransmission and neuroprotection. *Immunol Endocr Metab Agents Med Chem.* 2007;4:304–321.
24. Kitagawa H, Mori A, Shimada J, Mitsumoto Y, Kikuchi T. Intracerebral adenosine infusion improves neurological outcome after transient focal ischemia in rats. *Neurol Res.* 2002;24:317–323.
25. Latini S, Pedata F. Adenosine in the central nervous system: release mechanisms and extracellular concentrations. *J Neurochem.* 2001;79:463–484.
26. Corradetti R, Lo Conte G, Moroni F, Passani MB, Pepeu G. Adenosine decreases aspartate and glutamate release from rat hippocampal slices. *Eur J Pharmacol.* 1984;104:19–26.
27. Dunwiddie TV. Interactions between the effects of adenosine and calcium on synaptic responses in rat hippocampus *in vitro*. *J Physiol.* 1984;350:545–559.
28. Choi DW. Possible mechanisms limiting N-methyl-D-aspartate receptor overactivation and the therapeutic efficacy of N-methyl-D-aspartate antagonists. *Stroke.* 1990;21: III20–III22.
29. Stockwell J, Chen Z, Niazi M, Nosib S, Cayabyab FS. Protein phosphatase role in adenosine A_1 receptor-induced AMPA receptor trafficking and rat hippocampal neuronal damage in hypoxia/reperfusion injury. *Neuropharmacology.* 2016;102:254–265.
30. Greene RW, Haas HL. The electrophysiology of adenosine in the mammalian central nervous system. *Prog Neurobiol.* 1991;36:329–341.
31. Tupone D, Madden CJ, Morrison SF. Highlights in basic autonomic neurosciences: central adenosine A_1 receptor – the key to a hypometabolic state and therapeutic hypothermia? *Auton Neurosci.* 2013;176:1–2.
32. Jinka TR, Combs VM, Drew KL. Translating drug-induced hibernation to therapeutic hypothermia. *ACS Chem Neurosci.* 2015;6:899–904.
33. Marcoli M, Raiteri L, Bonfanti A, et al. Sensitivity to selective adenosine A_1 and A_{2A} receptor antagonists of the release of glutamate induced by ischemia in rat cerebrocortical slices. *Neuropharmacology.* 2003;45:201–210.
34. Sebastião AM, de Mendonca A, Moreira T, Ribeiro JA. Activation of synaptic NMDA receptors by action potential-dependent release of transmitter during hypoxia impairs recovery of synaptic transmission on reoxygenation. *J Neurosci.* 2001;21:8564–8571.

35. Lee KS, Lowenkopf T. Endogenous adenosine delays the onset of hypoxic depolarization in the rat hippocampus *in vitro* via an action at A_1 receptors. *Brain Res.* 1993;609:313–315.
36. Liang R, Pang ZP, Deng P, Xu ZC. Transient enhancement of inhibitory synaptic transmission in hippocampal CA1 pyramidal neurons after cerebral ischemia. *Neuroscience.* 2009;160:412–418.
37. Daval JL, Nicolas F. Opposite effects of cyclohexyladenosine and theophylline on hypoxic damage in cultured neurons. *Neurosci Lett.* 1994;175:114–116.
38. Mori M, Nishizaki T, Okada Y. Protective effect of adenosine on the anoxic damage of hippocampal slice. *Neuroscience.* 1992;46:301–307.
39. Milton SL, Nayak G, Kesaraju S, Kara L, Prentice HM. Suppression of reactive oxygen species production enhances neuronal survival *in vitro* and *in vivo* in the anoxia-tolerant turtle Trachemys scripta. *J Neurochem.* 2007;101:993–1001.
40. Johansson B, Halldner L, Dunwiddie TV, et al. Hyperalgesia, anxiety, and decreased hypoxic neuroprotection in mice lacking the adenosine A_1 receptor. *Proc Natl Acad Sci U S A.* 2001;98:9407–9412.
41. Bjorklund O, Shang M, Tonazzini I, Dare E, Fredholm BB. Adenosine A_1 and A_3 receptors protect astrocytes from hypoxic damage. *Eur J Pharmacol.* 2008;596:6–13.
42. Pedata F, Dettori I, Coppi E, et al. Purinergic signalling in brain ischemia. *Neuropharmacology.* 2016;104:105–130.
43. von Lubitz DK, Lin RC, Paul IA, et al. Postischemic administration of adenosine amine congener (ADAC): Analysis of recovery in gerbils. *Eur J Pharmacol.* 1996;316:171–179.
44. Sufianova GZ, Sufianov AA, Shapkin AG. Effect of cyclopentyladenosine on lipid peroxidation during focal cerebral ischemia. *Bull Exp Biol Med.* 2014;157:228–230.
45. Zamani M, Soleimani M, Golab F, Mohamadzadeh F, Mehdizadeh M, Katebi M. NeuroProtective effects of adenosine receptor agonist coadministration with ascorbic acid on CA1 hippocampus in a mouse model of ischemia reperfusion injury. *Metab Brain Dis.* 2013;28:367–374.
46. Gessi S, Merighi S, Stefanelli A, Fazzi D, Varani K, Borea PA. A(1) and A(3) adenosine receptors inhibit LPS-induced hypoxia-inducible factor-1 accumulation in murine astrocytes. *Pharmacol Res.* 2013;76:157–170.
47. Phillis JW. The effects of selective A_1 and A_{2A} adenosine receptor antagonists on cerebral ischemic injury in the gerbil. *Brain Res.* 1995;705:79–84.
48. Rudolphi KA, Keil M, Fastbom J, Fredholm BB. Ischaemic damage in gerbil hippocampus is reduced following upregulation of adenosine A_1 receptors by caffeine treatment. *Neurosci Lett.* 1989;103:275–280.
49. Jacobson KA, Von Lubitz DK, Daly JW, Fredholm BB. Adenosine receptor ligands: Differences with acute versus chronic treatment. *Trends Pharmacol Sci.* 1996;17:108–113.
50. Lee KS, Tetzlaff W, Kreutzberg GW. Rapid down regulation of hippocampal adenosine receptors following brief anoxia. *Brain Res.* 1986;380:155–158.
51. Coelho JE, Rebola N, Fragata I, Ribeiro JA, de Mendonça A, Cunha RA. Hypoxia-induced desensitization and internalization of adenosine A_1 receptors in the rat hippocampus. *Neuroscience.* 2006;138:1195–1203.
52. Olsson T, Cronberg T, Rytter A, et al. Deletion of the adenosine A_1 receptor gene does not alter neuronal damage following ischaemia *in vivo* or *in vitro*. *Eur J Neurosci.* 2004;20:1197–1204.
53. Bona E, Adén U, Gilland E, Fredholm BB, Hagberg H. Neonatal cerebral hypoxia-ischemia: the effect of adenosine receptor antagonists. *Neuropharmacology.* 1997;36:1327–1338.
54. Turner CP, Seli M, Ment L, et al. A_1 adenosine receptors mediate hypoxia-induced ventriculomegaly. *Proc Natl Acad Sci U S A.* 2003;100:11718–11722.

55. Rivkees SA, Zhao Z, Porter G, Turner C. Influences of adenosine on the fetus and newborn. *Mol Genet and Metab*. 2001;74:160–171.
56. Cheng P, Ren Y, Bai S, et al. Chronic cerebral ischemia induces downregulation of A_1 adenosine receptors during white matter damage in adult mice. *Cell Mol Neurobiol*. 2015;35:1149–1156.
57. Pimentel VC, Moretto MB, Oliveira MC, Zanini D, Sebastião AM, Schetinger MR. Neuroinflammation after neonatal hypoxia-ischemia is associated with alterations in the purinergic system: adenosine deaminase 1 isoenzyme is the most predominant after insult. *Mol Cell Biochem*. 2015;403:169–177.
58. Schmidt B, Roberts RS, Davis P, et al. Long-term effects of caffeine therapy for apnea of prematurity. *N Engl J Med*. 2007;357:1893–1902.
59. Pugliese AM, Latini S, Corradetti R, Pedata F. Brief, repeated, oxygen-glucose deprivation episodes protect neurotransmission from a longer ischemic episode in the *in vitro* hippocampus: Role of adenosine receptors. *Br J Pharmacol*. 2003;140:305–314.
60. Cui M, Bai X, Li T, et al. Decreased extracellular adenosine levels lead to loss of hypoxia-induced neuroprotection after repeated episodes of exposure to hypoxia. *PLoS One*. 2013;9:e57065.
61. Hu S, Dong H, Zhang H, et al. Noninvasive limb remote ischemic preconditioning contributes neuroprotective effects via activation of adenosine A_1 receptor and redox status after transient focal cerebral ischemia in rats. *Brain Res*. 2012;1459:81–90.
62. Baltos JA, Gregory KJ, White PJ, Sexton PM, Christopoulos A, May LT. Quantification of adenosine A_1 receptor biased agonism: implications for drug discovery. *Biochem Pharmacol*. 2016;99:101–112.
63. Martire A, Pepponi R, Tebano MT, Lambertucci K, Volpini R, Popoli P. Neuroprotective potential of adenosine A_1 receptor partial agonists in *ex vivo* and *in vitro* experimental models of ischemia. Communication to the Annual meeting of the Italian Purine Club. Rome, Italy, January 15, 2016.
64. Higashi H, Meno J, Marwaha AS, et al. Hippocampal injury and neurobehavioral deficits following hyperglycemic cerebral ischemia: Effect of theophylline and ZM241385. *J Neurosurg*. 2002;96:117–126.
65 Monopoli A, Lozza G, Forlani A, Mattavelli A, Ongini E. Blockade of adenosine A_{2A} receptors by SCH 58261 results in neuroprotective effects in cerebral ischaemia in rats. *Neuroreport*. 1998;9:3955–3959.
66. Melani A, Gianfriddo M, Vannucchi MG, et al. The selective A_{2A} receptor antagonist SCH58261 protects from neurological deficit, brain damage and activation of p38 MAPK in rat focal cerebral ischemia. *Brain Res*. 2006;1073–1074:470–480.
67. Melani A, Cipriani S, Vannucchi MG, et al. Selective adenosine A_{2A} receptor antagonism reduces JNK activation in oligodendrocytes after cerebral ischaemia. *Brain*. 2009;132:1480–1495.
68. Pedata F, Gianfriddo M, Turchi D, Melani A. The protective effect of adenosine A_{2A} receptor antagonism in cerebral ischemia. *Neurol Res*. 2005;27:169–174.
69. Yang ZJ, Wang B, Kwansa H, et al. Adenosine A_{2A} receptor contributes to ischemic brain damage in newborn piglet. *J Cereb Blood Flow Metab*. 2013;33:1612–1620.
70. Mohamed RA, Agha AM, Abdel-Rahman AA, Nassar NN. Role of adenosine A_{2A} receptor in cerebral ischemia reperfusion injury: Signaling to phosphorylated extracellular signal-regulated protein kinase (pERK1/2). *Neuroscience*. 2016;314:145–159.
71. Svenningsson P, Nomikos GG, Ongini E, Fredholm BB. Antagonism of adenosine A_{2A} receptors underlies the behavioural activating effect of caffeine and is associated with reduced expression of messenger RNA for NGFI-A and NGFI-B in caudate-putamen and nucleus accumbens. *Neuroscience*. 1997;79:753–764.
72. Gui L, Duan W, Tian H, et al. Adenosine A_{2A} receptor deficiency reduces striatal glutamate outflow and attenuates brain injury induced by transient focal cerebral ischemia in mice. *Brain Res*. 2009;1297:185–193.

73. Lopes LV, Sebastião AM, Ribeiro JA. Adenosine and related drugs in brain diseases: present and future in clinical trials. *Curr Top Med Chem.* 2011;11:1087–1101.
74. Latini S, Bordoni F, Corradetti R, Pepeu G, Pedata F. Effect of A_{2A} adenosine receptor stimulation and antagonism on synaptic depression induced by *in vitro* ischaemia in rat hippocampal slices. *Br J Pharmacol.* 1999;128:1035–1044.
75. Somjen GG. Mechanisms of spreading depression and hypoxic spreading depression-like depolarization. *Physiol Rev.* 2001;81:1065–1096.
76. Pugliese AM, Traini C, Cipriani S, et al. The adenosine A_{2A} receptor antagonist ZM241385 enhances neuronal survival after oxygen-glucose deprivation in rat CA1 hippocampal slices. *Br J Pharmacol.* 2009;157:818–830.
77. Maraula G, Traini C, Mello T, et al. Effects of oxygen and glucose deprivation on synaptic transmission in rat dentate gyrus: role of A_{2A} adenosine receptors. *Neuropharmacology.* 2013;67:511–520.
78. Lopes LV, Cunha RA, Kull B, Fredholm BB, Ribeiro JA. Adenosine A(2A) receptor facilitation of hippocampal synaptic transmission is dependent ontonic A(1) receptor inhibition. *Neuroscience.* 2002;112:319–329.
79. Pinto-Duarte A, Coelho JE, Cunha RA, Ribeiro JA, Sebastião AM. Adenosine A_{2A} receptors control the extracellular levels of adenosine through modulation of nucleoside transporters activity in the rat hippocampus. *J Neurochem.* 2005;93:595–604.
80. Dias RB, Ribeiro JA, Sebastião AM. Enhancement of AMPA currents and GluR1 membrane expression through PKA-coupled adenosine A(2A) receptors. *Hippocampus.* 2012;22:276–291.
81. Rebola N, Lujan R, Cunha RA, Mulle C. Adenosine A_{2A} receptors are essential for long-term potentiation of NMDA-EPSCs at hippocampal mossy fiber synapses. *Neuron.* 2008;57:121–134.
82. Trincavelli ML, Melani A, Guidi S, et al. Regulation of A(2A) adenosine receptor expression and functioning following permanent focal ischemia in rat brain. *J Neurochem.* 2008;104:479–490.
83. O'Regan MH, Simpson RE, Perkins LM, Phillis JW. The selective A_2 adenosine receptor agonist CGS 21680 enhances excitatory transmitter amino acid release from the ischemic rat cerebral cortex. *Neurosci Lett.* 1992;138:169–172.
84. Popoli P, Pintor A, Domenici MR, et al. Blockade of striatal adenosine A_{2A} receptor reduces, through a presynaptic mechanism, quinolinic acid-induced excitotoxicity: possible relevance to neuroprotective interventions in neurodegenerative diseases of the striatum. *J Neurosci.* 2002;22:1967–1975.
85. Cristóvão-Ferreira S, Vaz SH, Ribeiro JA, Sebastião AM. Adenosine A_{2A} receptors enhance GABA transport into nerve terminals by restraining PKC inhibition of GAT-1. *J Neurochem.* 2009;109:336–347.
86. Jones PA, Smith RA, Stone TW. Protection against kainate-induced excitotoxicity by adenosine A_{2A} receptor agonists and antagonists. *Neuroscience.* 1998;85:229–237.
87. Barone FC, Irving EA, Ray AM, et al. Inhibition of p38 mitogen-activated protein kinase provides neuroprotection in cerebral focal ischemia. *Med Res Rev.* 2001;21:129–145.
88. Jurewicz A, Matysiak M, Andrzejak S, Selmaj K. TRAIL-induced death of human adult oligodendrocytes is mediated by JNK pathway. *Glia.* 2006;53:158–166.
89. Kuan CY, Whitmarsh AJ, Yang DD, et al. A critical role of neural-specific JNK3 for ischemic apoptosis. *Proc Natl Acad Sci U S A.* 2003;100:15184–15189.
90. Coppi E, Cellai L, Maraula G, et al. Role of adenosine in oligodendrocyte precursor maturation. *Front Cell Neurosci.* 2015;9:155.
91. Coppi E, Maraula G, Fumagalli M, et al. UDP-glucose enhances outward K(+) currents necessary for cell differentiation and stimulates cell migration by activating the GPR17 receptor in oligodendrocyte precursors. *Glia.* 2013;61:1155–1171.

92. Kurino M, Fukunaga K, Ushio Y, Miyamoto E. Activation of mitogen-activated protein kinase in cultured rat hippocampal neurons by stimulation of glutamate receptors. *J Neurochem.* 1995;65:1282–1289.
93. Melani A, Dettori I, Corti F, Cellai L, Pedata F. Time-course of protection by the selective A_{2A} receptor antagonist SCH58261 after transient focal cerebral ischemia. *Neurol Sci.* 2015;36:1441–1448.
94. von Lubitz DK, Lin RC, Jacobson KA. Cerebral ischemia in gerbils: Effects of acute and chronic treatment with adenosine A_{2A} receptor agonist and antagonist. *Eur J Pharmacol.* 1995;287:295–302.
95. Sheardown MJ, Knutsen LJS. Unexpected neuroprotection observed with the adenosine A_{2A} receptor agonist CGS21680. *Drug Dev Res.* 1996;39:108–114.
96. Duan W, Gui L, Zhou Z, et al. Adenosine A_{2A} receptor deficiency exacerbates white matter lesions and cognitive deficits induced by chronic cerebral hypoperfusion in mice. *J Neurol Sci.* 2009;285:39–45.
97. Adén U, Halldner L, Lagercrantz H, Dalmau I, Ledent C, Fredholm BB. Aggravated brain damage after hypoxic ischemia in immature adenosine A_{2A} knockout mice. *Stroke.* 2003;34:739–744.
98. Melani A, Corti F, Cellai L, Vannucchi MG, Pedata F. Low doses of the selective adenosine A_{2A} receptor agonist CGS21680 are protective in a rat model of transient cerebral ischemia. *Brain Res.* 2014;1551:59–72.
99. Mayne M, Fotheringham J, Yan HJ, et al. Adenosine A_{2A} receptor activation reduces proinflammatory events and decreases cell death following intracerebral hemorrhage. *Ann Neurol.* 2001;49:727–735.
100. Sebastião AM, Ribeiro JA. Triggering neurotrophic factor actions through adenosine A_{2A} receptor activation: implications for neuroprotection. *Br J Pharmacol.* 2009;158: 15–22.
101. Zhang RL, Chopp M, Chen H, Garcia JH. Temporal profile of ischemic tissue damage, neutrophil response, and vascular plugging following permanent and transient (2H) middle cerebral artery occlusion in the rat. *J Neurol Sci.* 1994;125:3–10.
102. Pedata F, Pugliese AM, Coppi E, et al. Adenosine A_{2A} receptors modulate acute injury and neuroinflammation in brain ischemia. *Mediators Inflamm.* 2014;2014: 805198.
103. Li Y, Oskouian RJ, Day YJ, et al. Mouse spinal cord compression injury is reduced by either activation of the adenosine A_{2A} receptor on bone marrow-derived cells or deletion of the A_{2A} receptor on non-bone marrow-derived cells. *Neuroscience.* 2006;141:2029–2039.
104. Paterniti I, Melani A, Cipriani S, et al. Selective adenosine A_{2A} receptor agonists and antagonists protect against spinal cord injury through peripheral and central effects. *J Neuroinflammation.* 2011;8:31.
105. Day YJ, Marshall MA, Huang L, McDuffie MJ, Okusa MD, Linden J. Protection from ischemic liver injury by activation of A_{2A} adenosine receptors during reperfusion: inhibition of chemokine induction. *Am J Physiol Gastrointest Liver Physiol.* 2004;286: G285–G293.
106. Genovese T, Melani A, Esposito E, et al. Selective adenosine A(2a) receptor agonists reduce the apoptosis in an experimental model of spinal cord trauma. *J Biol Regul Homeost Agents.* 2010;24:73–86.
107. Mazzon E, Esposito E, Impellizzeri D, et al. CGS 21680, an agonist of the adenosine (A2A) receptor, reduces progression of murine type II collagen-induced arthritis. *J Rheumatol.* 2011;38:2119–2129.
108. Impellizzeri D, Di Paola R, Esposito E, et al. CGS 21680, an agonist of the adenosine (A2A) receptor, decreases acute lung inflammation. *Eur J Pharmacol.* 2011;668: 305–316.

109. Di Paola R, Melani A, Esposito E, et al. Adenosine A_{2A} receptor-selective stimulation reduces signaling pathways involved in the development of intestine ischemia and reperfusion injury. *Shock*. 2010;33:541–551.
110. Odashima M, Bamias G, Rivera-Nieves J, et al. Activation of A_{2A} adenosine receptor attenuates intestinal inflammation in animal models of inflammatory bowel disease. *Gastroenterology*. 2005;129:26–33.
111. Xu J, Guo S, Jia Z, Ma S, Li Z, Xue R. Additive effect of prostaglandin E2 and adenosine in mouse experimental autoimmune encephalomyelitis. *Prostaglandins Other Lipid Mediat*. 2013;100–101:30–35.
112. Popoli P, Pepponi R. Potential therapeutic relevance of adenosine A_{2B} and A_{2A} receptors in the central nervous system. *CNS Neurol Disord Drug Targets*. 2012;11:664–674.
113. Trincavelli ML, Marroni M, Tuscano D, et al. Regulation of A_{2B} adenosine receptor functioning by tumour necrosis factor a in human astroglial cells. *J Neurochem*. 2004;91:1180–1190.
114. Gu L, Huang B, Shen W, et al. Early activation of nSMase2/ceramide pathway in astrocytes is involved in ischemia-associated neuronal damage via inflammation in rat hippocampi. *J Neuroinflammation*. 2013;10:109.
115. Wei W, Du C, Lv J, et al. Blocking A_{2B} adenosine receptor alleviates pathogenesis of experimental autoimmune encephalomyelitis via inhibition of IL-6 production and Th17 differentiation. *J Immunol*. 2013;190:138–146.
116. Eusemann TN, Willmroth F, Fiebich B, Biber K, van Calker D. Adenosine receptors differentially regulate the expression of regulators of G-protein signalling (RGS) 2, 3 and 4 in astrocyte-like cells. *PLoS One*. 2015;11:e0134934.
117. Gessi S, Varani K, Merighi S, et al. Expression, pharmacological profile, and functional coupling of A_{2B} receptors in a recombinant system and in peripheral blood cells using a novel selective antagonist radioligand, [3H]MRE 2029-F20. *Mol Pharmacol*. 2005;67:2137–2147.
118. Hinz S, Lacher SK, Seibt BF, Müller CE. BAY60-6583 acts as a partial agonist at adenosine A_{2B} receptors. *J Pharmacol Exp Ther*. 2014;349:427–436.
119. Eckle T, Faigle M, Grenz A, Laucher S, Thompson LF, Eltzschig HK. $A2_B$ adenosine receptor dampens hypoxia-induced vascular leak. *Blood*. 2008;111:2024–2035.
120. Feoktistov I, Ryzhov S, Zhong H, et al. Hypoxia modulates adenosine receptors in human endothelial and smooth muscle cells toward an A_{2B} angiogenic phenotype. *Hypertension*. 2004;44:649–654.
121. Eltzschig HK, Thompson LF, Karhausen J, et al. Endogenous adenosine produced during hypoxia attenuates neutrophil accumulation: coordination by extracellular nucleotide metabolism. *Blood*. 2004;104:3986–3992.
122. Yang D, Zhang Y, Nguyen HG, et al. The A_{2B} adenosine receptor protects against inflammation and excessive vascular adhesion. *J Clin Invest*. 2006;116:1913–1923.
123. Eltzschig HK, Bonney SK, Eckle T. Attenuating myocardial ischemia by targeting A_{2B} adenosine receptors. *Trends Mol Med*. 2013;19:345–354.
124. Grenz A, Osswald H, Eckle T, et al. The reno-vascular A_{2B} adenosine receptor protects the kidney from ischemia. *PLoS Med*. 2008;5:e137.
125. Koeppen M, Eckle T, Eltzschig HK. Interplay of hypoxia and A_{2B} adenosine receptors in tissue protection. *Adv Pharmacol*. 2011;61:145–186.
126. Gessi S, Merighi S, Varani K, Leung E, Mac Lennan S, Borea PA. The A_3 adenosine receptor: An enigmatic player in cell biology. *Pharmacol Ther*. 2008;117:123–140.
127. Borea PA, Gessi S, Bar-Yehuda S, Fishman P. A_3 adenosine receptor: Pharmacology and role in disease. *Handb Exp Pharmacol*. 2009;(193):297–327.
128. Pedata F, Pugliese AM, Sebastiao AM, Ribeiro JA. Adenosine A_3 receptor signaling in the central nervous system. In: Borea PA, ed. *A_3 Adenosine Receptors From Cell Biology to Pharmacology and Therapeutics*. Netherlands: Springer; 2010:165–188.

129. Fedorova IM, Jacobson MA, Basile A, Jacobson KA. Behavioral characterization of mice lacking the A_3 adenosine receptor: Sensitivity to hypoxic neurodegeneration. *Cell Mol Neurobiol*. 2003;23:431–447.
130. Chen GJ, Harvey BK, Shen H, Chou J, Victor A, Wang Y. Activation of adenosine A_3 receptors reduces ischemic brain injury in rodents. *J Neurosci Res*. 2006;84:1848–1855.
131. von Lubitz DK, Lin RC, Popik P, Carter MF, Jacobson KA. Adenosine A_3 receptor stimulation and cerebral ischemia. *Eur J Pharmacol*. 1994;263:59–67.
132. Hentschel S, Lewerenz A, Nieber K. Activation of A(3) receptors by endogenous adenosine inhibits synaptic transmission during hypoxia in rat cortical neurons. *Restor Neurol Neurosci*. 2003;21:55–63.
133. Pugliese AM, Coppi E, Volpini R, et al. Role of adenosine A_3 receptors on CA1 hippocampal neurotransmission during oxygen-glucose deprivation episodes of different duration. *Biochem Pharmacol*. 2007;74:768–779.
134. Colotta V, Catarzi D, Varano F, et al. New 2-arylpyrazolo[3,4-c]quinoline derivatives as potent and selective human A_3 adenosine receptor antagonists. Synthesis, pharmacological evaluation, and ligand-receptor modeling studies. *J Med Chem*. 2007;50: 4061–4074.
135. Colotta V, Catarzi D, Varano F, et al. Synthesis, ligand-receptor modeling studies and pharmacological evaluation of novel 4-modified-2-aryl-1,2,4-triazolo[4,3-a]quinoxalin-1-one derivatives as potent and selective human A_3 adenosine receptor antagonists. *Bioorg Med Chem*. 2008;16:6086–6102.
136. Colotta V, Lenzi O, Catarzi D, et al. Pyrido[2,3-e]-1,2,4-triazolo[4,3-a]pyrazin-1-one as a new scaffold to develop potent and selective human A_3 adenosine receptor antagonists. Synthesis, pharmacological evaluation, and ligand-receptor modeling studies. *J Med Chem*. 2009;52:2407–2419.
137. Pugliese AM, Coppi E, Spalluto G, Corradetti R, Pedata F. A_3 adenosine receptor antagonists delay irreversible synaptic failure caused by oxygen and glucose deprivation in the rat CA1 hippocampus *in vitro*. *Br J Pharmacol*. 2006;147:524–532.
138. Dennis SH, Jaafari N, Cimarosti H, Hanley JG, Henley JM, Mellor JR. Oxygen/glucose deprivation induces a reduction in synaptic AMPA receptors on hippocampal CA3 neurons mediated by mGluR1 and adenosine A_3 receptors. *J Neurosci*. 2011;31:11941–11952.
139. Nieber K,Hentschel S. Signalling pathways of adenosine A_3 receptors in rat cortical neurons. Proceedings of the 8th international symposium on adenosine and adenine nucleotides. Ferrara, Italy, May 24–28, 2006.
140. Volpini R, Costanzi S, Lambertucci C, et al. N (6)-alkyl-2-alkynyl derivatives of adenosine as potent and selective agonists at the human adenosine A_3 receptor and a starting point for searching A_{2B} ligands. *J Med Chem*. 2002;45:3271–3279.
141. Volpini R, Dal Ben D, Lambertucci C, et al. N6-methoxy-2-alkynyladenosine derivatives as highly potent and selective ligands at the human A_3 adenosine receptor. *J Med Chem*. 2007;50:1222–1230.
142. Choi IY, Lee JC, Ju C, et al. A_3 adenosine receptor agonist reduces brain ischemic injury and inhibits inflammatory cell migration in rats. *Am J Pathol*. 2011;179: 2042–2052.
143. Butler M, Sanmugalingam D, Burton VJ, et al. Impairment of adenosine A_3 receptor activity disrupts neutrophil migratory capacity and impacts innate immune function *in vivo*. *Eur J Immunol*. 2012;42:3358–3368.
144. Fishman P, Bar-Yehuda S, Liang BT, Jacobson KA. Pharmacological and therapeutic effects of A_3 adenosine receptor agonists. *Drug Discov Today*. 2012;17:359–366.

CHAPTER

9

Adenosine Receptors and Memory Disorders

Alexandre de Mendonca[1], *David Blum*[2] *and Jonathan D. Geiger*[3]

[1]University of Lisbon, Lisbon, Portugal [2]University of Lille, Lille, France [3]University of North Dakota, Grand Forks, ND, United States

OUTLINE

ALZHEIMER'S DISEASE

Alzheimer's disease (AD) was first described more than a 100 years ago. With longer life expectancy, its incidence has increased dramatically and current forecasts speak in terms of a doubling of the number of persons affected every 20 years.[1] AD is the most frequently encountered form of dementia (about 70% of cases of dementia). The definitive diagnosis of AD is based on the observation of characteristic brain

Adenosine Receptors in Neurodegenerative Diseases.
DOI: http://dx.doi.org/10.1016/B978-0-12-803724-9.00009-0

lesions, usually found during a post-mortem examination—senile plaques and neurofibrillary tangles.

Neurofibrillary degeneration consists of the intraneuronal accumulation of proteinaceous fibrils forming flame-shaped neurons into paired helical filaments (PHF).[2,3] The major antigen of PHF was shown to correspond to Tau protein, which is phosphorylated. Tau is a neuronal protein essentially located within the axonal compartment. Its structure makes it essential for the organization, stabilization, and dynamics of microtubules,[2,3] but recent data also emphasize that Tau has other important neuronal functions at the dendritic and nuclear levels.[4–6] Post-translational modifications, such as phosphorylation of Tau, affect its physiologic and pathologic functions. Indeed, in AD and in other related neurodegenerative disorders, aggregated Tau proteins are almost always found to be hyperphosphorylated. Furthermore, Tau hyperphosphorylation appears to affect many different functions of Tau, including its ability to aggregate.[2,3] In terms of AD pathogenesis, the importance of Tau pathology in decreased cognitive abilities is underscored by findings that Tau pathology spreads to the basal forebrain and several cortical areas in an anatomically-defined pattern[7] along neuronal projections defining the Braak stages of Tau pathology[8] and that these stages are well correlated to the severity of dementia.[9,10]

Another prominent pathological feature of AD is the appearance of senile plaques extracellularly composed largely of amyloid-beta (Aβ) protein and the intraneuronal accumulation of Aβ peptides. Together, these features are referred to as amyloid pathology and increased levels and plaques continue to be linked to AD, because under normal conditions Aβ is present in low concentrations. Aβ peptides are derived from either a precursor called βAPP (β-Amyloid Precursor Protein) or result from the combined action of the two distinct proteolytic enzymatic activities of β- and γ-secretase.[11,12] The activity of β-secretase is now well characterized as an acid protease called BACE1 (β-site-APP Cleaving Enzyme 1), which releases the N-terminal fragment of the amyloid peptides. In comparison, the C-terminal fragment is released by γ-secretase. Numerous studies suggest that presenilins 1 and 2 (PS1 and PS2) are themselves carriers of the γ-secretase activity. Depending on the presenilins involved, the γ-secretase activity appears to be associated with a high molecular mass multi-protein complex with physical and functional associations between at least three other proteins; nicastrin (NCT), Aph-1 (Anterior pharynx defective 1 homolog), and Pen-2 (Presenilin enhancer 2 homolog).[13]

In addition to Tau and Aβ pathology, inflammatory processes have also been strongly implicated in the pathogenesis of AD.[14] This is substantiated by recent genome-wide association studies that variants located in genes notably involved in innate immunity regulation such as CR1, TREM2 or CD33 modulate AD risk.[15,16] Microglial cells are key mediators of brain inflammatory processes and while they have been implicated in the pathogenesis of AD it is not yet clear the extent to which microglial activation is protective or deleterious.[14] However, microglial activation has been linked strongly to Tau. For example, microglial activation has been found to precede neurofibrillary tangle formation,[17] and removal of the microglial CX3CR1 receptor leads to enhanced Tau pathology.[18] These observations fit well with the ability of pro-inflammatory mediators, known to be released by microglial cells, such as IL1β or TNFα, to promote Tau phosphorylation and even its neuritic aggregation.[19,20] Together, these observations fit with recent data demonstrating a good correlation between innate immunity and p-Tau levels in brain of AD patients.[21]

CAFFEINE MAY PROTECT AGAINST COGNITIVE DECLINE AND AD

Worldwide, caffeine is the most widely consumed psychoactive drug. The pharmacological actions of caffeine are complex and differ greatly depending on dosage and concentration. For example, at very high concentrations (millimolar) that are not obtainable under normal human consumption conditions caffeine is well known to mobilize stores of intracellular calcium through actions on ryanodine receptors, and to increase levels of cAMP by inhibiting cyclic nucleotide phosphodiesterase. In contrast, normal consumption leads to levels of caffeine clearly capable of blocking cell surface adenosine receptors, including all 4 known subtypes; A_1, A_{2A}, A_{2B}, and A_3.[22,23] In humans, caffeine is not the only methylxanthine capable of blocking adenosine receptors. Following consumption, caffeine is largely (80%) demethylated to its dimethyl metabolic intermediate paraxanthine (1,7-dimethyl-xanthine) and about 16% is converted to theophylline (3,7-dimethylxanthine) and theobromine (1,3-dimethylxanthine); all known antagonists of adenosine receptors.

Caffeine has long been used in attempts to enhance cognition. In healthy subjects, caffeine consumption increases alertness and improves cognition.[24–27] Caffeine also improves attention and information processing.[28] More recently, caffeine was also shown to enhance memory consolidation.[29] Accordingly, and because of its widespread consumption, increased attention has been paid to the study of the effects of caffeine consumption on AD pathogenesis and aging-associated cognitive decline.

A protective relationship between caffeine consumption and the pathophysiology of neurodegenerative diseases has been suspected for more than two decades. In one of the first retrospective case-controlled studies, this one performed in Australia, an association between coffee consumption and AD was not apparent.[30] However, about one decade later we showed an inverse relationship between caffeine intake and the occurrence of AD later in life. In our retrospective case-controlled study patients with AD had an average daily caffeine intake of about 74 ± 98 mg during the 20 years that preceded diagnosis of AD whereas age-matched controls had a larger average daily caffeine intake of 199 ± 136 mg during the corresponding 20 years of their lifetimes.[31] A few years later, we had the opportunity to participate in the largest population-based cohort study on the association between caffeine intake and cognitive decline so far performed; 7000 participants were enrolled in the study.[32] Women with the highest rates of caffeine consumption (over three cups per day) showed less decline in verbal retrieval (OR = 0.67, 95% CI = 0.53 to 0.85) and visuospatial memory (OR = 0.82, 95% CI = 0.65 to 1.03) over 4 years than women consuming one cup or less. A significant protective effect for caffeine was not found in men.

Reflecting the growing interest in the topic, in 2010 the *Journal of Alzheimer's Disease* devoted a whole issue to therapeutic opportunities for caffeine in AD and other neurodegenerative disorders. This issue included a systematic review and meta-analysis of 9 cohort and 2 case-controlled studies in which relationships between caffeine intake and cognitive decline or dementia were examined quantitatively. The summary relative risk (RR) for the association between caffeine intake and different measures of cognitive impairment or decline was 0.84 (95% CI = 0.72 to 0.99). This finding suggested that caffeine might be protective against age-related cognitive decline but that the large methodological heterogeneity across a still limited number of epidemiological studies precluded a more definite statement of findings.

Since then, other studies were published and summary information was included in a recent systematic review.[33] The authors found that several cross-sectional and longitudinal population-based studies suggest a protective effect of coffee, tea, and caffeine use against late-life cognitive impairment or decline, but again they were cautious that the association was not found in all cognitive domains investigated and there was a lack of a distinct dose-response association, with a stronger effect among women than men. Regarding dementia and AD prevention, the authors pointed out that, although some studies with baseline examination in midlife pointed to a lack of association, other case-controlled and longitudinal population-based studies with briefer follow-up periods supported favorable effects of coffee, tea, and caffeine consumption against AD.[33]

Certainly, it would be important to know the mechanisms whereby caffeine could exert the putative neuroprotective effects against cognitive decline and AD. It is noteworthy that the Honolulu-Asian Aging

Study could not find a significant association between caffeine intake and dementia risk.[34] However, the authors reported that, at autopsy, patients in the highest quartile of caffeine intake (>277.5 mg/d) were less likely to have any of the neuropathological lesions, such as AD-related lesions, microvascular ischemic lesions, cortical Lewy bodies, hippocampal sclerosis, or generalized atrophy. These results suggest that neuroprotective effects of caffeine might be relatively nonspecific, rather than target a specific pathological neurodegenerative disease process. To this regard, it is interesting that long-term caffeine consumption was hypothesized to exert protective effects against AD by facilitating cerebrospinal fluid (CSF) production, turnover, and clearance.[35]

Rather intriguingly, a recent study, performed in patients with mild cognitive impairment or AD, showed that theobromine, a methylxanthine contained in chocolate, but not caffeine, had a significant positive correlation between CSF and plasma levels and levels of the AD CSF biomarker $A\beta_{1-42}$. In other words, theobromine was associated with a healthier CSF biomarker profile suggesting that it eventually may have a protective effect against the development of AD. In addition to theobromine being ingested in the form of chocolate, as mentioned above, it can also originate from metabolism of caffeine to theobromine. Thus, caffeine consumption should lead to significantly increased levels of theobromine because it does not get metabolized into other dimethylxanthines (i.e., theophylline or paraxanthine) nor does it get converted back to the trimethylxanthine caffeine.[36] A recent prospective cohort study showed that chocolate intake was indeed associated with a lower risk of cognitive decline (unpublished data). Interestingly, this protective effect was observed mainly among subjects with a low daily consumption of caffeine. This further emphasizes what is now recognized to be a complex relationship between the types and amounts of methylxanthines contained in different types of ingested products and their possible protection against cognitive decline and AD.

Considering other neurodegenerative disorders, there is presently robust epidemiological evidence that caffeine consumption is protective against Parkinson's disease (see systematic review and meta-analysis by 37). Less is known about the role of caffeine in other neurodegenerative disorders. Caffeine intake was reported in a retrospective case-control study to be inversely associated with the risk for amyotrophic lateral sclerosis.[38] In contrast, caffeine was suggested to exhibit detrimental effects on Huntington's disease whereby it might accelerate the age of onset.[39]

Certainly, a large amount of epidemiological evidence suggests that caffeine and other associated methylxanthines may have protective effects against late-life cognitive impairment or decline. However, a definitive conclusion must await results from a yet-to-be-performed long-term interventional randomized controlled study. As discussed elsewhere, the planning and conduction of such a clinical trial with caffeine undoubtedly represents a difficult challenge.[40] Furthermore, it is noteworthy that although the beneficial effects of coffee on cognitive decline and decreased AD risk in humans have been mostly ascribed to caffeine, other coffee constituents may play an important role. Indeed, two recent studies support that non-caffeine components of coffee display neuroprotective effects in drosophila and *Caenorhabditis elegans* amyloid models though activation of the Nrf2 detoxification pathway.[41,42]

CAFFEINE, ADENOSINE RECEPTOR AND AD LESIONS

Caffeine has been shown to be protective in a number of animal and cell-based models of AD. In transgenic mice overexpressing mutated APP (APPsw), caffeine was shown to mitigate against cognitive decline induced by Aβ as well as to reduce amyloid burden in preventive as well as therapeutic paradigms. Indeed, APPsw mice chronically treated from 4 to 9.5 months of age with caffeine (300 mg/L by drinking water corresponding to 500 mg/d in humans) were cognitively improved in several behavioral tasks that evaluated working and spatial memories and exhibited reduction of hippocampal $A\beta_{1-40}$ and $A\beta_{1-42}$.[43] Importantly, a similar treatment of APPsw mice at late pathological stages (18–19 months) for 4–5 weeks reversed memory deficits and reduced amyloid deposition as well as soluble Aβ levels in both entorhinal cortex and hippocampus.[44] Such beneficial effects of caffeine against Aβ production have recently been confirmed by another group in an experimental model of sporadic AD based on feeding rabbits with cholesterol enriched-diet that elevates both Aβ levels and Tau phosphorylation in the brain.[45] In this study, rabbits fed with a cholesterol-enriched diet were treated with low doses of caffeine (0.5–30 mg/d) through drinking water, corresponding to a maximal 60 mg/d consumption in humans. In this paradigm, caffeine significantly decreased Aβ production in accordance with Arendash's results.[45] Furthermore, reduced production of $A\beta_{1-40}$ and $A\beta_{1-42}$ was also observed in a neuroblastoma model overexpressing mutated βAPP following treatment with concentrations of caffeine below 10 μM[43], further supporting that caffeine impacts on mechanisms underlying amyloid peptide production. More recently, we found that caffeine concentration-dependently blocked LDL-cholesterol internalization and the ability of LDL-cholesterol to increase Aβ levels through what appeared to be adenosine A_3 receptor-mediated actions,[46] and that caffeine effectively blocked HIV-1 Tat protein-induced Aβ production and Tau phosphorylation.[47] The effects of chronic caffeine intake (0.3 g/L through drinking water) given at an early pathological stage have also been examined in the THY-Tau22 transgenic mouse model of progressive AD-like Tau pathology.[48] We found that chronic caffeine intake prevented the development of spatial memory deficits in Tau mice and that the improved memory was associated with reduced hippocampal Tau phosphorylation and proteolytic fragments. Moreover, caffeine treatment mitigated several pro-inflammatory markers found to be upregulated in the hippocampus of THY-Tau22 animals. Thus, in models of sporadic and early onset (genetically-linked) AD, caffeine has been shown repeatedly to reduce behavioral and pathological features associated with AD.

Mechanistically, chronic caffeine treatment of APPsw mice has been associated with decreased PS1 and BACE1 protein expression as well as increased Insulin-Degrading Enzyme (IDE) levels, the latter presumably contributing to enhanced Aβ degradation.[43,45] The effect of caffeine on BACE1 expression could relate to its ability to reduce c-Raf1 activity, possibly through PKA activation.[44] In addition, caffeine would reduce GSK3 expression and/or activity and thereby influence Aβ production.[44] However, a direct effect of caffeine on γ-secretase activity remains elusive, and mechanisms linking caffeine and Aβ production/clearance deserve further evaluations. Some of these effects of caffeine might be due to reduced leakage of the blood brain barrier. In our cholesterol-fed rabbit model of sporadic AD, we demonstrated that caffeine (3 mg/day, a human equivalent of $\sim$1 cup of coffee per day) blocked cholesterol-induced increases in leakage of Evan's blue dye and decreases in levels of the tight junction proteins occludin and ZO-1.[49] Similarly, we found using a MPTP-induced mouse model of PD that caffeine blocked MPTP-induced decreases in dopaminergic neurons, increases in leakage of Evan's blue dye and FITC-albumin in striatum but not in cerebral cortex or hippocampus, decreases in levels of the tight junction proteins occludin and ZO-1, and increases in reactive gliosis.[50] Thus, caffeine might protect against AD as well as other neurodegenerative disorders including PD by stabilizing the BBB.[51]

Plasma and brain concentrations achieved following low to moderate caffeine consumption in humans and in animal models[48] are mostly compatible with adenosine receptor inhibition.[22,52] Therefore, it is reasonable simply on the basis of dosage and levels to ascribe the actions of caffeine to its effects upon adenosine receptors. Caffeine could notably mediate its long-term effects through A_{2A} receptor blockade. During aging, both we and others have found compelling evidence of cortical and hippocampal increases of A_{2A}R expression/function. Specifically, in the hippocampus of aged rats, A_{2A}R expression is nearly two fold higher compared to young rats.[53,54] More importantly, the A_{2A}R-dependent activation of glutamate release becomes more pronounced as aging progresses and shifts from protein kinase C-mediated signaling to cAMP-dependent effects.[54,55] This is accompanied by clear behavioral deficits in hippocampus-dependent tasks, such as spatial memory in rats.[56] Accordingly, rats overexpressing hippocampal A_{2A} receptors also exhibit behavioral deficits including spatial memory defects as well as LTP impairments.[57] Interestingly, other detrimental conditions associated with cognitive impairment, such as hypoxia, diabetes, or epilepsy share similar $A_{2A}R$ overactivation.[54,58,59] Recently, decreased adult hippocampal long-term potentiation (LTP), cognitive/memory impairment, and increased hippocampal $A_{2A}R$ expression were observed in a chronic stress induced ageing-like model generated by maternal separation during the early post-natal period.[60]

Adenosine A_{2A} receptor antagonists might be a useful strategy against AD. Adenosine A_{2A} receptor blockade with the selective antagonist KW6002 normalized synaptic and cognitive dysfunctions in aged rats thus supporting the important role that A_{2A} receptors play in the genesis of synaptic dysfunction and cognitive impairment. With regards to AD, adenosine A_{2A} receptors are dysregulated[61–63] and the blockade of $A_{2A}R$ appears to be a rational strategy to mitigate several detrimental components in AD. Convergent data indicate that caffeine protects against the synapto-neurotoxicity induced by Aβ through blockade of A_{2A} receptors.[64] Caffeine was indeed shown to be protective against the death of rat primary hippocampal neurons induced by an adenovirus carrying mutated βAPP gene.[65] These data are in accordance with former works showing that, in primary cultures of cerebellar granule cells, low doses of caffeine (1–25 μM) were able to counteract $A\beta_{25-35}$ toxicity, an effect mimicked by ZM241385, an A_{2A} receptor antagonist, but not CPT, a selective A_1 receptor antagonist.[66] These protective effects were confirmed in vivo. A_{2A} receptor blockade, through intraperitoneal injection of SCH58261 and KW6002 or genetic knock-out, was also shown to prevent working memory impairment as well as synaptic loss induced by icv injection of $A\beta_{1-42}$.[64,67] Working memory improvement observed following A_{2A} receptor blockade was thought to be related to prevention of synaptotoxicity promoted by Aβ through modulation of p38 MAPK and mitochondrial function.[64] Even though A_{2A} receptor activation has been shown to favor amyloidogenesis,[68] it remains so far unclear the extent to which A_{2A} modulation impacts amyloidogenesis in vivo.

Caffeine has also been shown to protect against Tau pathology, and the effects of caffeine appear to be mediated through blockade of A_{2A} receptors because caffeine's effects were mimicked with specific A_{2A} receptor antagonists in a Tau transgenic mouse model.[69] Furthermore, we removed the gene encoding A_{2A} receptors from THY-Tau22 mice and analyzed the subsequent effects on both pathological (Tau phosphorylation and aggregation, neuroinflammation) and functional impairments (spatial learning and memory, hippocampal plasticity, neurotransmitter profile). We found that deleting A_{2A} receptors protected against Tau pathology-induced deficits in spatial memory and hippocampal long-term depression. These effects were concomitant with a normalization of the hippocampal glutamate/GABA ratio together with a global reduction in neuroinflammatory markers and a decrease in Tau hyperphosphorylation. Additionally, oral therapy using a specific A_{2A} receptor antagonist (MSX-3) significantly improved memory and reduced Tau hyperphosphorylation in Tau mice. Together, these results suggest that A_{2A} receptor modulation is capable of decreasing both amyloid and Tau mediated pathologies and might therefore be a path forward in developing A_{2A} receptor-based strategies in AD.

CONCLUSION

AD is a debilitating neurodegenerative disease the incidence of which continues to increase as our populations become more elderly. Caffeine is a commonly ingested and generally recognized as safe drug that can decrease the incidence of, as well as reduce behavioral and pathological features associated with, AD. Involvement of adenosine receptors appears to be a root mechanism for caffeine but clearly additional work is warranted to understand better the extent to which and mechanisms by which caffeine exerts its protective actions against AD.

Acknowledgments

David Blum lab is supported by France Alzheimer, LECMA/Alzheimer Forschung Initiative, LabEx (excellence laboratory) DISTALZ (Development of Innovative Strategies for a Transdisciplinary approach to ALZheimer's disease), Inserm, CNRS, Université Lille, RégionHauts de France (PartnAIRR), ANR (ADORATAU, SPREADTAU), ERA-Net (ABeta-ID) and FUIMEDIALZ. Work in the Geiger lab is supported by 3 grants from the National Institutes of Health (USA); P30GM103329, R01MH100972 and R01MH105329.

References

1. Wittchen HU, Jacobi F. Size and burden of mental disorders in Europe—a critical review and appraisal of 27 studies. *Eur Neuropsychopharmacol*. 2005;15:357–376.
2. Buée L, Bussière T, Buée-Scherrer V, Delacourte A, Hof PR. Tau protein isoforms, phosphorylation and role in neurodegenerative disorders. *Brain Res Brain Res Rev*. 2000;33:95–130.
3. Sergeant N, Bretteville A, Hamdane M, et al. Biochemistry of Tau in Alzheimer's disease and related neurological disorders. *Expert Rev Proteomics*. 2008;5:207–224.
4. Ittner LM, Ke YD, Delerue F, et al. Dendritic function of tau mediates amyloid-beta toxicity in Alzheimer's disease mouse models. *Cell*. 2010;142:387–397.

5. Morris M, Maeda S, Vossel K, Mucke L. The many faces of tau. *Neuron*. 2011;70:410–426.
6. Sultan A, Nesslany F, Violet M, et al. Nuclear tau, a key player in neuronal DNA protection. *J Biol Chem*. 2011;286:4566–4575.
7. Dujardin S, Colin M, Buée L. Invited review: Animal models of tauopathies and their implications for research/translation into the clinic. *Neuropathol Appl Neurobiol*. 2015;41:59–80.
8. Braak H, Braak E. Neuropathological stageing of Alzheimer-related changes. *Acta Neuropathol*. 1991;82:239–359.
9. Duyckaerts C, Bennecib M, Grignon Y, et al. Modeling the relation between neurofibrillary tangles and intellectual status. *Neurobiol Aging*. 1997;18:267–273.
10. Grober E, Dickson D, Sliwinski MJ, et al. Memory and mental status correlates of modified Braak staging. *Neurobiol Aging*. 1999;20(6):573–579.
11. Checler F. Processing of the beta-amyloid precursor protein and its regulation in Alzheimer's disease. *J Neurochem*. 1995;65:1431–1444.
12. De Strooper B. Proteases and proteolysis in Alzheimer disease: a multifactorial view on the disease process. *Physiol Rev*. 2010;90:465–494.
13. De Strooper B, Vassar R, Golde T. The secretases: enzymes with therapeutic potential in Alzheimer disease. *Nat Rev Neurol*. 2010;6:99–107.
14. Heneka MT, Carson MJ, El Khoury J, et al. Neuroinflammation in Alzheimer's disease. *Lancet Neurol*. 2015;14:388–405.
15. Lambert JC, Heath S, Even G, et al. Genome-wide association study identifies variants at CLU and CR1 associated with Alzheimer's disease. *Nat Genet*. 2009;41:1094–1099.
16. Guerreiro R, Brás J, Hardy J. SnapShot: genetics of Alzheimer's disease. *Cell*. 2013;155:968.
17. Yoshiyama Y, Higuchi M, Zhang B, et al. Synapse loss and microglial activation precede tangles in a P301S tauopathy mouse model. *Neuron*. 2007;53:337–351.
18. Bhaskar K, Konerth M, Kokiko-Cochran ON, Cardona A, Ransohoff RM, Lamb BT. Regulation of tau pathology by the microglial fractalkine receptor. *Neuron*. 2010;68:19–31.
19. Gorlovoy P, Larionov S, Pham TT, Neumann H. Accumulation of tau induced in neurites by microglial proinflammatory mediators. *FASEB J*. 2009;23:2502–2513.
20. Li Y, Liu L, Barger SW, Griffin WS. Interleukin-1 mediates pathological effects of microglia on tau phosphorylation and on synaptophysin synthesis in cortical neurons through a p38-MAPK pathway. *J Neurosci*. 2003;23:1605–1611.
21. Zotova E, Bharambe V, Cheaveau M, et al. Inflammatory components in human Alzheimer's disease and after active amyloid-β42 immunization. *Brain*. 2013;136:2677–2696.
22. Fredholm BB, Bättig K, Holmén J, Nehlig A, Zvartau EE. Actions of caffeine in the brain with special reference to factors that contribute to its widespread use. *Pharmacol Rev*. 1999;51:83–133.
23. Fredholm BB, Chen JF, Cunha RA, Svenningsson P, Vaugeois JM. Adenosine and brain function. *Int Rev Neurobiol*. 2005;63:191–270.
24. Fisone G, Borgkvist A, Usiello A. Caffeine as a psychomotor stimulant: mechanism of action. *Cell Mol Life Sci*. 2004;61:857–872.
25. Haskell CF, Kennedy DO, Wesnes KA, Scholey AB. Cognitive and mood improvements of caffeine in habitual consumers and habitual non-consumers of caffeine. *Psychopharmacology (Berl)*. 2005;179:813–825.
26. Sawyer DA, Julia HL, Turin AC. Caffeine and human behavior: arousal, anxiety, and performance effects. *J Behav Med*. 1982;5:415–439.
27. Smit HJ, Rogers PJ. Effects of low doses of caffeine on cognitive performance, mood and thirst in low and higher caffeine consumers. *Psychopharmacology (Berl)*. 2000;152:167–173.
28. Lorist MM, Tops M. Caffeine, fatigue, and cognition. *Brain Cogn*. 2003;53:82–94.
29. Borota D, Murray E, Keceli G, et al. Post-study caffeine administration enhances memory consolidation in humans. *Nat Neurosci*. 2014;17:201–203.

30. Broe GA, Henderson AS, Creasey H, et al. A case-control study of Alzheimer's disease in Australia. *Neurology*. 1990;40:1698–1707.
31. Maia L, de Mendonça A. Does caffeine intake protect from Alzheimer's disease? *Eur J Neurol*. 2002;9:377–382.
32. Ritchie K, Carrière I, de Mendonca A, et al. The neuroprotective effects of caffeine: a prospective population study (the Three City Study). *Neurology*. 2007;69:536–545.
33. Panza F, Solfrizzi V, Barulli MR, et al. Coffee, tea, and caffeine consumption and prevention of late-life cognitive decline and dementia: a systematic review. *J Nutr Health Aging*. 2015;19:313–328.
34. Gelber RP, Petrovitch H, Masaki KH, Ross GW, White LR. Coffee intake in midlife and risk of dementia and its neuropathologic correlates. *J Alzheimers Dis*. 2011;23:607–615.
35. Wostyn P, Van Dam D, Audenaert K, De Deyn PP. Increased cerebrospinal fluid production as a possible mechanism underlying caffeine's protective effect against Alzheimer's disease. *Int J Alzheimers Dis*. 2011;2011:617420.
36. Perera V, Gross AS, McLachlan AJ. Measurement of CYP1A2 activity: a focus on caffeine as a probe. *Curr Drug Metab*. 2012;13:667–678.
37. Costa J, Lunet N, Santos C, Santos J, Vaz-Carneiro A. Caffeine exposure and the risk of Parkinson's disease: a systematic review and meta-analysis of observational studies. *J Alzheimers Dis*. 2010;20(Suppl 1):S221–S238.
38. Beghi E, Pupillo E, Messina P, et al. EURALS Group. Coffee and amyotrophic lateral sclerosis: a possible preventive role. *Am J Epidemiol*. 2011;174:1002–1008.
39. Simonin C, Duru C, Salleron J, et al. Association between caffeine intake and age at onset in Huntington's disease. *Neurobiol Dis*. 2013;58:179–182.
40. de Mendonça A, Cunha RA. Therapeutic opportunities for caffeine in Alzheimer's disease and other neurodegenerative disorders. *J Alzheimers Dis*. 2010;20(Suppl 1):S1–S2.
41. Dostal V, Roberts CM, Link CD. Genetic mechanisms of coffee extract protection in a Caenorhabditis elegans model of β-amyloid peptide toxicity. *Genetics*. 2010;186:857–866.
42. Trinh K, Andrews L, Krause J, et al. Decaffeinated coffee and nicotine-free tobacco provide neuroprotection in Drosophila models of Parkinson's disease through an NRF2-dependent mechanism. *J Neurosci*. 2010;30:5525–5532.
43. Arendash GW, Schleif W, Rezai-Zadeh K, et al. Caffeine protects Alzheimer's mice against cognitive impairment and reduces brain beta-amyloid production. *Neuroscience*. 2006;142:941–952.
44. Arendash GW, Mori T, Cao C, et al. Caffeine reverses cognitive impairment and decreases brain amyloid-beta levels in aged Alzheimer's disease mice. *J Alzheimers Dis*. 2009;17:661–680.
45. Prasanthi JR, Dasari B, Marwarha G, et al. Caffeine protects against oxidative stress and Alzheimer's disease-like pathology in rabbit hippocampus induced by cholesterol-enriched diet. *Free Radic Biol Med*. 2010;49:1212–1220.
46. Li S, Geiger NH, Soliman ML, Hui L, Geiger JD, Chen X. Caffeine, Through adenosine A3 receptor-mediated actions, suppresses amyloid-β protein precursor internalization and amyloid-β generation. *J Alzheimers Dis*. 2015;47:73–83.
47. Soliman ML, Geiger JD, Chen X. Caffeine blocks HIV-1 Tat-induced amyloid beta production and tau phosphorylation. *J Neuroimmune Pharmacol*. 2016 (in press).
48. Laurent C, Eddarkaoui S, Derisbourg M, et al. Beneficial effects of caffeine in a transgenic model of Alzheimer's disease-like tau pathology. *Neurobiol Aging*. 2014;35:2079–2090.
49. Chen X, Lan X, Roche I, Liu R, Geiger JD. Caffeine protects against MPTP-induced blood-brain barrier dysfunction in mouse striatum. *J Neurochem*. 2008;107:1147–1157.
50. Chen X, Gawryluk JW, Wagener JF, Ghribi O, Geiger JD. Caffeine blocks disruption of blood brain barrier in a rabbit model of Alzheimer's disease. *J Neuroinflammation*. 2008;5:12.

51. Chen X, Ghribi O, Geiger JD. Caffeine protects against disruptions of the blood-brain barrier in animal models of Alzheimer's and Parkinson's diseases. *J Alzheimers Dis.* 2010;20:S127–S141.
52. Müller CE, Jacobson KA. Recent developments in adenosine receptor ligands and their potential as novel drugs. *Biochim Biophys Acta.* 2011;1808:1290–1308.
53. Cunha RA, Constantino MC, Sebastião AM, Ribeiro JA. Modification of A1 and A2a adenosine receptor binding in aged striatum, hippocampus and cortex of the rat. *Neuroreport.* 1995;6:1583–1588.
54. Lopes LV, Cunha RA, Ribeiro JA. Cross talk between A(1) and A(2A) adenosine receptors in the hippocampus and cortex of young adult and old rats. *J Neurophysiol.* 1999;82:3196–3203.
55. Rebola N, Sebastião AM, de Mendonca A, Oliveira CR, Ribeiro JA, Cunha RA. Enhanced adenosine A2A receptor facilitation of synaptic transmission in the hippocampus of aged rats. *J Neurophysiol.* 2003;90:1295–1303.
56. Diógenes MJ, Costenla AR, Lopes LV, et al. Enhancement of LTP in aged rats is dependent on endogenous BDNF. *Neuropsychopharmacology.* 2011;36:1823–1836.
57. Batalha VL, Ferreira DG, Coelho JE, et al. The caffeine-binding adenosine A2A receptor induces age-like HPA-axis dysfunction by targeting glucocorticoid receptor function. *Sci Rep.* 2016;6:31493.
58. Barros-Barbosa AR, Ferreirinha F, Oliveira A, et al. Adenosine A(2A), receptor and ecto-5′-nucleotidase/CD73 are upregulated in hippocampal astrocytes of human patients with mesial temporal lobe epilepsy (MTLE). *Purinergic Signal.* 2016;12(4):719-734.
59. Lopes LV, Sebastião AM, Ribeiro JA. Adenosine and related drugs in brain diseases: present and future in clinical trials. *Curr Top Med Chem.* 2011;11:1087–1101.
60. Batalha VL, Pego JM, Fontinha BM, et al. Adenosine A(2A) receptor blockade reverts hippocampal stress-induced deficits and restores corticosterone circadian oscillation. *Mol Psychiatry.* 2013;18:320–331.
61. Albasanz JL, Perez S, Barrachina M, Ferrer I, Martín M. Up-regulation of adenosine receptors in the frontal cortex in Alzheimer's disease. *Brain Pathol.* 2008;18:211–219.
62. Orr AG, Hsiao EC, Wang MM, et al. Astrocytic adenosine receptor A2A and Gs-coupled signaling regulate memory. *Nat Neurosci.* 2015;18:423–434.
63. Viana da Silva S, Haberl MG, Zhang P, et al. Early synaptic deficits in the APP/PS1 mouse model of Alzheimer's disease involve neuronal adenosine A2A receptors. *Nat Commun.* 2016;7:11915.
64. Canas PM, Porciúncula LO, Cunha GM, et al. Adenosine A2A receptor blockade prevents synaptotoxicity and memory dysfunction caused by beta-amyloid peptides via p38 mitogen-activated protein kinase pathway. *J Neurosci.* 2009;29:14741–14751.
65. Stoppelkamp S, Bell HS, Palacios-Filardo J, Shewan DA, Riedel G, Platt B. In vitro modelling of Alzheimer's disease: degeneration and cell death induced by viral delivery of amyloid and tau. *Exp Neurol.* 2011;229:226–237.
66. Dall'Igna OP, Porciúncula LO, Souza DO, Cunha RA, Lara DR. Neuroprotection by caffeine and adenosine A2A receptor blockade of beta-amyloid neurotoxicity. *Br J Pharmacol.* 2003;138:1207–1209.
67. Cunha GM, Canas PM, Melo CS, et al. Adenosine A2A receptor blockade prevents memory dysfunction caused by beta-amyloid peptides but not by scopolamine or MK-801. *Exp Neurol.* 2008;210:776–781.
68. Nagpure BV, Bian JS. Hydrogen sulfide inhibits A2A adenosine receptor agonist induced β-amyloid production in SH-SY5Y neuroblastoma cells via a cAMP dependent pathway. *PLoS One.* 2014;9:e88508.
69. Laurent C, Burnouf S, Ferry B, et al. A2A adenosine receptor deletion is protective in a mouse model of Tauopathy. *Mol Psychiatry.* 2016;21:149.

CHAPTER

10

Control of Motor Function by Adenosine A_{2A} Receptors in Parkinson's and Huntington's Disease

Annalisa Pinna[1], Jadwiga Wardas[2], Maria R. Domenici[3], Patrizia Popoli[3], Giovanni Cossu[4] and Micaela Morelli[1,5]

[1]Neuroscience Institute, Cagliari, Italy [2]Polish Academy of Sciences, Krakow, Poland [3]Italian National Institute of Health, Rome, Italy [4]AOB S. Michele General Hospital, Cagliari, Italy [5]University of Cagliari, Cagliari, Italy

OUTLINE

Adenosine Receptors in Neurodegenerative Diseases.
DOI: http://dx.doi.org/10.1016/B978-0-12-803724-9.00010-7

INTRODUCTION

Adenosine A_{2A} receptors are predominantly expressed in the basal ganglia (BG) circuitry, which is responsible for the integration of sensorimotor information that controls the planning and initiation of voluntary movement.[1] Moreover, the caudate-putamen (CPu) or striatum and nucleus accumbens contain higher concentrations of these receptors. Of great importance for adenosine A_{2A} receptor control of motor functions is their colocalization with the dopamine (DA) D_2 receptors.[2,3]

BASAL GANGLIA CIRCUITRY

The motor BG circuitry, involved in the pathophysiology of movement disorders, consists of several subcortical structures, including the CPu, the globus pallidus (internal (GPi) and external (GPe) divisions), the substantia nigra (pars reticulata (SNr), pars compacta (SNc)), and the subthalamic nucleus (STN) (Fig. 10.1).[4,5] All BG-related nuclei are connected through well-established neurochemical circuits, and with the specific cortical areas. The CPu receives and integrates glutamatergic input from the thalamus (centromedian, ventral anterior, and ventrolateral thalamic nuclei) and cerebral cortex, and transmits to the output nuclei, such as the SNr and GPi, which then project back the information to the CPu through the thalamus and cerebral cortex (Fig. 10.1).[4,5] Other BG output nuclei connect with the tegmental pedunculopontine nucleus and with

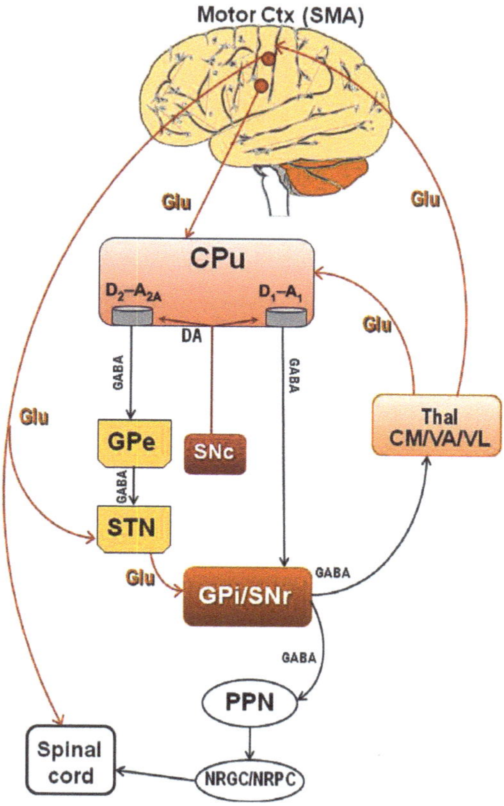

FIGURE 10.1 Scheme of neuronal circuitry (connections) between Basal Ganglia, Cerebral Cortex and other brainstem nuclei important for motor dysfunction in PD.
Cerebral cortex (the supplementary motor area, SMA) sends excitatory glutamatergic (Glu) projections to the CPu, spinal cord and STN. CPu receives dopamine (DA) ergic projections from the SNc and Glu afferents from cerebral cortex (Ctx) and intralaminar nuclei of the thalamus (Thal).

Output structures of the basal ganglia are the GPi (in primates) and SNr (in rodents), which send GABAergic projections to the thalamus and then Glu from the thalamus back to the cortical areas and CPu.

The GPi/SNr send also the GABAergic projections to midbrain tegmentum and pedunculopontine nucleus (PPN) which project to the nucleus reticularis giganto-cellularis (NRPG) and nucleus reticularis pontis caudalis (NRPC) whose efferents reach spinal motor neurons.

CM, centromedian thalamic nuclei; CPu, caudate-putamen; DA, dopamine; D_1, dopamine D_1 receptor; D_2, dopamine D_2 receptor; GABA, γ-aminobutyric acid; Glu, glutamic acid; GPe, globus pallidus pars externa; GPi, globus pallidus pars interna; NRGC, nucleus reticularis giganto-cellularis; NRPC, nucleus reticularis pontis caudalis; PPN, pedunculopontine tegmental nucleus; SMA, supplementary motor areas; SNc, substantia nigra pars compacta; SNr, substantia nigra pars reticulata; STN, subthalamic nucleus; Thal, thalamic nuclei; VA, ventral anterior nucleus; VL, ventrolateral nucleus.

the caudal intralaminar nuclei.[4,5] The neural striatal population is characterized by 95% of spiny γ-aminobutyric acid (GABA)ergic neurons and by 5% of aspiny interneurons, including GABAergic and cholinergic interneurons. The striatal population of spiny GABAergic neurons is divided into two neuronal pathways: The "striatonigral direct projection" which connects the CPu with the SNr/GPi and the "striatopallidal indirect projection" which connects the CPu with the GP/GPe (Fig. 10.1). The striatonigral neurons mainly express DA D_1 receptors and the neuropeptides substance P and dynorphin, whereas the striatopallidal neurons express mostly DA D_2 receptors and the neuropeptide enkephalin. DArgic input to the CPu arises primarily from the mesencephalon from the SNc, and plays a critical modulatory role in neuronal signaling at this level by exerting a dual effect, depending on the type of postsynaptic DA receptor stimulated. Specifically, DA modulates motor coordination and fine movements by facilitating the direct pathway, through acting on excitatory DA D_1 receptors and by inhibiting the indirect pathway, through acting on inhibitory DA D_2 receptors.[5] Although this classical model of BG function and dysfunction provides an excellent starting point,[5,6] it is important to highlight that BG organization is far more sophisticated than supposed in this model (reviewed in Refs. 1,7,8).

ADENOSINE A_{2A} RECEPTOR ANTAGONISTS

The restricted adenosine A_{2A} receptors expression in the BG circuitry and their close interaction with DA, especially with DA D_2 receptors, rendered adenosine A_{2A} receptors very attractive as a non-DArgic target for motor disease therapy, including Parkinson's disease (PD) and Huntington's disease (HD) therapy (Fig. 10.2). Indeed, adenosine A_{2A} receptors are localized in areas of the BG receiving the DArgic nigrostriatal and mesolimbic pathways, such as the CPu, GP, nucleus accumbens, and olfactory tubercle.[9] Specifically, in the CPu, adenosine A_{2A} receptors are predominantly restricted on the dendritic spines of GABAergic striatopallidal neurons, where they are colocalized with DA D_2 receptors,[10] whereas striatonigral neurons do not contain appreciable levels of adenosine A_{2A} receptors.[10] This colocalization of adenosine A_{2A} and DA D_2 receptors in the striatopallidal neurons leads to a functional antagonistic interaction between these receptors.[3,10,11] Specifically, stimulation of the DA D_2 receptors enhances motor activity, whereas stimulation of the adenosine A_{2A} receptors reduces this effect.[3,11] At the biochemical level, this antagonistic functional interaction between adenosine A_{2A} and DA D_2 receptors takes place both directly, with an intramembrane receptor–receptor interaction, in which the activation of adenosine A_{2A} receptors decreases the binding affinity of D_2 receptors

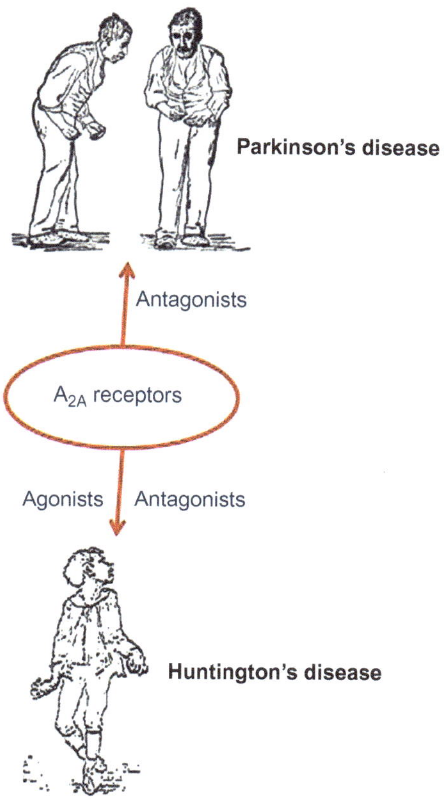

FIGURE 10.2 Influence of adenosine A_{2A} receptors in Parkinson's disease and in Huntington disease.

for DA,[3] and at the level of second messengers, such as adenylyl cyclase, in which stimulation of adenosine A_{2A} receptors counteracts the DA D_2 receptor-mediated inhibition of 3′,5′-cyclic adenosine monophosphate (cAMP) formation and D_2 receptor-induced intracellular Ca^{2+} responses.[12,13]

The functionally opposing roles of the adenosine A_{2A} and DA D_2 receptors on the indirect pathway neurons offers a rationale for the extensive investigation of the activity of A_{2A} receptor antagonists on counteracting motor deficits in pharmacological and toxicological animal models of PD and HD.

PARKINSON'S DISEASE

PD is the second most common chronic neurodegenerative disease, with a progressive course, affecting over 5 million individuals worldwide.[14]

The prevalence of PD is expected to rise dramatically over the next 20 years as the population gets older.[15] The motor symptoms that characterize PD, i.e., bradykinesia, muscle rigidity, resting tremor, gait disorders, and postural instability, are primarily due to the degeneration of the DArgic nigrostriatal neurons; therefore, PD therapy is mainly based on the replacement of impaired DArgic transmission.[16] The principal effect of DArgic neurodegeneration in parkinsonian patients leads to a disruption of processing in the BG circuitry (Fig. 10.1).[1,16] Indeed, in PD, the nigrostriatal DArgic neurodegeneration causes DA depletion in the CPu, consequently reducing activation of both DA D_1 and D_2 receptors (Fig. 10.1). This lack of striatal DA generates an imbalance in the activity of striatal output pathways, characterized by reduced excitation of the striatonigral direct pathway, which leads to a decrease in inhibitory control of the GPi/SNr, and a concomitant disinhibition of the striatopallidal indirect pathway to the STN and increases stimulation of the GPi/SNr neurons (Fig. 10.1). Taken together, this sequence of events exacerbates the activation of GABAergic BG output neurons, finally leading to excessive inhibition of thalamocortical projections of the motor systems, causing parkinsonian motor symptoms (Fig. 10.1).[1,6] Moreover, neuronal loss has also been observed in brain areas other than the BG, producing changes in neurotransmitters, such as noradrenaline, serotonin, glutamate, acetylcholine, and adenosine, which contribute to the symptomatology of PD.[17]

To date, DA replacement with the DA precursor L-dopa (in combination with a peripheral decarboxylase inhibitor) remains the most efficacious treatment to counteract PD motor symptoms, thereby improving survival.[18] However, a major limiting factor in chronic and pulsatile L-dopa therapy is the development of motor fluctuations that appear after several years of treatment.[19,20] Indeed, the duration of L-dopa's efficacy shortens over time, an effect known as *wearing-off*; responses become less predictable with rapid switching between time spent by patients in a state of mobility *on state* and severe immobility *off state*.[20] Moreover, patients affected by these motor response swings often show a range of types of choreic or dystonic drug-induced involuntary movements known as dyskinesia, which, in themselves, could become a major source of disability when the drug produces its maximal effect.[19,20] Motor fluctuations and dyskinesia, however, are not only due to the duration, dosage, and pulsatile administration of the pharmacologic treatment, but are also linked to intrinsic factors, such as the degree of loss of DArgic innervation in the BG (Fig. 10.1).[21]

Even though a few pharmacological and surgical strategies exist to ameliorate L-dopa-induced motor complications, they do not completely solve this problem.[22] This urgent topic has prompted research to find non-DArgic adjunctive treatments that modulate DArgic transmission, rather than stimulating it directly, in order to prevent and reduce the above-mentioned side effects.[23] Among the non-DArgic pharmacologic approaches to minimize motor dysfunction in PD, antagonists of adenosine A_{2A} receptors have emerged as leading candidates.[2,23–25]

EFFECT OF A_{2A} RECEPTOR ANTAGONISTS ON MOTOR AND SENSORIMOTOR DEFICITS

The efficacy of A_{2A} receptor antagonists against the symptomatic parkinsonian akinesia, bradykinesia, and motor activity impairment have been demonstrated using a wide range of pharmacological and/or toxicological rodent and primate models of PD, including counteraction of hypomotility or catalepsy induced by haloperidol or reserpine, and modulation of rotational behavior in rodents, as well as a reduction of motor impairment in nonhuman primates (Table 10.1).[26–28] The majority of A_{2A} receptor antagonists were able to counteract, in a dose-dependent manner, catalepsy and/or hypolocomotion induced by haloperidol or reserpine in rodents, and to strengthen the anticataleptic effect induced by L-dopa, suggesting that there may be a synergism between the adenosine A_{2A} receptor antagonists and the DArgic agents (Table 10.1).[28–30] Interestingly, Varty and collaborators have demonstrated that adenosine A_{2A} receptor blockade can attenuate haloperidol-induced cataleptic motor impairment also in monkeys (Table 10.1).[31]

To verify the antiparkinsonian effects of A_{2A} receptor antagonists, these compounds have been evaluated in the most frequently used PD model of hemiparkinsonian rats, characterized by a unilateral intracerebral infusion of the DArgic neurotoxin 6-hydroxydopamine (6-OHDA), which produces massive degeneration of the nigrostriatal DArgic neurons, similar to that occurring in idiopathic PD.[59,60] In this model, the ability of a specific drug to induce contralateral rotational behavior, as well as to potentiate the rotational behavior stimulated by DA receptor agonists, can be assumed as a parameter reflecting its antiparkinsonian activity.[59,60]

A_{2A} receptor antagonists clearly showed a motor facilitative activity in hemiparkinsonian rats. In particular, acute administration of adenosine A_{2A} receptor antagonists induced no contralateral rotations per se, but significantly potentiated rotational behavior induced by L-dopa or apomorphine and by either DA D_1 or D_2 receptor agonists in hemiparkinsonian rodents (Table 10.1; for review see Ref. 28).

Moreover, in hemiparkinsonian rats, more sophisticated measurements of akinesia, bradykinesia, and gait impairment induced by the

TABLE 10.1 Main Effects Exerted by A_{2A} Antagonists on Cardinal Parkinsonian-Like Symptoms and Motor Complications in Rodent and Primate Models of PD

Parkinsonian-like symptoms	**Effects of A_{2A} antagonists**
Parkinsonian-like akinesia, bradykinesia, motor impairment, sensorimotor integration deficit and tremor	Antagonism of hypolocomotion and catalepsy induced by reserpine or haloperidol (alone or in combination with L-dopa) in rodents and primates[26–34]
	Potentiation of contralateral rotational behavior induced by DAergic antiparkinsonian drugs in the hemiparkinsonian rats[26–28,33,35–37]
	Restoration of the impaired initiation time of stepping and adjusting steps of the forelimb contralateral to the lesion in hemiparkinsonian rats[28,35,38]
	Restoration of the placement of the contralateral forelimb after vibrissae brushing in hemiparkinsonian rats[28,35,38]
	Reinstatement of hindlimb bradykinesia and performance in rotarod test of MitoPark mice[39]
	Reduction of motor impairment in MPTP-treated primates (alone or in combination with DAergic antiparkinsonian drugs)[40–44]
	Reversal of parkinsonian-like tremor elicited by several tremorigenic agents in rodents[26,28,45–50]
Parkinsonian-like muscle rigidity	Amelioration of parkinsonian-like muscle rigidity produced by either haloperidol or reserpine (alone or in combination with L-dopa) in rodents[32,51] and primate[31]
Parkinsonian-like motor complications induced by L-dopa	**Effects of A_{2A} antagonists on motor complication**
	Attenuation of the *wearing-off* phenomenon induced by L-dopa in hemiparkinsonian rodents[28,36,52–54]
	Lack of sensitization to the intensity of contralateral turning in hemiparkinsonian rodents chronically treated with L-dopa[26,28,33,37,55,56]
	Lack of exacerbation of the intensity of AIMs in hemiparkinsonian rodents chronically treated with L-dopa[26,28,34,37,57,58]
	No onset of dyskinesia when given as monotherapy to MPTP-treated primate[40–43]
	Reduction of dyskinesia elicited by L-dopa in MPTP-treated primates[42,43]
	Efficacy in delaying apomorphine-induced dyskinesia in MPTP-treated primates[53]

unilateral 6-OHDA lesion have been assessed. Different strategies, such as adjusting step counting (index of gait impairment) and initiation time of stepping (index of akinesia/bradykinesia), have been developed in order to evaluate and quantify these parkinsonian-like symptoms and their relief by drugs.[61] In addition, similar to parkinsonian patients, hemiparkinsonian rats showed marked sensorimotor integration deficits correlated with a unilateral lesion of the DArgic nigrostriatal pathway and evaluated by vibrissae test.[62] Administration of the A_{2A} receptor antagonists, similarly to L-dopa, significantly counteracted forelimb akinesia/hypokinesia and motor initiation deficit, and restored the sensorimotor integration deficit, supporting a potential efficacy of these compounds to ameliorate these impairments in PD patients (Table 10.1).[28,35,38] Recently, the antiakinetic/bradykinetic effects of A_{2A} receptor antagonists have been demonstrated, also, in a genetic mouse model of PD that displays a progressive loss of DA neurons, such as in the MitoPark mouse (Table 10.1).[39]

The antiparkinsonian activity of A_{2A} receptor antagonists against bradykinesia, akinesia, and motor disability shown in the rodent model of PD has been confirmed in a neurotoxic primate model of PD (Table 10.1). Primates treated with the neurotoxin 1-methyl-4-phenyl-1,2,3,6-tetrahydropyridine (MPTP) is a model that closely mimics the clinical features of PD in humans and in which all currently used antiparkinsonian medications have been shown to be effective and highly predictive of their effect in humans. Moreover, the MPTP-treated primate develops clear dyskinesia when repeatedly exposed to L-dopa.[63]

Acute administration of A_{2A} receptor antagonists increased locomotor activity and reversed motor disability in a dose-dependent manner in parkinsonian MPTP-treated primates (Table 10.1).[40–43] Furthermore, when coadministered with L-dopa or DA D_1 and D_2 receptor agonists, A_{2A} receptor antagonists enhanced the intensity and duration of the efficacy of L-dopa in reversing motor disabilities and increasing locomotor activity (Table 10.1).[42–44] Interestingly, despite producing an enhanced antiparkinsonian response, acute A_{2A} receptor antagonists did not exacerbate the dyskinesia induced by L-dopa in MPTP-treated primates previously rendered dyskinetic by L-dopa exposure.[40,41,43]

Notably, chronic administration of A_{2A} receptor antagonists has been demonstrated to effectively improve motor deficits in rodent and primate models of PD, and does not produce tolerance to their motor-stimulant effects.[36,41,55] Recently, the long-lasting efficacy of A_{2A} receptor antagonists in preventing the reduction of spontaneous locomotor activity has been demonstrated in both early and mild to severe parkinsonian genetic MitoPark mice.[39,64]

Parkinsonian Tremor and Involvement of A_{2A} Receptors

Another important antiparkinsonian effect exerted by A_{2A} receptor antagonists is the antitremorigenic effect (Table 10.1). Acute administration of several A_{2A} receptor antagonists significantly reversed jaw tremor induced by several different drugs including tacrine, pilocarpine, haloperidol, reserpine, and pimozide in rats, suggesting a beneficial use of these compounds as specific drugs against this parkinsonian symptom (Table 10.1).[35,45–49] Moreover, A_{2A} receptor antagonism or genetic deletion of the adenosine A_{2A} receptor significantly attenuated the tremulous jaw movements induced by pilocarpine in mice (Table 10.1).[50] Considering the important role played by an increase in striatal acetylcholine in tremor development, and the reduction of the evoked release of this neurotransmitter exerted by A_{2A} receptor antagonists, it might be suggested that the anticholinergic effect exerted by blockade of the adenosine A_{2A} receptors may explain the antitremorigenic effect.[46,47]

Parkinsonian Muscle Rigidity and Involvement of A_{2A} Receptors

Another cardinal motor feature of PD besides bradykinesia, tremor at rest, and postural instability is muscle rigidity. Rigidity is typically tested in PD patients at rest by assessing passive movement of the wrists, elbows, neck, and ankles. Parkinsonian rigidity is characterized by increased resistance to passive movement of a joint (stretching and shortening), with both flexor and extensor muscles involved.[65] Moreover, the tonic electromyographic (EMG) activity at rest shows no changes in short-latency, monosynaptic spinal components, but displays increased long-latency EMG components based mainly on polysynaptic and spino-supraspinal reflexes.[65–67] The underlying mechanism of parkinsonian rigidity is poorly understood, however contribution of spinal cord, brainstem, and higher cortical circuits was suggested to be important in the pathophysiology of this symptom.[65–68]

Experimental studies showed that the lesion of nigrostriatal neurons by the 6-OHDA, injection of reserpine or blockade of DA D_2 receptors by haloperidol were associated with an increased muscle tone, measured by combined mechano (MMG) and EMG (MMG/EMG) methods in rats.[69–72] Moreover, such treatments modeled well the parkinsonian muscle rigidity since all the above-mentioned features (resistance of the hind foot to passive movements, enhanced EMG activity at rest and increased long-latency EMG components) were observed.[32,51,69,71,72] Additionally, the rigidity was antagonized by antiparkinsonian drugs such as L-dopa and pramipexole.[32,51,73]

Previous studies have shown that stimulation of adenosine A_{2A} receptors by very high doses of CGS 21680, injected either peripherally[74] or directly into the rostral CPu,[75] evokes parkinsonian muscle rigidity measured as an increased resistance in response to passive movements in the hind limb and the EMG activity of the antagonistic muscles, gastrocnemius (GS) and tibialis anterior (Tib). Furthermore, a tonic EMG activity develops at rest, which, similarly to in PD patients, reflects some difficulty in relaxing the muscles.[66,74,75] Studies by Wu et al.[76] examining the influence of peripheral injection of CGS 21680 on joint stiffness and viscosity, analyzed separately, showed that stimulation of adenosine A_{2A} receptors, similarly to raclopride, increased stiffness measured at the hind paw. However, the effect on viscosity was less clear, as only a nonsignificant trend after CGS 21680 was found, which suggests that this component of joint resistance seems to be less consistently affected than stiffness.

Accordingly, selective A_{2A} receptor antagonist SCH 58261 counteracted muscle rigidity evoked by haloperidol or reserpine. It decreased both the muscle resistance and tonic and reflex EMG activities in the GS and Tib muscles (Table 10.1).[51] Moreover, blockade of A_{2A} receptors was shown to potentiate the efficacy of a low, ineffective dose of L-dopa in both models of parkinsonian-like rigidity, an effect which might be attributed to the synergistic action of adenosine A_{2A} receptor blockade with DA D_1 and D_2 stimulation (Table 10.1).[32,51]

A_{2A} Receptor Antagonists on L-Dopa-Induced Motor Complications

The most important limitation of long-term therapy with L-dopa in PD patients is characterized by motor fluctuations consistent with the progressive reduction of the drug's efficacy in preventing parkinsonian motor symptoms, usually known as *wearing-off* and *on–off* phenomena.[19,20] During *wearing-off*, L-dopa counteracts PD motor deficits for a shorter period of time, after which akinesia and rigidity become manifest again. In the *on–off* phenomenon, the patient fluctuates from *on state* in which the parkinsonian impairments are counteracted, to *off state* in which the patient shows bradykinesia and rigidity.

In hemiparkinsonian rats, the duration of rotational behavior induced by L-dopa progressively decreases during the long-term treatment with this drug, a phenomenon that mimics *wearing-off* of L-dopa observed in parkinsonian patients.[77] Consistent with the acute effect of A_{2A} receptor antagonists producing an increased duration of rotational behavior induced by L-dopa or apomorphine,[36,52] the coadministration of A_{2A} receptor antagonists with L-dopa reversed the shortening of rotational behavior, supporting a potential beneficial influence of adenosine A_{2A} receptor blockade on L-dopa-induced *wearing-off* (Table 10.1).[36,52–54]

Chronic therapy with L-dopa is associated with the development of dyskinesia, characterized by dystonic-choreic abnormal involuntary movements (AIMs), which are highly disabling for parkinsonian patients.[19,20] The influence of adenosine A_{2A} receptor blockade on dyskinesia has been investigated by means of validated experimental paradigms in which dyskinetic movements induced by chronic L-dopa are expressed both in hemiparkinsonian rodents (sensitization of rotational behavior and/or AIMs affecting parts of the body contralateral to the lesion)[37,55,57,78–80] and in parkinsonian MPTP-treated primates (dyskinetic movements affecting several parts of the body, similar to those observed in parkinsonian patients).[81]

Chronic administration of A_{2A} receptor antagonists together with L-dopa, at dosages that induced the same number of contralateral rotations of a full effective dose of L-dopa alone, did not lead to a modification of the intensity of rotational behavior or AIMs during treatment, while L-dopa alone produced a sensitization in rotational behavior intensity (index of dyskinesia), and increased the intensity of AIMs (Table 10.1),[33,37,55] strongly indicating the lack of dyskinetic potential of A_{2A} receptor antagonists. Parallel studies demonstrated that different A_{2A} receptor antagonists, given in combination with a full effective dose of L-dopa, did not affect the severity of AIMs in dyskinetic hemiparkinsonian rats (Table 10.1).[34,57] Interestingly, this hypothesis has been supported by studies showing that genetic deletion of the adenosine A_{2A} receptor prevents the sensitization of contralateral rotational behavior and AIMs stimulated by L-dopa in hemiparkinsonian mice.[56,58]

Findings in dyskinetic parkinsonian primates confirmed that A_{2A} receptor antagonists associated with a low nondyskinetic dose of L-dopa may satisfactorily ameliorate motor deficits, limiting the severity of L-dopa-induced dyskinesia (Table 10.1).[42,43] A further interesting study was performed by the research group of Bibbiani et al.[53] who showed that istradefylline delayed the shortening of the motor response after chronic administration of the mixed DA D_1–D_2 receptor agonist apomorphine, suggesting that A_{2A} receptors may play an important role in the development of dyskinesia, rather than its expression, in nonhuman primates (Table 10.1). Taken together, these results suggested that, although no study has yet demonstrated the ability of A_{2A} receptor antagonists to revert an already established dyskinesia in both rodents and primates, the association of A_{2A} receptor antagonists with a low nondyskinetic dose of L-dopa might produce an efficient improvement of motor symptoms, contrasting the increased intensity of dyskinetic movements that develop during a chronic treatment with L-dopa. Moreover, in hemiparkinsonian rodent studies, a prolongation of L-dopa efficacy was observed, suggesting a potential therapeutic efficacy of A_{2A} receptor antagonists on *wearing-off* and *on–off* phenomena when administered in association with L-dopa.[33,55]

Clinical Studies on A_{2A} Receptor Antagonists in Parkinson's Disease

The development of new highly selective adenosine A_{2A} receptor antagonists, and their encouraging antiparkinsonian responses in animal models of PD, have provided a rationale for clinical trials to evaluate the therapeutic potential and the safety of these agents in parkinsonian patients. Several clinical trials with A_{2A} receptor antagonists have so far been performed in PD patients with advanced disease and motor complications. In the majority of those studies, A_{2A} receptor antagonists were utilized as an adjunct to L-dopa, and results generally showed that the A_{2A} receptor antagonists tested were safe and well tolerated.[24,25] Moreover, numerous clinical trials in advanced PD patients have demonstrated the efficacy of A_{2A} receptor antagonists in reducing the *wearing-off* phenomenon and in reducing the waking time spent in the *off state* and increasing the *on state*. During the *on* state, an increase in dyskinesia classified as nontroublesome was observed.[24,25] The commercialization of the A_{2A} receptor antagonist istradefylline as a drug to counteract *wearing-off* in PD patients,[25,82] and the clinical use of istradefylline (NOURIAST tablets 20 mg) was approved in Japan in 2013, but not in the United States.[82] To date, clinical development of new adenosine A_{2A} antagonists as a treatment for advanced PD is in progress (Fig. 10.2).[25]

MODULATION OF ADENOSINE A_{2A} RECEPTORS AND INTERACTION WITH THE OTHER NEUROTRANSMITTER SYSTEMS

Considering the neuromodulatory role of adenosine, adenosine A_{2A} receptors have been shown to interact either directly or indirectly with various receptors, such as the DA D_3, *N*-methyl-D-aspartate (NMDA), cannabinoid CB_1, serotonin 5-HT_{1A}, metabotropic glutamate 4 ($mGlu_4$) and 5 ($mGlu_5$) receptors, and to form heteromeric complexes with some of them, suggesting a complex mechanism through which they influence PD motor functions.[27,83–85] This receptor heteromer concept postulates that receptors from different families combine to generate complexes with distinctive biochemical and functional characteristics, thus generating unique functional entities and novel potential targets for therapy.[27,86]

Interaction Between A_{2A} Receptor Antagonists and Dopaminergic Receptors

To better understand the antiparkinsonian efficacy of A_{2A} receptor antagonists on the cardinal symptoms of PD, it is necessary to illustrate the main role played by adenosine A_{2A} receptors in the motor BG circuitry involved in the pathophysiology of movement disorders (Fig. 10.1).

As described above, the colocalization of adenosine A_{2A} and DA D_2 receptors in the striatopallidal neurons provides the anatomical basis for the existence of a functional antagonistic interaction between these receptors (Fig. 10.1).[11] Adenosine A_{2A} receptor blockade reduces the excessive inhibitory output of the striatal indirect pathway, similar to DA D_2 receptor activation (Fig. 10.1).[11] In addition, activation or blockade of the adenosine A_{2A} receptors in the indirect striatopallidal pathway indirectly impairs or facilitates DArgic D_1-mediated responses as well.[11,87] Thus, A_{2A} receptor antagonists seem to restore some balance between the striatonigral and striatopallidal neurons, reinstating thalamocortical activity (Fig. 10.1). Moreover, an important function of adenosine A_{2A} receptors has been showed in the GP (Fig. 10.1). Indeed, in PD, the blockade of pallidal adenosine A_{2A} receptors, by reducing extracellular GABA, may contribute to restoring GP and, in turn, STN activity.[88] Furthermore, when associated with DArgic receptor agonists, they induce an enhancement of their motor-stimulating effects.[46,47,88,89] In addition, stimulation of the postsynaptic adenosine A_{2A} receptors antagonizes the inhibitory modulation of the NMDA receptor activity mediated by DA D_2 receptors, including modulation of Ca^{2+} influx, transition to the firing "up" state and modulation of neuronal firing in the "up" state.[90,91] This interaction appears to be responsible for most of the locomotor activation and depression induced by A_{2A} receptor antagonists and agonists, respectively.[86] Further contribution to the antiparkinsonian effects—in particular the antitremorigenic effect—of adenosine A_{2A} receptor antagonists may be related to a cholinergic mechanism.[26,27,92] Indeed, functional antagonism between the adenosine A_{2A} and DA D_2 receptors has been reported in striatal cholinergic interneurons.[93]

Interaction Between A_{2A} Receptor Antagonists and Glutamate Receptors

While the DArgic nigrostriatal projection, which modulates medium spiny neurons, is at the basis of the DA–A_{2A} receptor interaction, the cortical glutamatergic projection modulates these projecting neurons through multiple glutamate receptor subtypes. Neuroanatomical ultrastructural studies have shown that striatal A_{2A} receptors are highly expressed in the dendrites and dendritic spines of GABAergic postsynaptic neurons in asymmetric excitatory synapses[9,10] and, therefore, these receptors may not only interact with the DA receptors, but also with the glutamatergic ionotropic and metabotropic receptors. Since excessive increase in glutamate plays an important role in the neuroplasticity taking place in the BG, A_{2A} receptor modulation of glutamatergic transmission at the extracellular level can influence neuronal plasticity in a significant way.[94] As several lines of preclinical evidence have demonstrated, glutamate plays a most important role in neuroplasticity and in the abnormal modifications related to L-dopa therapy, in particular dyskinesia.[95] Consequently, multiple presynaptic as well as postsynaptic mechanisms could contribute to the modulatory role played by the A_{2A} receptors on glutamatergic transmission. Indeed, several glutamatergic receptors, such as α-amino-3-hydroxy-5-methyl-4-isoxazole-propionic acid (AMPA), NMDA and mGlu receptors, have been involved in these events.

In hemiparkinsonian rats, chronic administration of L-dopa was reported to induce a hyperphosphorylation of the AMPA receptor, an effect that was significantly attenuated when L-dopa was administered in combination with an A_{2A} receptor antagonist.[53] Since A_{2A} receptors activate protein kinase A (PKA) and C (PKC),[89,96] it is possible that A_{2A} receptor antagonists, by inhibiting these kinases, might attenuate hyperphosphorylation of these glutamatergic receptors.

In addition, A_{2A} receptors regulate the conductance[97] and phosphorylation of NMDA receptors.[98] A role of the striatal A_{2A} and NMDA receptors in dyskinesia is also suggested by a study by Ekonomou et al.,[99] who showed that A_{2A} receptor stimulation upregulates a subunit of the NMDA receptor of weaver mice, in which overactivity of these receptors worsens dyskinesia.

Besides the ionotropic receptors, the mGluR are present in the BG and interaction with the A_{2A} receptors of some of them has been described.[83]

The mGlu5 receptor antagonists were considered as a therapeutic approach for PD. It was, in fact, reported that chronic treatment with mGlu5 receptor antagonists normalized glutamate neurotransmission and reduced the development of dyskinesia.[95] Moreover, a positive interaction between the A_{2A} receptors and the mGlu5 receptor was described in models of PD,[100,101] the basis of which might be the modulatory role of A_{2A} receptors and mGlu5 receptor on striatal electrical activity.[102] Moreover, a recent study in hemiparkinsonian rats has demonstrated that the combined treatment with A_{2A} and mGlu5 receptor antagonists results in a therapeutic benefit and provides better results than those produced by either drug given alone, both in terms of motor performance and neuroprotection, shedding some light on the efficacy and advantages of combined non-DArgic PD treatments using low drug concentrations.[103]

A different example of an A_{2A} and mGlu receptor interaction in preclinical models of PD is the mGlu4 receptor. Agonists and positive allosteric modulators (PAM) of the mGlu4 receptors have been suggested to be efficacious in reducing L-dopa dose, while maintaining the same benefit on PD motor impairment.[84,104,105] Interestingly, the selective PAM of mGlu4 receptor, VU0364770, produces synergistic effects when administered with either L-dopa or the A_{2A} receptor antagonist preladenant.[84]

Interaction Between A_{2A} Receptor Antagonists and Cannabinoid Receptors

Additional mechanisms involved in the modulation of L-dopa-induced abnormal motor responses include the cannabinoid CB_1 receptors.[106] Selective CB_1 antagonists may enhance the antiparkinsonian action of L-dopa and allow the use of lower doses of DA receptor agonists, possibly reducing side effects,[107] whereas dyskinesia may be alleviated by activation of the CB_1 receptors.[108,109] $A_{2A}-CB_1$ receptor heteromeric complexes in cotransfected HEK-293T cells and rat CPu have been demonstrated.[110,111] Moreover, adenosine A_{2A}, cannabinoid CB_1 and DA D_2 receptors may interact to form $A_{2A}-CB_1-D_2$ receptor heteromers in cotransfected cells as well as in the rat CPu.[110,111] These heteromers are postsynaptically located in the spines of GABAergic enkephalinergic neurons.[110] In rats, the motor effects induced by the intrastriatal administration of CB_1 receptor agonists could be counteracted by A_{2A} receptor antagonists,[110] whereas the inhibitory effect of CB_1 receptor agonists on D_2 receptor agonist-induced hyperlocomotion could be counteracted by CB_1 receptor antagonists and A_{2A} receptor antagonists,[111] providing evidence for the existence of functional receptor–receptor interactions. $A_{2A}-CB_1-D_2$ receptor heteromers are also present in the CPu of 6-OHDA-lesioned rats; however, following acute or chronic treatment with L-dopa, the heteromer crosstalk is lost.[112] Similar results were obtained in nonhuman primates in which the expression of $A_{2A}-CB_1$, $A_{2A}-D_2$, and CB_1-D_2 heteromers was reduced in the caudate nucleus of monkeys that received chronic L-dopa treatment compared with control, supporting the evidence that L-dopa alters heteromer expression in models of PD.[113] As reported by Bonaventura et al.,[113] the L-dopa-induced disruption of $A_{2A}-CB_1-D_2$ receptor heteromers contributed to the alteration of the balance between striatal direct and indirect efferent pathways by eliminating the "brake" that A_{2A} or CB_1 receptor activation exerts on D_2 receptor-mediated motor behavior.

Therefore, drugs targeting A_{2A}–CB_1–D_2 receptor heteromers might balance striatal efferent pathways and prevent L-dopa-induced motor complications.

Interaction Between A_{2A} Receptor Antagonists and 5-HT Receptors

While in the early stages of PD L-dopa is transformed into DA, released and reuptaken steadily in the remaining presynaptic DArgic terminals, in later stages, when DArgic innervation is totally lost, L-dopa is mainly transformed into DA in non-DArgic neurons, such as serotoninergic neurons. In these neurons, neither DA reuptake occurs nor are DA autoreceptors present and, therefore, the duration of the effect of L-dopa seems to reflect its plasma half-life rise and fall.[114,115] In this context, long-term aberrant changes due to pulsatile stimulation of DA receptors take place. According to this view, toxin lesions or pharmacological silencing of serotonin neurons have been shown to suppress dyskinetic movements in dyskinetic rats and macaques.[116,117] Moreover, dampening of serotonin neuron release by the partial 5-HT_{1A} receptor agonist buspirone was shown to reduce both dyskinesia and synaptic DA levels.[118] Whereas targeting of serotonin autoreceptors has been shown to be extremely effective in counteracting AIMs in animal models, concerns have been raised about the preservation of the therapeutic effect of L-dopa. Indeed, excessive dampening of serotonin neuron release led, unavoidably, not only to reduction of AIMs but also of L-dopa therapeutic efficacy in dyskinetic rats and MPTP-treated macaques.[119] As mentioned above, A_{2A} receptor antagonists do not counteract dyskinesia; however, when administered chronically with a low dose of L-dopa, they potentiate the motor efficacy of the drug and extend its duration without exacerbating dyskinesia compared with a full dose of L-dopa.[37,55] Interestingly, a recent study demonstrated that the combination of the 5-$HT_{1A/1B}$ receptor agonist, eltoprazine, and the A_{2A} receptor antagonist, preladenant, reduces AIMs induced by L-dopa in both dyskinetic hemiparkinsonian rats and MPTP-treated monkeys, without impairing motor performance.[52] As for the mechanisms that underlie these findings, although a study by Lukasiewicz and coworkers[85] demonstrated a direct adenosine A_{2A}/5-HT_{1A} receptor interaction in cotransfected cells, the lack of evidence of receptor–receptor interaction *in vivo* suggests that two complementary mechanisms may take place: A presynaptic one that reduces dyskinesia through an inhibitory control of DA release by eltoprazine from serotonergic neurons[120] and another that potentiates motor activation through a direct postsynaptic interaction between A_{2A} and DA receptors.[2]

ADENOSINE A_{2A} RECEPTORS AND HUNTINGTON'S DISEASE

Chorea is a primary feature of HD, a genetic degenerative disorder characterized by motor disability, psychiatric symptoms, dementia, and early death. Chorea is a movement disorder described as brief, abrupt, and irregular involuntary movements, which seriously impair everyday life. The most common symptomatic treatment of chorea includes the use of neuroleptic agents, DA depleters, such as tetrabenazine, and sometimes benzodiazepines. None of these drugs, however, are able to efficiently control this hyperkinetic disorder and new therapeutic strategies are greatly needed. In the latest years, several research groups focused their interest on the role of adenosine A_{2A} receptors in the pathogenesis and—possibly—on the treatment of HD, on the basis that neurons that degenerate early in HD (i.e., the medium size spiny neurons of the CPu) are those where adenosine A_{2A} receptors are more abundantly expressed. Pharmacological and genetic approaches have been used in different animal models of HD to verify whether the stimulation or the blockade of A_{2A} receptors could prevent or slow down the progression of the disease. The results, however, are quite controversial, with some studies suggesting a beneficial effect of the treatment with the A_{2A} receptor antagonists while others propose the A_{2A}R agonists as potential therapeutic drugs (Fig. 10.2; for review see Refs. 121,122). Most of these studies have been performed in the R6/2 mice, a transgenic mouse model that displays some of the HD symptoms such as motor impairment, loss of body weight, and early death. These mice also show aggregates of mutant huntingtin.[123] In these animals, it has been demonstrated that the A_{2A} receptor antagonist SCH 58261 normalized the alteration in the emotional responses and reduced NMDA-induced toxicity, but had no effect on motor impairment (Table 10.2).[124] On the contrary, a chronic treatment with A_{2A} receptor agonists (CGS 21680 and T1-11), besides normalizing some neurochemical parameters and reducing the formation of mutant huntingtin aggregates, also improves motor functions (Table 10.2).[125,126] This improvement in motor ability is probably the result of the neuroprotective effect exerted by the chronic treatment with the agonist, rather than a direct effect on the mechanisms that control the movement. In this context, it is important to underline that transgenic mouse models of HD do not satisfactorily parallel the motor disability of the human disease, since they are mainly hypokinetic models, in contrast with the hyperkinesia that characterizes the human disease. In this view, since the blockade of A_{2A}Rs promotes the movement while their stimulation depresses it (on the basis of the well-known A_{2A}/D_2 antagonistic receptor interaction), the treatment with the antagonists should revert hypokinesia in mice, while the agonists should be proposed for the treatment of hyperkinesia in humans, making clear that transgenic mice do not represent an ideal model for investigating the effects of A_{2A} receptor ligands in the control of motor functions in HD.

Before the discovery of the mutation underlying HD, animal models mainly relied on neurotoxin-induced lesioning, on the basis that the primary site of neurodegeneration in HD is the CPu. Intrastriatal injections of toxins, such as quinolinic acid, resulted in selective loss of the GABAergic projection neurons that are most severely affected in this disorder.[129] Rats bilaterally injected with quinolinic acid in the CPu show an exacerbated motor response to D-amphetamine,[127] thus providing an HD model to test antihyperkinetic drugs. In this model, the intrastriatal injection of the selective A_{2A} receptor agonist CGS 21680, 1 and 2 weeks after quinolinic acid lesion, antagonized the motor hyperactivity induced by parentelar administration of D-amphetamine,

TABLE 10.2 Effects Exerted by A_{2A}R Agonists and Antagonists in Animal Models of HD

Animal model	Agonist	Antagonist	Dosage	Effect
Transgenic mice R6/2		SCH 58261	0.01 mg/kg (daily i.p. 1 week)	↓ NMDA-induced toxicity[124] ↓ Behavioral alterations[124]
Transgenic mice R6/2	CGS 21680		0.5, 2.5, and 5 mg/kg (daily i.p. 5 weeks)	↑ Motor coordination[125] ↓ Brain atrophy[125] mhtt aggregates[125]
Transgenic mice R6/2	T1-11		0.05 mg/mL (drinking water 5 weeks)	↑ Motor coordination[126] BDNF levels[126] ↓ mhtt aggregates[126]
Rat intrastriatally injected with quinolinic acid	CGS 21680		3 μg/2 μl (intrastriatal, single injection, 1–2 weeks after quinolinic acid)	↓ D-amphetamine-induced motor hyperactivity[127]
Rat intrastriatally injected with quinolinic acid		SCH 58261	0.01 mg/kg (intrastriatal, single injection, 20 min before quinolinic acid)	↓ D-amphetamine-induced motor hyperactivity[128]

evaluated 4 weeks after the lesion (Table 10.2).[127] However, in the same model, the A_{2A} receptor antagonist SCH 58261, intrastriatally injected 20 min before the quinolinic acid, was also able to reduce the motor response to D-amphetamine, evaluated 4 weeks after the lesion (Table 10.2).[128] In both cases, it could be assumed that the normalization of the motor response to D-amphetamine is the result of the neuroprotective effects. Interestingly enough, it is the time of the treatment (soon before or 1–2 weeks after the lesion) which dictates the effect achieved by A_{2A} receptor blockade or activation, respectively.

Such a time-related response is one of the reasons for which, despite great efforts and numerous studies demonstrating a role of A_{2A} receptors in the pathogenesis of HD, exploitation of these receptors as therapeutic targets for HD is far from being achieved. Many other reasons stand for this: (1) in the CPu A_{2A} receptors are expressed by different cell populations (neuronal, astroglial, microglial, and peripheral-blood cells) that differently contribute to the HD pathogenesis; (2) A_{2A} receptors mediate both positive and negative effects (they possess trophic and antiinflammatory effects, but also stimulate glutamate outflow and excessive glial activation); (3) the function and expression of A_{2A} receptors change during the course of the disease, making it difficult to establish when the stimulation or the blockade of the receptor is advantageous.

CONCLUSIONS

In conclusion, we have reviewed the interaction of the A_{2A} and DArgic, glutamatergic, cannabinoid, and serotonin receptors, since these are the principal receptors implicated in the motor complications induced by L-dopa in PD. Moreover, data reported in the present chapter describe A_{2A} receptor antagonists as being promising compounds to be used in PD motor control therapy. Their potential is largely represented by the efficacy demonstrated in alleviating every cardinal PD motor symptom observed in pharmacological and toxicological animal models of PD. The findings achieved in both rodent and primate models of PD suggested that A_{2A} receptor antagonist agents might have symptomatic therapeutic effectiveness in the early stages of PD, when motor complications have not yet appeared. In addition, clinical trials show the efficacy of A_{2A} receptor antagonists in reducing the *wearing-off* and in increasing the *on state* with no exacerbation of dyskinesia. Furthermore, expectation of potential antidepressant, cognitive enhancement, and neuroprotective effects of A_{2A} receptor antagonists in PD is growing.

Regarding HD, although adenosine A_{2A} receptors still represent an interesting target for the development of therapeutic treatments for HD, the complexity of both A_{2A} receptor pharmacology and HD pathogenesis dampened the initial enthusiasm, and further studies are needed to understand the real therapeutic potential of these receptors in the control of motor disability in HD.

References

1. Obeso JA, Rodríguez-Oroz MC, Rodríguez M, et al. Pathophysiology of the basal ganglia in Parkinson's disease. *Trends Neurosci.* 2000;23:S8–19.
2. Morelli M, Di Paolo T, Wardas J, Calon F, Xiao D, Schwarzschild MA. Role of adenosine A2A receptors in parkinsonian motor impairment and l-DOPA-induced motor complications. *Prog Neurobiol.* 2007;83:293–309.
3. Fredholm BB, IJzerman AP, Jacobson KA, Linden J, Müller CE. International Union of Basic and Clinical Pharmacology. LXXXI. Nomenclature and classification of adenosine receptors-an update. *Pharmacol Rev.* 2011;63:1–34.
4. DeLong MR, Wichmann T. Circuits and circuit disorders of the basal ganglia. *Arch Neurol.* 2007;64:20–24.
5. Gerfen CR, Bolam JP. The neuroanatomical organization of the basal ganglia part A. In: Steiner H, Tseng KY, eds. *Handbook of Basal Ganglia Structure and Function: A Decade of Progress.* 1. San Diego, CA: Elsevier Academic Press; 2010:3–28.
6. Albin RL, Young AB, Penney JB. The functional anatomy of basal ganglia disorders. *Trends Neurosci.* 1989;12:366–375.
7. Bar-Gad I, Bergman H. Stepping out of the box: information processing in the neural networks of the basal ganglia. *Curr Opin Neurobiol.* 2001;11:689–695.
8. Steiner H, Tseng KY. *Handbook of Basal Ganglia Structure and Function: A Decade of Progress.* San Diego, CA: Elsevier Academic Press; 2010.
9. Rosin DL, Robeva A, Woodard RL, Guyenet PG, Linden J. Immunohistochemical localization of adenosine A2A receptors in the rat central nervous system. *J Comp Neurol.* 1998;401:163–186.
10. Hettinger BD, Lee A, Linden J, Rosin DL. Ultrastructural localization of adenosine A2A receptors suggests multiple cellular sites for modulation of GABAergic neurons in rat CPu. *J Comp Neurol.* 2001;431:331–346.
11. Ferré S, Fredholm BB, Morelli M, Popoli P, Fuxe K. Adenosine-dopamine receptor-receptor interactions as an integrative mechanism in the basal ganglia. *Trends Neurosci.* 1997;20:482–487.
12. Hillion J, Canals M, Torvinen M, et al. Coaggregation, cointernalization, and codesensitization of adenosine A2A receptors and dopamine D2 receptors. *J Biol Chem.* 2002;277:18091–18097.
13. Fredholm BB, Chen JF, Cunha RA, Svenningsson P, Vaugeois JM. Adenosine and brain function. *Int Rev Neurobiol.* 2005;63:191–270.
14. Van Den Eeden SK, Tanner CM, Bernstein AL, et al. Incidence of Parkinson's disease: variation by age, gender, and race/ethnicity. *Am J Epidemiol.* 2003;157:1015–1022.
15. Dorsey ER, Constantinescu R, Thompson JP, et al. Projected number of people with Parkinson disease in the most populous nations, 2005 through 2030. *Neurology.* 2007;68:384–386.
16. Schneider SA, Obeso JA. Clinical and pathological features of Parkinson's disease. *Curr Top Behav Neurosci.* 2015;22:205–220.

17. Jellinger KA. Recent developments in the pathology of Parkinson's disease. *J Neural Transm Suppl*. 2002;62:347–376.
18. Olanow CW, Stern MB, Sethi K. The scientific and clinical basis for the treatment of Parkinson disease. *Neurology*. 2009;72:S1–136.
19. Jenner P. Molecular mechanisms of L-DOPA-induced dyskinesia. *Nat Rev Neurosci*. 2008;9:665–677.
20. Ondo WG. Motor complications in Parkinson's disease. *Int J Neurosci*. 2011;121:37–44.
21. Obeso JA, Rodriguez-Oroz M, Marin C, et al. The origin of motor fluctuations in Parkinson's disease: importance of dopaminergic innervation and basal ganglia circuits. *Neurology*. 2004;62:S17–S30.
22. Horstink M, Tolosa E, Bonuccelli U, et al. European Federation of Neurological Societies; Movement Disorder Society-European Section. Review of the therapeutic management of Parkinson's disease. Report of a joint task force of the European Federation of Neurological Societies (EFNS) and the Movement Disorder Society-European Section (MDS-ES). Part II: late (complicated) Parkinson's disease. *Eur J Neurol*. 2006;13:1186–1202.
23. Rascol O, Perez-Lloret S, Ferreira JJ. New treatments for levodopa-induced motor complications. *Mov Disord*. 2015;30:1451–1460.
24. Kanda T, Uchida S. Clinical/pharmacological aspect of adenosine A2A receptor antagonist for dyskinesia. *Int Rev Neurobiol*. 2014;119:127–150.
25. Pinna A. Adenosine A2A receptor antagonists in Parkinson's disease: progress in clinical trials from the newly approved istradefylline to drugs in early development and those already discontinued. *CNS Drugs*. 2014;28:455–474.
26. Simola N, Morelli M, Pinna A. Adenosine A2A receptor antagonists and Parkinson's disease: state of the art and future directions. *Curr Pharm Des*. 2008;14:1475–1489.
27. Armentero MT, Pinna A, Ferré S, Lanciego JL, Müller CE, Franco R. Past, present and future of A(2A) adenosine receptor antagonists in the therapy of Parkinson's disease. *Pharmacol Ther*. 2011;132:280–299.
28. Pinna A, Morelli M. A critical evaluation of behavioral rodent models of motor impairment used for screening of antiparkinsonian activity: the case of adenosine A (2A) receptor antagonists. *Neurotox Res*. 2014;25:392–401.
29. Kanda T, Shiozaki S, Shimada J, Suzuki F, Nakamura J. KF17837: a novel selective adenosine A2A receptor antagonist with anticataleptic activity. *Eur J Pharmacol*. 1994;256:263–268.
30. Stasi MA, Borsini F, Varani K, et al. ST 1535: a preferential A2A adenosine receptor antagonist. *Int J Neuropsychopharmacol*. 2006;9:575–584.
31. Varty GB, Hodgson RA, Pond AJ, Grzelak ME, Parker EM, Hunter JC. The effects of adenosine A2A receptor antagonists on haloperidol-induced movement disorders in primates. *Psychopharmacology (Berl)*. 2008;200:393–401.
32. Wardas J. Synergistic effect of SCH 58261, an adenosine A2A receptor antagonist, and L-DOPA on the reserpine-induced muscle rigidity in rats. *Pol J Pharmacol*. 2003;55: 155–164.
33. Hodgson RA, Bertorelli R, Varty GB, et al. Characterization of the potent and highly selective A2A receptor antagonists preladenant and SCH 412348 [7-[2-[4-2,4-difluorophenyl]-1-piperazinyl]ethyl]-2-(2-furanyl)-7H-pyrazolo[4,3-][1,2,4]triazolo[1,5-c]pyrimidin-5-amine] in rodent models of movement disorders and depression. *J Pharmacol Exp Ther*. 2009;330:294–303.
34. Jones N, Bleickardt C, Mullins D, Parker E, Hodgson R. A2A receptor antagonists do not induce dyskinesias in drug-naive or L-dopa sensitized rats. *Brain Res Bull*. 2013;98:163–169.
35. Pinna A, Tronci E, Schintu N, et al. A new ethyladenine antagonist of adenosine A (2A) receptors: behavioral and biochemical characterization as an antiparkinsonian drug. *Neuropharmacology*. 2010;58:613–623.

36. Koga K, Kurokawa M, Ochi M, Nakamura J, Kuwana Y. Adenosine A(2A) receptor antagonists KF17837 and KW-6002 potentiate rotation induced by dopaminergic drugs in hemi-Parkinsonian rats. *Eur J Pharmacol.* 2000;408(3):249–255.
37. Tronci E, Simola N, Borsini F, et al. Characterization of the antiparkinsonian effects of the new adenosine A2A receptor antagonist ST1535: acute and subchronic studies in rats. *Eur J Pharmacol.* 2007;566:94–102.
38. Pinna A, Pontis S, Borsini F, Morelli M. Adenosine A2A receptor antagonists improve deficits in initiation of movement and sensory motor integration in the unilateral 6-hydroxydopamine rat model of Parkinson's disease. *Synapse.* 2007;61:606–614.
39. Smith KM, Browne SE, Jayaraman S, et al. Effects of the selective adenosine A2A receptor antagonist, SCH 412348, on the parkinsonian phenotype of MitoPark mice. *Eur J Pharmacol.* 2014;728:31–38.
40. Grondin R, Bédard PJ, Hadj Tahar A, Grégoire L, Mori A, Kase H. Antiparkinsonian effect of a new selective adenosine A2A receptor antagonist in MPTP-treated monkeys. *Neurology.* 1999;52:1673–1677.
41. Kanda T, Jackson MJ, Smith LA, et al. Adenosine A2A antagonist: a novel antiparkinsonian agent that does not provoke dyskinesia in parkinsonian monkeys. *Ann Neurol.* 1998;43:507–513.
42. Kanda T, Jackson MJ, Smith LA, et al. Combined use of the adenosine A(2A) antagonist KW-6002 with L-DOPA or with selective D1 or D2 dopamine agonists increases antiparkinsonian activity but not dyskinesia in MPTP-treated monkeys. *Exp Neurol.* 2000;162:321–327.
43. Hodgson RA, Bedard PJ, Varty GB, et al. Preladenant, a selective A(2A) receptor antagonist, is active in primate models of movement disorders. *Exp Neurol.* 2010;225:384–390.
44. Rose S, Jackson MJ, Smith LA, et al. The novel adenosine A2a receptor antagonist ST1535 potentiates the effects of a threshold dose of L-DOPA in MPTP treated common marmosets. *Eur J Pharmacol.* 2006;546:82–87.
45. Collins LE, Galtieri DJ, Brennum LT, et al. Oral tremor induced by the muscarinic agonist pilocarpine is suppressed by the adenosine A2A antagonists MSX-3 and SCH58261, but not the adenosine A1 antagonist DPCPX. *Pharmacol Biochem Behav.* 2010;94:561–569.
46. Simola N, Fenu S, Baraldi PG, Tabrizi MA, Morelli M. Blockade of adenosine A2A receptors antagonizes parkinsonian tremor in the rat tacrine model by an action on specific striatal regions. *Exp Neurol.* 2004;189:182–188.
47. Simola N, Fenu S, Baraldi PG, Tabrizi MA, Morelli M. Dopamine and adenosine receptor interaction as basis for the treatment of Parkinson's disease. *J Neurol Sci.* 2006;248:48–52.
48. Salamone JD, Betz AJ, Ishiwari K, et al. Tremorolytic effects of adenosine A2A antagonists: implications for parkinsonism. *Front Biosci.* 2008;(13):3594–3605.
49. Collins-Praino LE, Paul NE, Rychalsky KL, et al. Pharmacological and physiological characterization of the tremulous jaw movement model of parkinsonian tremor: potential insights into the pathophysiology of tremor. *Front Syst Neurosci.* 2011;5:49.
50. Salamone JD, Collins-Praino LE, Pardo M, et al. Conditional neural knockout of the adenosine A(2A) receptor and pharmacological A(2A) antagonism reduce pilocarpine-induced tremulous jaw movements: studies with a mouse model of parkinsonian tremor. *Eur Neuropsychopharmacol.* 2013;23:972–977.
51. Wardas J, Konieczny J, Lorenc-Koci E. SCH 58261, an A(2A) adenosine receptor antagonist, counteracts parkinsonian-like muscle rigidity in rats. *Synapse.* 2001;41:160–171.
52. Pinna A, Ko WK, Costa G, et al. Antidyskinetic effect of A(2A) and 5HT(1A/1B) receptor ligands in two animal models of Parkinson's disease. *Mov Disord.* 2016; Available from: http://dx.doi.org/10.1002/mds.26475.
53. Bibbiani F, Oh JD, Petzer JP, et al. A2A antagonist prevents dopamine agonist-induced motor complications in animal models of Parkinson's disease. *Exp Neurol.* 2003;184:285–294.

54. Bové J, Marin C, Bonastre M, Tolosa E. Adenosine A2A antagonism reverses levodopa-induced motor alterations in hemiparkinsonian rats. *Synapse*. 2002;46:251–257.
55. Pinna A, Fenu S, Morelli M. Motor stimulant effects of the adenosine A2A receptor antagonist SCH 58261 do not develop tolerance after repeated treatments in 6-hydroxydopamine-lesioned rats. *Synapse*. 2001;39(3):233–238.
56. Fredduzzi S, Moratalla R, Monopoli A, et al. Persistent behavioral sensitization to chronic L-DOPA requires A2A adenosine receptors. *J Neurosci*. 2002;22:1054–1062.
57. Lundblad M, Vaudano E, Cenci MA. Cellular and behavioural effects of the adenosine A2a receptor antagonist KW-6002 in a rat model of l-DOPA-induced dyskinesia. *J Neurochem*. 2003;84:1398–1410.
58. Xiao D, Bastia E, Xu YH, et al. Forebrain adenosine A2A receptors contribute to L-3,4-dihydroxyphenylalanine-induced dyskinesia in hemiparkinsonian mice. *J Neurosci*. 2006;26:13548–13555.
59. Schwarting RK, Huston JP. The unilateral 6-hydroxydopamine lesion model in behavioral brain research. Analysis of functional deficits, recovery and treatments. *Prog Neurobiol*. 1996;50(2–3):275–331.
60. Simola N, Morelli M, Carta AR. The 6-hydroxydopamine model of Parkinson's disease. *Neurotox Res*. 2007;11(3–4):151–167.
61. Meredith GE, Kang UJ. Behavioral models of Parkinson's disease in rodents: a new look at an old problem. *Mov Disord*. 2006;21:1595–1606.
62. Schallert T, Fleming SM, Leasure JL, Tillerson JL, Bland ST. CNS plasticity and assessment of forelimb sensorimotor outcome in unilateral rat models of stroke, cortical ablation, parkinsonism and spinal cord injury. *Neuropharmacology*. 2000;39:777–787.
63. Duty S, Jenner P. Animal models of Parkinson's disease: a source of novel treatments and clues to the cause of the disease. *Br J Pharmacol*. 2011;164:1357–1391.
64. Marcellino D, Lindqvist E, Schneider M, et al. Chronic A2A antagonist treatment alleviates parkinsonian locomotor deficiency in MitoPark mice. *Neurobiol Dis*. 2010;40:460–466.
65. Rodriguez-Oroz MC, Jahanshahi M, Krack P, et al. Initial clinical manifestations of Parkinson's disease: features and pathophysiological mechanisms. *Lancet Neurol*. 2009;8:1128–1139.
66. Lee RG. Pathophysiology of rigidity and akinesia in Parkinson's disease. *Eur Neurol*. 1989;29(Suppl 1):13–18.
67. Delwaide PJ. Parkinsonian rigidity. *Funct Neurol*. 2001;16:147–156.
68. Baradaran N, Tan SN, Liu A, et al. Parkinson's disease rigidity: relation to brain connectivity and motor performance. *Front Neurol*. 2013;4:67.
69. Wolfarth S, Konieczny J, Smiałowska M, Schulze G, Ossowska K. Influence of 6-hydroxydopamine lesion of the dopaminergic nigrostriatal pathway on the muscle tone and electromyographic activity measured during passive movements. *Neuroscience*. 1996;74:985–996.
70. Johnels B. Reserpine-induced rigidity in rats: drug effects on muscle tone from corpus CPu and nucleus accumbens. *Pharmacol Biochem Behav*. 1983;19:463–470.
71. Lorenc-Koci E, Ossowska K, Wardas J, Wolfarth S. Does reserpine induce parkinsonian rigidity? *J Neural Transm Park Dis Dement Sect*. 1995;9:211–223.
72. Lorenc-Koci E, Wolfarth S, Ossowska K. Haloperidol-increased muscle tone in rats as a model of parkinsonian rigidity. *Exp Brain Res*. 1996;109:268–276.
73. Lorenc-Koci E, Wolfarth S. Efficacy of pramipexole, a new dopamine receptor agonist, to relieve the parkinsonian-like muscle rigidity in rats. *Eur J Pharmacol*. 1999;385:39–46.
74. Wardas J, Konieczny J, Pietraszek M. Influence of CGS 21680, a selective adenosine A (2A) agonist, on the phencyclidine-induced sensorimotor gating deficit and motor behaviour in rats. *Psychopharmacology (Berl)*. 2003;168:299–306.
75. Wardas J, Konieczny J, Lorenc-Koci E. The role of striatal adenosine A2A receptors in regulation of the muscle tone in rats. *Neurosci Lett*. 1999;276:79–82.

76. Wu YN, Chen JJ, Zhang LQ, Hyland BI. Regulation of hind-limb tone by adenosine A2A receptor in rats. *Neuroscience*. 2009;159:1408–1413.
77. Marin C, Aguilar E, Bonastre M, Tolosa E, Obeso JA. Early administration of entacapone prevents levodopa-induced motor fluctuations in hemiparkinsonian rats. *Exp Neurol*. 2005;192:184–193.
78. Henry B, Crossman AR, Brotchie JM. Characterization of enhanced behavioral responses to L-DOPA following repeated administration in the 6-hydroxydopamine-lesioned rat model of Parkinson's disease. *Exp Neurol*. 1998;151:334–342.
79. Carta M, Lindgren HS, Lundblad M, Stancampiano R, Fadda F, Cenci MA. Role of striatal L-DOPA in the production of dyskinesia in 6-hydroxydopamine lesioned rats. *J Neurochem*. 2006;96:1718–1727.
80. Lundblad M, Andersson M, Winkler C, Kirik D, Wierup N, Cenci MA. Pharmacological validation of behavioural measures of akinesia and dyskinesia in a rat model of Parkinson's disease. *Eur J Neurosci*. 2002;15:120–132.
81. Bibbiani F, Costantini LC, Patel R, Chase TN. Continuous dopaminergic stimulation reduces risk of motor complications in parkinsonian primates. *Exp Neurol*. 2005;192 (1):73–78.
82. Dungo R, Deeks ED. Istradefylline: first global approval. *Drugs*. 2013;73:875–882.
83. Bogenpohl JW, Ritter SL, Hall RA, Wichmann T, Smith Y. Adenosine A2A receptor in the monkey basal ganglia: ultrastructural localization and colocalization with the metabotropic glutamate receptor 5 in the CPu. *J Comp Neurol*. 2012;520:570–589.
84. Jones CK, Bubser M, Thompson AD, et al. The metabotropic glutamate receptor 4-positive allosteric modulator VU0364770 produces efficacy alone and in combination with L-DOPA or an adenosine 2A antagonist in preclinical rodent models of Parkinson's disease. *J Pharmacol Exp Ther*. 2012;340:404–421.
85. Łukasiewicz S, Błasiak E, Faron-Górecka A, et al. Fluorescence studies of homooligomerization of adenosine A2A and serotonin 5-HT1A receptors reveal the specificity of receptor interactions in the plasma membrane. *Pharmacol Rep*. 2007;59(4):379–392.
86. Ferré S, Quiroz C, Woods AS, et al. An update on adenosine A2A-dopamine D2 receptor interactions. Implications for the function of G protein-coupled receptors. *Curr Pharm Des*. 2008;14:1468–1474.
87. Pinna A, di Chiara G, Wardas J, Morelli M. Blockade of A2a adenosine receptors positively modulates turning behaviour and c-Fos expression induced by D1 agonists in dopamine-denervated rats. *Eur J Neurosci*. 1996;8(6):1176–1181.
88. Ochi M, Shiozaki S, Kase H. Adenosine A(2A) receptor-mediated modulation of GABA and glutamate release in the output regions of the basal ganglia in a rodent model of Parkinson's disease. *Neuroscience*. 2004;127(1):223–231.
89. Shindou T, Nonaka H, Richardson PJ, Mori A, Kase H, Ichimura M. Presynaptic adenosine A2A receptors enhance GABAergic synaptic transmission via a cyclic AMP dependent mechanism in the rat globus pallidus. *Br J Pharmacol*. 2002;136 (2):296–302.
90. Azdad K, Gall D, Woods AS, Ledent C, Ferré S, Schiffmann SN. Dopamine D2 and adenosine A2A receptors regulate NMDA-mediated excitation in accumbens neurons through A2A-D2 receptor heteromerization. Neuropsychopharmacology. 34:972–986.
91. Higley MJ, Sabatini BL. Competitive regulation of synaptic Ca2+ influx by D2 dopamine and A2A adenosine receptors. *Nat Neurosci*. 2010;13:958–966.
92. Kurokawa M, Koga K, Kase H, Nakamura J, Kuwana Y. Adenosine A2a receptor-mediated modulation of striatal acetylcholine release in vivo. *J Neurochem*. 1996;66:1882–1888.
93. Tozzi A, de Iure A, Di Filippo M, et al. The distinct role of medium spiny neurons and cholinergic interneurons in the D_2/A_2A receptor interaction in the CPu: implications for Parkinson's disease. *J Neurosci*. 2011;31:1850–1862.

94. Popoli P, Frank C, Tebano MT, et al. Modulation of glutamate release and excitotoxicity by adenosine A2A receptors. *Neurology*. 2003;61:S69–S71.
95. Morin N, Di Paolo T. Interaction of adenosine receptors with other receptors from therapeutic perspective in Parkinson's disease. *Int Rev Neurobiol*. 2014;119:151–167.
96. Cheng HC, Shih HM, Chern Y. Essential role of cAMP-response element-binding protein activation by A2A adenosine receptors in rescuing the nerve growth factor-induced neurite outgrowth impaired by blockage of the MAPK cascade. *J Biol Chem*. 2002;277:33930–33942.
97. Wirkner K, Gerevich Z, Krause T, et al. Adenosine A2A receptor-induced inhibition of NMDA and GABAA receptor-mediated synaptic currents in a subpopulation of rat striatal neurons. *Neuropharmacology*. 2004;46:994–1007.
98. Köles L, Wirkner K, Illes P. Modulation of ionotropic glutamate receptor channels. *Neurochem Res*. 2001;26:925–932.
99. Ekonomou A, Poulou PD, Matsokis N, Angelatou F. Stimulation of adenosine A2A receptors elicits zif/268 and NMDA epsilon2 subunit mRNA expression in cortex and striatum of the "weaver" mutant mouse, a genetic model of nigrostriatal dopamine deficiency. *Neuroscience*. 2004;123:1025–1036.
100. Lopez S, Turle-Lorenzo N, Johnston TH, et al. Functional interaction between adenosine A2A and group III metabotropic glutamate receptors to reduce parkinsonian symptoms in rats. *Neuropharmacology*. 2008;55:483–490.
101. Kachroo A, Orlando LR, Grandy DK, Chen JF, Young AB, Schwarzschild MA. Interactions between metabotropic glutamate 5 and adenosine A2A receptors in normal and parkinsonian mice. *J Neurosci*. 2005;25:10414–10419.
102. Domenici MR, Pepponi R, Martire A, Tebano MT, Potenza RL, Popoli P. Permissive role of adenosine A2A receptors on metabotropic glutamate receptor 5 (mGluR5)-mediated effects in the striatum. *J Neurochem*. 2004;90:1276–1279.
103. Fuzzati-Armentero MT, Cerri S, Levandis G, et al. Dual target strategy: combining distinct non-dopaminergic treatments reduces neuronal cell loss and synergistically modulates L-DOPA-induced rotational behavior in a rodent model of Parkinson's disease. *J Neurochem*. 2015;134:740–747.
104. Amalric M, Lopez S, Goudet C, et al. Group III and subtype 4 metabotropic glutamate receptor agonists: discovery and pathophysiological applications in Parkinson's disease. *Neuropharmacology*. 2013;66:53–64.
105. Bennouar KE, Uberti MA, Melon C, et al. Synergy between L-DOPA and a novel positive allosteric modulator of metabotropic glutamate receptor 4: implications for Parkinson's disease treatment and dyskinesia. *Neuropharmacology*. 2013;66:158–169.
106. Fernández-Ruiz J. The endocannabinoid system as a target for the treatment of motor dysfunction. *Br J Pharmacol*. 2009;156:1029–1040.
107. Cao X, Liang L, Hadcock JR, et al. Blockade of cannabinoid type 1 receptors augments the antiparkinsonian action of levodopa without affecting dyskinesias in 1-methyl-4-phenyl-1,2,3,6-tetrahydropyridine-treated rhesus monkeys. *J Pharmacol Exp Ther*. 2007;323:318–326.
108. Ferrer B, Asbrock N, Kathuria S, Piomelli D, Giuffrida A. Effects of levodopa on endocannabinoid levels in rat basal ganglia: implications for the treatment of levodopa-induced dyskinesias. *Eur Neurosci*. 2003;18(6):1607–1614.
109. Morgese MG, Cassano T, Cuomo V, Giuffrida A. Anti-dyskinetic effects of cannabinoids in a rat model of Parkinson's disease: role of CB(1) and TRPV1 receptors. *Exp Neurol*. 2007;208:110–119.
110. Carriba P, Ortiz O, Patkar K, et al. Striatal adenosine A2A and cannabinoid CB1 receptors form functional heteromeric complexes that mediate the motor effects of cannabinoids. *Neuropsychopharmacology*. 2007;32:2249–2259.

111. Marcellino D, Carriba P, Filip M, et al. Antagonistic cannabinoid CB1/dopamine D2 receptor interactions in striatal CB1/D2 heteromers. A combined neurochemical and behavioral analysis. *Neuropharmacology*. 2008;54(5):815–823.
112. Pinna A, Bonaventura J, Farré D, et al. L-DOPA disrupts adenosine A(2A)-cannabinoid CB(1)-dopamine D(2) receptor heteromer cross-talk in the CPu of hemiparkinsonian rats: biochemical and behavioral studies. *Exp Neurol*. 2014;253:180–191.
113. Bonaventura J, Rico AJ, Moreno E, et al. L-DOPA-treatment in primates disrupts the expression of A(2A) adenosine-CB(1) cannabinoid-D(2) dopamine receptor heteromers in the caudate nucleus. *Neuropharmacology*. 2014;79:90–100.
114. Carta M, Tronci E. Serotonin system implication in l-DOPA-induced dyskinesia: from animal models to clinical investigations. *Front Neurol*. 2014;5:78.
115. Cenci MA, Lundblad M. Post- versus presynaptic plasticity in L-DOPA-induced dyskinesia. *J Neurochem*. 2006;99:381–392.
116. Carta M, Carlsson T, Kirik D, Björklund A. Dopamine released from 5-HT terminals is the cause of L-DOPA-induced dyskinesia in parkinsonian rats. *Brain*. 2007;130:1819–1833.
117. Muñoz A, Li Q, Gardoni F, et al. Combined 5-HT1A and 5-HT1B receptor agonists for the treatment of L-DOPA-induced dyskinesia. *Brain*. 2008;131:3380–3394.
118. Politis M, Wu K, Loane C, et al. Serotonergic mechanisms responsible for levodopa-induced dyskinesias in Parkinson's disease patients. *J Clin Invest*. 2014;124:1340–1349.
119. Bezard E, Tronci E, Pioli EY, et al. Study of the antidyskinetic effect of eltoprazine in animal models of levodopa-induced dyskinesia. *Mov Disord*. 2013;28:1088–1096.
120. Bezard E, Carta M. Could the serotonin theory give rise to a treatment for levodopa-induced dyskinesia in Parkinson's disease? *Brain*. 2015;138:829–830.
121. Popoli P, Blum D, Domenici MR, Burnouf S, Chern Y. A critical evaluation of adenosine A2A receptors as potentially "druggable" targets in Huntington's disease. *Curr Pharm Des*. 2008;14:1500–1511.
122. Lee CF, Chern Y. Adenosine receptors and Huntington's disease. *Int Rev Neurobiol*. 2014;119:195–232.
123. Mangiarini L, Sathasivam K, Seller M, et al. Exon 1 of the HD gene with an expanded CAG repeat is sufficient to cause a progressive neurological phenotype in transgenic mice. *Cell*. 1996;87:493–506.
124. Domenici MR, Scattoni ML, Martire A, et al. Behavioral and electrophysiological effects of the adenosine A2A receptor antagonist SCH 58261 in R6/2 Huntington's disease mice. *Neurobiol Dis*. 2007;28:197–205.
125. Chou SY, Lee YC, Chen HM, et al. CGS21680 attenuates symptoms of Huntington's disease in a transgenic mouse model. *J Neurochem*. 2005;93:310–320.
126. Huang NK, Lin JH, Lin JT, et al. A new drug design targeting the adenosinergic system for Huntington's disease. *PLoS One*. 2011;(6):e20934.
127. Popoli P, Pèzzola A, Reggio R, Caporali MG, Scotti de Carolis A. CGS 21680 antagonizes motor hyperactivity in a rat model of Huntington's disease. *Eur J Pharmacol*. 1994;257:R5–R6.
128. Popoli P, Pintor A, Domenici MR, et al. Blockade of striatal adenosine A2A receptor reduces, through a presynaptic mechanism, quinolinic acid-induced excitotoxicity: possible relevance to neuroprotective interventions in neurodegenerative diseases of the CPu. *J Neurosci*. 2002;22:1967–1975.
129. Beal MF, Kowall NW, Ellison DW, Mazurek MF, Swartz KJ, Martin JB. Replication of the neurochemical characteristics of Huntington's disease by quinolinic acid. *Nature*. 1986;321:168–171.

CHAPTER

11

Adenosine Receptors Oligomers in Parkinson's Disease

Víctor Fernández-Dueñas and Francisco Ciruela

University of Barcelona, Barcelona, Spain

OUTLINE

INTRODUCTION

The role of adenosine as an extracellular signaling molecule begun with the seminal work of Drury and Szent-Györgyi in 1929 describing the profound adenosine-mediated hypotension, bradycardia, and kidney regulation produced by adenosine in mammals.[1] Since then, the physiological roles of adenosine have been widely investigated.[2] Adenosine is a nucleoside mostly generated by the catabolism of ATP

Adenosine Receptors in Neurodegenerative Diseases.
DOI: http://dx.doi.org/10.1016/B978-0-12-803724-9.00011-9

both at the intracellular and extracellular level (Fig. 11.1). In addition, small amounts of adenosine can be originated intracellularly by S-adenosyl-L-homocysteine (SAH) catabolism (Fig. 11.1). Once produced, intracellular adenosine can be: (1) secreted outside the cell, (2) phosphorylated forming AMP, (3) reacted with L-homocysteine to form SAH, and (4) deaminated to form inosine by the action of adenosine deaminase (Fig. 11.1). On the other hand, extracellular adenosine can be also deaminated by the action of ecto-adenosine deaminase (ADA, Fig. 11.1). Extracellular adenosine, either ubiquitously released by means of Na^+-dependent transporters and/or formed extracellularly by the breakdown of ATP, is considered a "retaliatory metabolite," thus mediating the increase of oxygen supply and decrease of oxygen consumption.[3] In addition, adenosine is involved in modulating a large array of physiological processes: respiratory function,[4] neural activity,[5] platelet aggregation,[6] neutrophil function,[7] lymphocyte differentiation,[8] and vascular tone.[9] Furthermore, adenosine is able to provoke both coronary arteries dilatation and kidney blood vessels contraction, thus reducing renal filtration.[10] Also, it exerts a negative chronotropic and dromotropic effect on the heart,[11] as well as mediating the inhibition of neurotransmitters release[12] and lipolysis.[13] Overall, this purine nucleoside, postulated as a mediator of metabolic distress, has a considerable impact on a large variety of physiological functions.

ADENOSINE RECEPTORS IN THE BRAIN

Adenosine, by acting at pre-, post- and/or extra-synaptic G protein–coupled adenosine receptors (ARs), plays an important neuromodulatory role in the brain.[14] Interestingly, the concentration of adenosine at the extracellular milieu ranges the high nM upon basal conditions, and it is related to the intracellular concentration of both adenosine and nucleotides (i.e., ATP, AMP, and cAMP).[15] Indeed, intracellular adenosine concentration depends on ATP catabolism rate,[15] thus once adenosine is generated as a consequence of ATP breakdown it is readily released to the extracellular milieu where it acts as a neuromodulator[16] in response to increased cellular metabolic demand.[17] Interestingly, it has been postulated that the main source of striatal extracellular adenosine is the intracellular cAMP accumulation.[18] Thus, cAMP generated at the striatum is readily metabolized to AMP by means of phosphodiesterases (PDEs) and then to adenosine, which is secreted (Fig. 11.1). Of note, since cAMP is generated by the activity of adenylyl cyclase, it seems likely that striatal extracellular adenosine

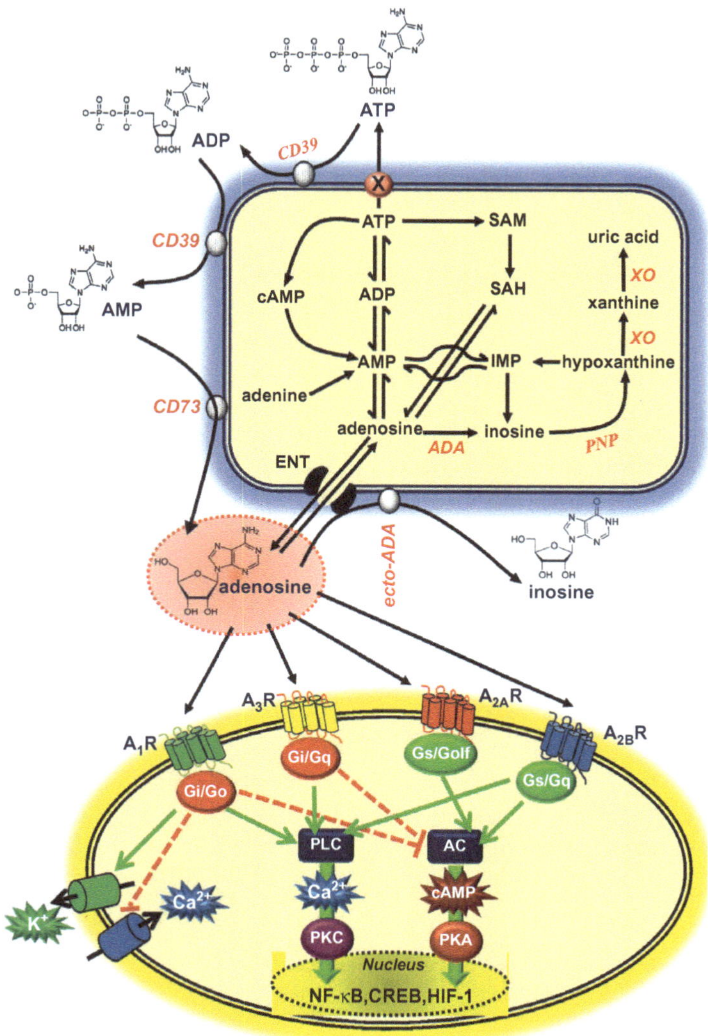

FIGURE 11.1 General overview of purine metabolism and adenosine receptors signaling.
The upper panel illustrated the purinergic metabolism and the different sources of extracellular adenosine. Thus, the intracellular ATP could be catabolized into adenosine, which can be further metabolized into inosine and hypoxanthine by intracellular adenosine deaminase (ADA) and purine nucleoside phosphorylase (PNP), respectively. Hypoxanthine can be either salvaged into inosine monophosphate (IMP) or further metabolized to xanthine and uric acid by xanthine oxidase (XO). In addition, adenosine could also be generated intracellularly through the hydrolysis of the *S*-adenosyl-homo-cysteine (SAH) by an SAH hydrolase. Interestingly, the intracellularly generated adenosine and ATP can be released to the extracellular milieu through an equilibrative nucleoside transporter (ENT) or open-ended systems (i.e., nucleotide-permeable channels, exocytosis,

may correlate well with the activity of GPCRs that are positively coupled to adenylyl cyclase (i.e., $A_{2A}R$) (Fig. 11.1). Interestingly, early in the 1970s, it was demonstrated that electrical stimulation of brain slices promoted adenosine release,[19] which mirrored intracellular cAMP accumulation, a phenomenon that was precluded by methylxanthines (i.e., caffeine and theophylline),[20] and that was also observed in other tissues (i.e., heart).[21] Indeed, these experimental observations constituted the first evidence suggesting that extracellular adenosine exerted its effects via specific plasma membrane receptors. Subsequently, it was demonstrated that the adenosine-dependent antilipolytic effect observed on fat cells was mediated by cAMP reduction.[22] Interestingly, the dual effect on cAMP formation was further demonstrated when the adenosine-dependent inhibition or stimulation of adenylyl cyclase activity was demonstrated. Overall, these observations ended with the first subclassification of ARs into R_i and R_a,[23] or alternatively, A_1 and A_2 ARs.[24]

Presently, it is well accepted that adenosine mediates its physiological actions throughout specific G protein–coupled receptors (GPCRs), for which four subtypes (A_1R, $A_{2A}R$, $A_{2B}R$, and A_3R) have been identified so far (Fig. 11.1). Interestingly, these receptors have distinctive pharmacological profile, tissue distribution, and effector coupling[25] (Fig. 11.1), and their functioning has been largely assessed in the brain (Table 11.1). ARs belong to the class A of GPCRs or rhodopsin-like.[47] This class of GPCRs share some common molecular features, namely the presence of the conserved NPxxY(x)5,6F and the DRY motifs,[48,49] amongst others. Accordingly, adenosine-mediated conformational rearrangement of ARs determines the binding and activation of specific G proteins (Table 11.1), which are responsible for

injury or lysis, transport vesicles, lysosomes), respectively. Subsequently, the ATP is dephosphorylated into adenosine by the ectonucleoside triphosphate diphosphohydrolase CD39 and the 5′-nucleotidase CD73, thus constituting the main mechanism behind high extracellular adenosine levels. Finally, extracellular adenosine can be either transported back into the cell through ENTs or transformed into inosine through ecto-adenosine deaminase ADA.

The lower panel illustrated the different type adenosine receptors and the main signal transduction associated to. Extracellular adenosine can bind to A_1R, $A_{2A}R$, $A_{2B}R$ and A_3R. The A_1R and A_3R subtypes inhibit adenylate cyclase (AC) activity and induce stimulation of phospholipase C (PLC). In addition, A_1R also modulates K^+ and Ca^{2+} channels. $A_{2A}R$ and $A_{2B}R$ stimulate AC, with a consequent increase in cAMP levels. Also, $A_{2B}R$ can activate PLC, raising intracellular Ca^{2+} levels. Interestingly, the activation of PKA and PKC through cAMP and Ca^{2+}, respectively, triggers the downstream signaling modulation of several transcription factors like the nuclear factor κB (NF-κB), the cAMP response element-binding protein (CREB), and the hypoxia-inducible factor 1 (HIF1), within others, which can be related to processes like inflammation and cell regulation. Enzymes are in red italics.

TABLE 11.1 Adenosine Receptors

Receptor	Adenosine affinity	G protein	Transduction mechanisms[b]	Physiological actions
A_1	~70 nM	$G_{i/o}$[a] $G_{q/11}$ G_s	• *Inhibits*: AC[a], • *Activates*: PLC, AC	Vasoconstriction[26]; hypothermia and sedation[27]; analgesia[28]; neurotransmitter release[29,30]; chemotaxis[31]; Neuroprotection[32]
A_{2A}	~150 nM	G_s[a] G_{olf} $G_{15,16}$ §	• *Activates*: AC[a], PLC • *Inhibits*: Ca^{b+} channels	Platelet aggregation inhibition[33]; vasodilation[34]; neurotransmitter release[35]; regulation of sensorimotor integration in basal ganglia[36]; sleep promotion[37]
A_{2B}	~5000 nM	G_s[a] $G_{q/11}$	• *Activates*: AC[a], PLC	Vasodilation[38]; vasoconstriction[39]; cytokine production[40]; inhibition of cell proliferation[41]
A_3	~6500 nM	$G_{i/o}$[a]	• *Inhibits*: AC[a] • *Activates*: PLC	Mast cell activation[42]; preconditioning[43]; coronary vasodilation[44]; regulation of intraocular pressure[45]; hypotension[46]

[a] *Main mechanism of coupling.*
[b] *AC, adenylyl cyclase; PLC, phospholipase C; PLA2, phospholipase A2; PLD, phospholipase D; GIRKs, G protein-dependent inwardly rectifying K^+ channels.*

activation of different intracellular signaling pathways associated with adenosine function (Table 11.1). A_1Rs and A_{2A}Rs are mainly responsible for the brain effects of adenosine (Table 11.1).[50] The A_1R is the most abundant and homogeneously distributed AR within the brain. This AR couples to pertussis toxin-sensitive G proteins (G_{i1}, G_{i2}, G_{i3}, and G_o), whose activation regulates several intracellular effector molecules (adenylyl cyclase, Ca^{2+} channels, K^+ channels, and phospholipase C; Table 11.1).[51] Conversely, the A_{2A}R is expressed at high levels only in certain brain regions, for instance the striatum, the olfactory tubercle and the nucleus accumbens.[18,52] This AR mainly couples to G_s/G_{olf} proteins,[53] thus prompting adenylyl cyclase activation and cAMP accumulation (Table 11.1). In addition, A_{2A}R may also signal through a G-protein independent pathway eventually associated to mitogen-activated protein kinase (MAPK) signaling cascade activation.[54] Next, the A_{2B}R, thought to be fairly ubiquitous in the brain, is positively coupled to adenylyl cyclase and PLC through a G_s and G_q protein, respectively (Table 11.1).[2] Finally, the A_3R has been shown to be coupled to $G_{i/o}$ proteins (Table 11.1),[2] and it is basically expressed peripherally and involved in immune and inflammatory processes.

ADENOSINE RECEPTOR-CONTAINING OLIGOMERS

The idea that receptors expressed at the cell surface interact with other receptors, forming receptors oligomers, appeared in the early 1980s, when it was proposed that the establishment of direct receptor–receptor interactions in the central nervous system would explain the functional cross-talk occurring between neuropeptide and glutamate receptors with serotonin or dopamine receptors.[55,56] The first results confirming such a theory consisted of radioligand binding experiments, in which cooperativity between receptors would lead to changes on selective ligands affinity (for review see Ref. 57). Later on, other techniques were included on the palette in order to sustain the concept of GPCR oligomerization, such as photo-affinity labeling, radiation inactivation experiments, or hydrodynamic and cross-linking analysis.[58–61] More recently, a number of techniques have been considered to demonstrate a direct receptor–receptor interaction at the plasma membrane. Thus, coimmunoprecipitation, bioluminescence, and time-resolved fluorescence resonance energy transfer techniques have been extensively used to show the occurrence of GPCR oligomerization in the cell surface of transfected cells.[62] On the other hand, evidences demonstrating GPCR oligomerization in native tissue have been scarce, and only indirect measurements were initially possible; from which yeast two hybrid system and coimmunoprecipitation techniques have been the most used.[63] Finally, a few years ago, by means of time-resolved FRET, it was possible to undeniably show the formation of oxytocin oligomers in rat mammary glands,[64] a technique that has been reproduced to reveal dopamine D_2 receptor (D_2R)-Ghrelin oligomers in mice hippocampus[65] or D_2R-$A_{2A}R$ oligomers in rat striatum.[66] Importantly, in this last study, it was implemented as a multimodal approach, thus using immunogold and proximity ligation assays, which permitted to validate these two techniques in order to confirm the existence of direct receptor–receptor interactions.[66]

As mentioned above, in recent years numerous studies have focused on elucidating the existence of direct receptor–receptor interactions. Concerning ARs, it has been possible to characterize a number of them (for review see Ref. 62). To start with, the A_1R was firstly shown to feasibly oligomerize with the dopamine D_1 receptor (D_1R),[67] with the metabotropic glutamate type 1α($mGlu_{1\alpha}$),[68] and also with the purinergic $P2Y_1$ receptor ($P2Y_1R$).[69] Thus, these receptor–receptor interactions would be important for desensitization, receptor trafficking, or in the case of the glutamate receptor, to prevent glutamate excitotoxicity.[68] In addition, the A_1R was shown to heterodimerize with the $A_{2A}R$[70] (Fig. 11.2). Interestingly, within the $A_1R/A_{2A}R$ oligomer low adenosine

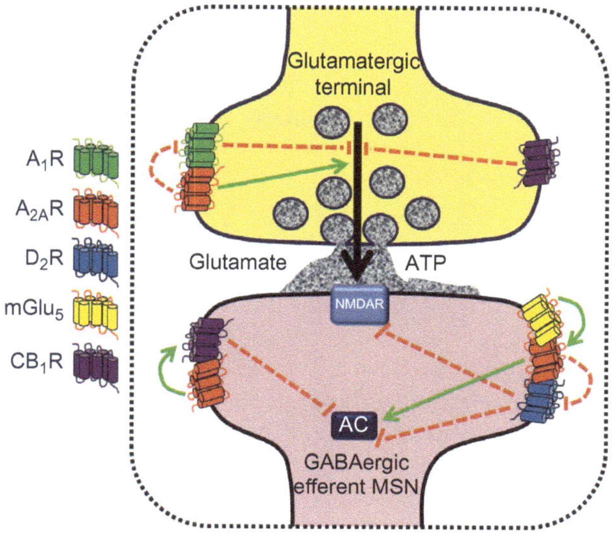

FIGURE 11.2 Scheme showing putative subsynaptic localization of striatal adenosine receptor-containing oligomers.
A_{2A}Rs are localized in glutamatergic terminals that contact striatal medium spiny neuron (MSN) where they can heterodimerize with A_1Rs. In addition, the $A_{2A}R$ is also localized postsynaptically where they can heterodimerize with the $mGlu_5$ receptor, D_2R and/or CB_1R. Indeed, the formation of higher-order oligomers (i.e., $mGlu_5/A_{2A}R/D_2R$) has been also described.[71] The inhibitory modulation of NMDA receptor (NMDAR) by the D_2R, the fine-tuning modulation of adenylate cyclase (AC) or the control of glutamate release by the receptors is also depicted.

concentrations activate mostly the A_1R, which reduces the probability of glutamate release to the synaptic space (Fig. 11.2). However, when the adenosine concentration raises until a certain level the $A_{2A}R$ could be activated and trigger a transinhibition of the A_1R, thus overcoming the A_1R-mediated inhibition of glutamate release[70] (Fig. 11.2). Overall, the $A_1R/A_{2A}R$ oligomer allows adenosine to exert a fine-tuning modulation of striatal glutamatergic neurotransmission, thus providing a switch mechanism by which low and high concentrations of adenosine inhibit and stimulate, respectively, glutamate release.[72] Interestingly, in recent years novel A_1R-containing oligomers have been studied, such as the one formed with the thromboxane A[73] and the β1- and β2-adrenergic receptors.[74] Of note, in the last study, the formation of the GPCR oligomer would lead to changes on the transduction pathways coupling, which in fact is a matter of intense study in recent years, i.e., the existence of functional selectivity and biased ligands (for review see Ref. 75). On the other hand, the $A_{2A}R$ has been proposed to oligomerize with the metabotropic glutamate type 5 ($mGlu_5$) receptor[54] and, as aforementioned, with the D_2R[76–78] (Fig. 11.2). In fact, in the central nervous system it has been postulated that a higher-order oligomer would exist, in which these three receptors would form a multimeric complex, which would permit the fine-tuning modulation of neurotransmitter release[71] (Fig. 11.2). More recently, it has been proposed that the $A_{2A}R$ may also oligomerize with the cannabinoid type 1 receptor (CB_1R) (for review see Ref. 79) (Fig. 11.2). Similarly to that described above, it has been suggested that a multimeric complex including $A_{2A}R$, D_2R, and CB_1R would be found in the striatum,[79] thus it seems clear that the role of these kinds of direct interactions between two or more receptors would be relevant for the proper functioning of these structures. In such a way, taking into account the most studied AR-containing oligomer, the $A_{2A}R/D_2R$ oligomer, findings revealing the relevance of these complexes have prompted the targeting of them for development of novel therapeutic strategies for Parkinson's disease (PD).[80–82] Below, we will deal with such approaches, which are expected to improve the treatment of the pathology.

PARKINSON'S DISEASE AND ADENOSINE RECEPTOR OLIGOMERS: IMPACT ON PHARMACOTHERAPY

PD is a progressive neurodegenerative disorder that affects approximately 1% of individuals older than 60 years, thus being one of the most common neurological disorders. This pathology was described by James Parkinson in 1817, who named it as the "shalking palsy" due to its symptomatology; which consists of resting tremor, rigidity, bradykinesia, and postural instability.[83] On the other hand, PD is characterized by the progressive loss of pigmented dopaminergic neurons in the substantia nigra pars compacta (SNc) and the presence of Lewy bodies (LB). This neurodegenerative process leads not only to more severe motor symptoms but also to cognitive dysfunction and dementia, since cortical regions are also affected.[84] It is worthy of note that before the appearance of any motor symptoms and a diagnostic is done, around 3/4 of dopaminergic neurons are lost. Accordingly, nowadays, research is focused not only on improving PD treatment but also on finding out potential biomarkers allowing an early diagnostic.

The SNc is a nucleus that forms part of the basal ganglia, which is the main brain area responsible of controlling cortical-related normal movements.[85] There exist two different pathways (direct and indirect) that are interrelated and permit that the basal ganglia exert a fine-tuning modulation on the thalamocortical motor circuit. In brief, dopamine released from the SNc reaches the striatum, where two different populations of GABAergic neurons exist: (1) dopamine D_1 receptors (D_1R)-containing neurons, and (2) neurons that express the D_2R. Thus, the striatal GABAergic neurons containing D_1R constitute the direct pathway and project to the globus pallidus pars interna (GPi) and the substantia nigra pars reticulata (SNr) to inhibit the thalamus, thus suppressing motor cortex stimulation. Conversely, the striatal GABAergic neurons containing D_2R are part of the indirect pathway and project to the globus pallidus externa (GPe), which tonically inhibits the subthalamic nucleus (STN). The STN, in turn, projects excitatory inputs to the GPi/SNr complex. The end-result of this circuitry consists of the inhibition of the thalamus, since dopamine released by the SNc activates the direct pathway and inhibits the indirect pathway, leading to the inhibition of the GPi/SNr. Importantly, the above described circuitry is disrupted in PD. Thus, the loss of dopaminergic neurons in the SNc leads to decreased dopamine release, and consequently the fine-control of the GPi/SNr is lost. Hence, GPi/SNr inhibition is diminished and the inhibitory output to the thalamus increased. As a result, the central role that basal ganglia play in processes such as motor activity control, motor learning, and some forms of associative and visual learning, is altered.[86]

As mentioned above, the current PD research trend consists of exploring possible biomarkers providing an early diagnostic and also of assaying drugs with a neuroprotective profile. Nevertheless, at present, the treatment of PD is mainly symptomatic, and it is primarily based on pharmacotherapy.[86] In addition, surgical interventions (i.e., deep brain stimulation) are possible.[87] However, since surgical procedures have associated risks and are not suitable for everyone, they represent an infrequent approach in clinics. Similarly, research around gene therapy and stem cell therapy is promising but still remains underdeveloped.[88] Regarding drug-based treatments, the most effective consist of using drugs mimicking dopamine, thus attempting to restore their normal levels. From them, the first drug used was L-DOPA, which is a dopamine precursor, and later on other agents (i.e., pramipexole) have been introduced into clinics.[89] Importantly, L-DOPA is a very efficacious drug on the treatment of PD symptoms. However, upon long-term treatment the efficacy is reduced, and furthermore severe undesired effects may appear, such as dyskinesia (i.e., twisting, turning movements), rigidity, and even psychiatric problems.[90] These undesired effects are mainly based on the fact that fewer dopamine neurons are able to store and release L-DOPA-derived dopamine when the disease advances, and plasma L-DOPA levels dramatically fluctuate. In clinics, the symptoms appearing upon fluctuating L-DOPA concentrations are defined as L-DOPA induced dyskinesias (LID).[90] Overall, a major objective of current research consists of increasing the benefit/risk profile of pharmacological therapies; i.e., providing an efficacious treatment of the symptoms together with lowering undesired effects; particularly for younger patients, where the development of LID is more likely.

In order to solve classical problems in therapeutics, researchers have incorporated in recent years the ultimate pharmacological findings. Among them, GPCR oligomerization appears as a very attractive drug target, since it seems likely that acting on these kinds of receptors' complexes may permit to tackle the disorder in an integrative way, better than using a single drug for a single receptor. As mentioned above, a clear-cut demonstration of adenosine-containing oligomers in the striatum (the main area affected in PD) has only been achieved for the $D_2R/A_{2A}R$,[66] but a number of evidences support the existence of other promising drug-target receptor–receptor interacting partners, such as the $A_1R/A_{2A}R$[70] or the $CB_1R/A_{2A}R$ oligomer[79] (Fig. 11.2). Regarding the $D_2R/A_{2A}R$ oligomer several studies dissected the exact mechanisms and/or domains supporting the direct interaction.[77,91,92] Interestingly, all this molecular information has been translated into a more clinical framework, thus new drugs or multimodal approaches are being explored to achieve a better management of PD. In this sense, the increased knowledge about $D_2R/A_{2A}R$ oligomerization has propelled the development of $A_{2A}R$ antagonists for the treatment of PD.[93] Interestingly, preclinical data suggests that these $A_{2A}R$ antagonists might have major applications in the treatment of PD by: (1) providing motor benefit as monotherapy, (2) potentiating the benefit of dopamine-based drugs, (3) preventing the development of LID, and (4) allowing the maintenance of the motor response with less dyskinesia using a lower L-DOPA dose. Indeed, $A_{2A}R$ antagonists have been shown to possess antiparkinsonian properties in human PD patients with LID.[94] This antiparkinsonian effect would be probably due to the action of the $A_{2A}R$ antagonist within the $D_2R/A_{2A}R$ oligomer, thereby reducing the $A_{2A}R$-mediated tonic inhibition of D_2R functioning. Furthermore, the status of some oligomers might be altered in pathological situations such as PD or LID, as demonstrated in our preclinical studies.[66] In line with this, the appearance of LID has been suggested to be associated to the prevalence of $A_{2A}R/A_{2A}R$ homomers versus $D_2R/A_{2A}R$ heteromers and D_2R/D_2R homomers, thus providing an abnormal increased $A_{2A}R$ signaling.[95] Interestingly, a similar situation would occur for other receptor–receptor interacting proteins, such as the $CB_1R/A_{2A}R$. Thus, it has been suggested that upon $A_{2A}R$-overexpression, the balance of $CB_1R/A_{2A}R$ and $A_{2A}R/A_{2A}R$ complexes may be altered, thus resulting in a different integration of receptors-mediated signaling,[96] which may be attempted to be tackled pharmacologically. Overall, a major goal in pharmacology for the forthcoming years may consist of establishing a precise knowledge of AR-containing oligomers' roles in the striatum. Accordingly, it would be possible to provide novel pharmacological approaches based on restoring oligomers' physiological functioning in order to improve PD management.

CONCLUDING REMARKS

PD treatment has been based for a long time on single-handed drugs, such as L-DOPA and other dopamine-based drugs. However, increased knowledge about GPCR oligomerization in general, and on AR heteromerization in particular, points to the use of multimodal approaches in which receptors' complex functioning may be finely modulated. From the AR-containing oligomers present in the striatum, the main area affected in PD, the $D_2R/A_{2A}R$ currently constitutes the main drug target. Accordingly, $A_{2A}R$ antagonists have been proposed to be used together with D_2R-based drugs in order to achieve a better management of the disease. The evidences existing support the antiparkinsonian properties of these kinds of drugs in human PD patients with LID,[94] thus several clinical trials are currently on the way,[93] may lead to these novel and more integrative manners of managing patients' diseases. Overall, $A_{2A}R$ antagonists are nowadays considered, based on their action through $D_2R/A_{2A}R$ oligomers, the most attractive potential non-dopaminergic drugs for the management of PD.

Acknowledgments

This work was supported by MINECO/ISC III (SAF2014-55700-P, PCIN-2013-019-C03-03, and PIE14/00034), IWT (SBO-140028) and Fundació la Marató de TV3 (Grant 20152031) to FC. VF-D and FC belong to the "Neuropharmacology and Pain" accredited research group (Generalitat de Catalunya, 2014 SGR 1251).

References

1. Drury AN, Szent-Gyorgyi A. The physiological activity of adenine compounds with especial reference to their action upon the mammalian heart. *J Physiol.* 1929;68:213–237.
2. Jacobson KA, Gao ZG. Adenosine receptors as therapeutic targets. *Nat Rev Discov.* 2006;5:247–264.
3. Newby AC. Adenosine and the concept of "retaliatory metabolites." *Trends Biochem.* 1984;9:42–44.
4. Brown RA, Spina D, Page CP. Adenosine receptors and asthma. *Br J Pharmacol.* 2008;153(suppl):S446–S456.
5. Burnstock G, Fredholm BB, Verkhratsky A. Adenosine and ATP receptors in the brain. *Curr Top Med Chem.* 2011;11:973–1011.
6. Johnston-Cox HA, Ravid K. Adenosine and blood platelets. *Purinergic Signal.* 2011;7:357–365.
7. Barletta KE, Ley K, Mehrad B. Regulation of neutrophil function by adenosine. *Arterioscler Thromb Vasc Biol.* 2012;32:856–864.

8. Ernst PB, Garrison JC, Thompson LF. Much ado about adenosine: adenosine synthesis and function in regulatory T cell biology. *J Immunol*. 2010;185:1993–1998.
9. Burnstock G. Purinergic regulation of vascular tone and remodelling. *Auton Autacoid Pharmacol*. 2009;29:63–72.
10. Vallon V, Mühlbauer B, Osswald H. Adenosine and kidney function. *Physiol Rev*. 2006;86:901–940.
11. Fredholm BB, Sollevi A. Cardiovascular effects of adenosine. *Clin Physiol*. 1986;6:1–21.
12. Sebastião AM, Ribeiro JA. Tuning and fine-tuning of synapses with adenosine. *Curr Neuropharmacol*. 2009;7:180–194.
13. Fredholm BB. Adenosine and lipolysis. *Int J Obes*. 1981;5:643–649.
14. Sebastiao AM, Ribeiro JA. Fine-tuning neuromodulation by adenosine. *Trends Pharmacol Sci*. 2000;21:341–346.
15. Latini S, Pedata F. Adenosine in the central nervous system: release mechanisms and extracellular concentrations. *J Neurochem*. 2001;79:463–484.
16. Snyder SH. Adenosine as a neuromodulator. *Annu Rev Neurosci*. 1985;8:103–124.
17. Ferre S, Fuxe K. Adenosine as a volume transmission signal. A feedback detector of neuronal activation. *Prog Brain Res*. 2000;125:353–361.
18. Fredholm BB. Purinoceptors in the nervous system. *Pharmacol Toxicol*. 1995;76:228–239.
19. Pull I, McIlwain H. Adenine derivatives as neurohumoral agents in the brain. The quantities liberated on excitation of superfused cerebral tissues. *Biochem J*. 1972;130:975–981.
20. Sattin A, Rall TW. The effect of adenosine and adenine nucleotides on the cyclic adenosine 3′, 5′-phosphate content of guinea pig cerebral cortex slices. *Mol Pharmacol*. 1970;6:13–23.
21. Degubareff T, Sleator Jr W. Effects of caffeine on mammalian atrial muscle, and its interaction with adenosine and calcium. *J Pharmacol Exp Ther*. 1965;148:202–214.
22. Trost T, Stock K. Effects of adenosine derivatives on cAMP accumulation and lipolysis in rat adipocytes and on adenylate cyclase in adipocyte plasma membranes. *Naunyn Schmiedebergs Arch Pharmacol*. 1977;299:33–40.
23. Londos C, Cooper DM, Wolff J. Subclasses of external adenosine receptors. *Proc Natl Acad Sci U S A*. 1980;77:2551–2554.
24. van Calker D, Muller M, Hamprecht B. Adenosine regulates via two different types of receptors, the accumulation of cyclic AMP in cultured brain cells. *J Neurochem*. 1979;33:999–1005.
25. Olah ME, Stiles GL. Adenosine receptor subtypes: characterization and therapeutic regulation. *Annu Rev Pharmacol Toxicol*. 1995;35:581–606.
26. Murray RD, Churchill PC. Effects of adenosine receptor agonists in the isolated, perfused rat kidney. *Am J Physiol*. 1984;247:H343–H348.
27. Anderson R, Sheehan MJ, Strong P. Characterization of the adenosine receptors mediating hypothermia in the conscious mouse. *Br J Pharmacol*. 1994;113:1386–1390.
28. Yamamoto S, Nakanishi O, Matsui T, et al. Intrathecal adenosine A1 receptor agonist attenuates hyperalgesia without inhibiting spinal glutamate release in the rat. *Cell Mol Neurobiol*. 2003;23:175–185.
29. De Lorenzo S, Veggetti M, Muchnik S, Losavio A. Presynaptic inhibition of spontaneous acetylcholine release induced by adenosine at the mouse neuromuscular junction. *Br J Pharmacol*. 2004;142:113–124.
30. Scholz KP, Miller RJ. Inhibition of quantal transmitter release in the absence of calcium influx by a G protein-linked adenosine receptor at hippocampal synapses. *Neuron*. 1992;8:1139–1150.
31. Schnurr M, Toy T, Shin A, et al. Role of adenosine receptors in regulating chemotaxis and cytokine production of plasmacytoid dendritic cells. *Blood*. 2004;103:1391–1397.

32. MacGregor DG, Miller WJ, Stone TW. Mediation of the neuroprotective action of R-phenylisopropyl-adenosine through a centrally located adenosine A1 receptor. *Br J Pharmacol.* 1993;110:470–476.
33. Varani K, Portaluppi F, Gessi S, et al. Dose and time effects of caffeine intake on human platelet adenosine A(2A) receptors: functional and biochemical aspects. *Circulation.* 2000;102:285–289.
34. Carroll MA, Doumad AB, Li J, Cheng MK, Falck JR, McGiff JC. Adenosine2A receptor vasodilation of rat preglomerular microvessels is mediated by EETs that activate the cAMP/PKA pathway. *Am J Physiol Renal Physiol.* 2006;291:F155–F161.
35. Popoli P, Betto P, Reggio R, Ricciarello G. Adenosine A2A receptor stimulation enhances striatal extracellular glutamate levels in rats. *Eur J Pharmacol.* 1995;287:215–217.
36. Nagel J, Schladebach H, Koch M, Schwienbacher I, Müller CE, Hauber W. Effects of an adenosine A2A receptor blockade in the nucleus accumbens on locomotion, feeding, and prepulse inhibition in rats. *Synapse.* 2003;49:279–286.
37. Scammell TE, Gerashchenko DY, Mochizuki T, et al. An adenosine A2a agonist increases sleep and induces Fos in ventrolateral preoptic neurons. *Neuroscience.* 2001;107:653–663.
38. Kemp BK, Cocks TM. Adenosine mediates relaxation of human small resistance-like coronary arteries via A2B receptors. *Br J Pharmacol.* 1999;126:1796–1800.
39. Donoso MV, López R, Miranda R, Briones R, Huidobro-Toro JP. A2B adenosine receptor mediates human chorionic vasoconstriction and signals through arachidonic acid cascade. *Am J Physiol Heart Circ Physiol.* 2005;288:H2439–H2449.
40. Zhong H, Belardinelli L, Maa T, Feoktistov I, Biaggioni I, Zeng D. A(2B) adenosine receptors increase cytokine release by bronchial smooth muscle cells. *Am J Respir Cell Mol Biol.* 2004;30:118–125.
41. Dubey RK, Gillespie DG, Mi Z, Jackson EK. Adenosine inhibits PDGF-induced growth of human glomerular mesangial cells via A(2B) receptors. *Hypertension.* 2005;46:628–634.
42. Zhong H, Shlykov SG, Molina JG, et al. Activation of murine lung mast cells by the adenosine A3 receptor. *J Immunol.* 2003;171:338–345.
43. Das S, Cordis GA, Maulik N, Das DK. Pharmacological preconditioning with resveratrol: role of CREB-dependent Bcl-2 signaling via adenosine A3 receptor activation. *Am J Physiol Heart Circ Physiol.* 2005;288:H328–H335.
44. Hinschen AK, Rose'Meyer RB, Headrick JP. Adenosine receptor subtypes mediating coronary vasodilation in rat hearts. *J Cardiovasc Pharmacol.* 2003;41:73–80.
45. Avila MY, Stone RA, Civan MM. Knockout of A3 adenosine receptors reduces mouse intraocular pressure. *Invest Ophthalmol Vis Sci.* 2002;43:3021–3026.
46. Stella L, de Novellis V, Marabese I, et al. The role of A3 adenosine receptors in central regulation of arterial blood pressure. *Br J Pharmacol.* 1998;125:437–440.
47. Kolakowski Jr LF. GCRDb: a G-protein-coupled receptor database. *Receptors Channels.* 1994;2:1–7.
48. Fritze O, Filipek S, Kuksa V, Palczewski K, Hofmann KP, Ernst OP. Role of the conserved NPxxY(x)5,6F motif in the rhodopsin ground state and during activation. *Proc Natl Acad Sci U S A.* 2003;100:2290–2295.
49. Rovati GE, Capra V, Neubig RR. The highly conserved DRY motif of class A G protein-coupled receptors: beyond the ground state. *Mol Pharmacol.* 2007;71:959–964.
50. Dunwiddie TV, Masino SA. The role and regulation of adenosine in the central nervous system. *Annu Rev Neurosci.* 2001;24:31–55.
51. Palmer TM, Stiles GL. Adenosine receptors. *Neuropharmacology.* 1995;34:683–694.
52. Rosin DL, Hettinger BD, Lee A, Linden J. Anatomy of adenosine A2A receptors in brain: morphological substrates for integration of striatal function. *Neurology.* 2003;61:S12–S18.

53. Marala RB, Mustafa SJ. Direct evidence for the coupling of A2-adenosine receptor to stimulatory guanine nucleotide-binding-protein in bovine brain striatum. *J Pharmacol Exp Ther*. 1993;266:294–300.
54. Ferre S, Karcz-Kubicha M, Hope BT, et al. Synergistic interaction between adenosine A2A and glutamate mGlu5 receptors: implications for striatal neuronal function. *Proc Natl Acad Sci U S A*. 2002;99:11940–11945.
55. Agnati LF, Fuxe K, Zini I, Lenzi P, Hokfelt T. Aspects on receptor regulation and isoreceptor identification. *Med Biol*. 1980;58:182–187.
56. Fuxe K, Agnati LF, Benfenati F, et al. Evidence for the existence of receptor–receptor interactions in the central nervous system. Studies on the regulation of monoamine receptors by neuropeptides. *J Neural Transm*. 1983;18:165–179.
57. Agnati LF, Ferre S, Lluis C, Franco R, Fuxe K. Molecular mechanisms and therapeutical implications of intramembrane receptor/receptor interactions among heptahelical receptors with examples from the striatopallidal GABA neurons. *Pharmacol Rev*. 2003;55:509–550.
58. Avissar S, Amitai G, Sokolovsky M. Oligomeric structure of muscarinic receptors is shown by photoaffinity labeling: subunit assembly may explain high- and low-affinity agonist states. *Proc Natl Acad Sci U S A*. 1983;80:156–159.
59. Fraser CM, Venter JC. The size of the mammalian lung beta 2-adrenergic receptor as determined by target size analysis and immunoaffinity chromatography. *Biochem Biophys Res Commun*. 1982;109:21–29.
60. Peterson GL, Rosenbaum LC, Broderick DJ, Schimerlik MI. Physical properties of the purified cardiac muscarinic acetylcholine receptor. *Biochemistry*. 1986;25:3189–3202.
61. Herberg JT, Codina J, Rich KA, Rojas FJ, Iyengar R. The hepatic glucagon receptor. Solubilization, characterization, and development of an affinity adsorption assay for the soluble receptor. *J Biol Chem*. 1984;259:9285–9294.
62. Ciruela F, Gomez-Soler M, Guidolin D, et al. Adenosine receptor containing oligomers: their role in the control of dopamine and glutamate neurotransmission in the brain. *Biochim Biophys Acta*. 2011;1808:1245–1255.
63. Berggård T, Linse S, James P. Methods for the detection and analysis of protein-protein interactions. *Proteomics*. 2007;7:2833–2842.
64. Albizu L, Cottet M, Kralikova M, et al. Time-resolved FRET between GPCR ligands reveals oligomers in native tissues. *Nat Chem Biol*. 2010;6:587–594.
65. Kern A, Albarran-Zeckler R, Walsh HE, Smith RG. Apo-ghrelin receptor forms heteromers with DRD2 in hypothalamic neurons and is essential for anorexigenic effects of DRD2 agonism. *Neuron*. 2012;73:317–332.
66. Fernández-Dueñas V, Taura JJ, Cottet M, et al. Untangling dopamine-adenosine receptor assembly in experimental parkinsonism. *Dis Model Mech*. 2015;8:57–63.
67. Torvinen M, Gines S, Hillion J, et al. Interactions among adenosine deaminase, adenosine A(1) receptors and dopamine D(1) receptors in stably cotransfected fibroblast cells and neurons. *Neuroscience*. 2002;113:709–719.
68. Ciruela F, Escriche M, Burgueno J, et al. Metabotropic glutamate 1alpha and adenosine A1 receptors assemble into functionally interacting complexes. *J Biol Chem*. 2001;276:18345–18351.
69. Nakata H, Yoshioka K, Kamiya T, Tsuga H, Oyanagi K. Functions of heteromeric association between adenosine and P2Y receptors. *J Mol Neurosci*. 2005;26:233–238.
70. Ciruela F, Casado V, Rodrigues RJ, et al. Presynaptic control of striatal glutamatergic neurotransmission by adenosine A1-A2A receptor heteromers. *J Neurosci*. 2006;26:2080–2087.
71. Cabello N, Gandia J, Bertarelli DC, et al. Metabotropic glutamate type 5, dopamine D (2) and adenosine A(2a) receptors form higher-order oligomers in living cells. *J Neurochem*. 2009;109:1497–1507.

72. Ciruela F, Ferre S, Casado V, et al. Heterodimeric adenosine receptors: a device to regulate neurotransmitter release. *Cell Mol Life Sci*. 2006;63:2427–2431.
73. Mizuno N, Suzuki T, Hirasawa N, Nakahata N. Hetero-oligomerization between adenosine A_1 and thromboxane A_2 receptors and cellular signal transduction on stimulation with high and low concentrations of agonists for both receptors. *Eur J Pharmacol*. 2012;677:5–14.
74. Chandrasekera PC, Wan TC, Gizewski ET, Auchampach JA, Lasley RD. Adenosine A1 receptors heterodimerize with β1- and β2-adrenergic receptors creating novel receptor complexes with altered G protein coupling and signaling. *Cell Signal*. 2013;25:736–742.
75. Onaran HO, Rajagopal S, Costa T. What is biased efficacy? Defining the relationship between intrinsic efficacy and free energy coupling. *Trends Pharmacol Sci*. 2014;35:639–647.
76. Hillion J, Canals M, Torvinen M, et al. Coaggregation, cointernalization, and codesensitization of adenosine A2A receptors and dopamine D2 receptors. *J Biol Chem*. 2002;277:18091–18097.
77. Ciruela F, Burgueno J, Casado V, et al. Combining mass spectrometry and pull-down techniques for the study of receptor heteromerization. Direct epitope-epitope electrostatic interactions between adenosine A2A and dopamine D2 receptors. *Anal Chem*. 2004;76:5354–5363.
78. Canals M, Marcellino D, Fanelli F, et al. Adenosine A2A-dopamine D2 receptor-receptor heteromerization: qualitative and quantitative assessment by fluorescence and bioluminescence energy transfer. *J Biol Chem*. 2003;278:46741–46749.
79. Ferré S, Lluís C, Justinova Z, et al. Adenosine-cannabinoid receptor interactions. Implications for striatal function. *Br J Pharmacol*. 2010;160:443–453.
80. Fuxe K, Agnati LF, Jacobsen K, et al. Receptor heteromerization in adenosine A2A receptor signaling: relevance for striatal function and Parkinson's disease. *Neurology*. 2003;61:S19–S23.
81. Fuxe K, Marcellino D, Borroto-Escuela DO, et al. Adenosine-dopamine interactions in the pathophysiology and treatment of CNS disorders. *CNS Neurosci Ther*. 2010;16:e18–e42.
82. Fuxe K, Borroto-Escuela DO, Romero-Fernandez W, et al. Moonlighting proteins and protein-protein interactions as neurotherapeutic targets in the G protein-coupled receptor field. *Neuropsychopharmacology*. 2014;39:131–155.
83. Lang AE, Lozano AM. Parkinson's disease. First of two parts. *N Engl J Med*. 1998;339:1044–1053.
84. Poewe W, Mahlknecht P. The clinical progression of Parkinson's disease. *Parkinsonism Relat Disord*. 2009;15(suppl 4):S28–S32.
85. Gerfen CR. Basal ganglia. In: Paxinos G, ed. *The Rat Nervous System*. Amsterdam: Elsevier Academic Press; 2004:445–508.
86. Meissner WG, Frasier M, Gasser T, et al. Priorities in Parkinson's disease research. *Nat Rev Discov*. 2011;10:377–393.
87. Volkmann J. Deep brain stimulation for the treatment of Parkinson's disease. *J Clin Neurophysiol*. 2004;21:6–17.
88. Mochizuki H, Choong C-J, Yasuda T. The promises of stem cells: stem cell therapy for movement disorders. *Parkinsonism Relat Disord*. 2014;20(suppl 1):S128–S131.
89. Cotzias GC, Van Woert MH, Schiffer LM. Aromatic amino acids and modification of parkinsonism. *N Engl J Med*. 1967;276:374–379.
90. Huot P, Johnston TH, Koprich JB, Fox SH, Brotchie JM. The pharmacology of L-DOPA-induced dyskinesia in Parkinson's disease. *Pharmacol Rev*. 2013;65:171–222.
91. Borroto-Escuela DO, Romero-Fernandez W, Tarakanov AO, et al. Characterization of the A2AR-D2R interface: focus on the role of the C-terminal tail and the transmembrane helices. *Biochem Biophys Res Commun*. 2010;402:801–807.

92. Fernandez-Duenas V, Gomez-Soler M, Jacobson KA, et al. Molecular determinants of A(2A) R-D(2) R allosterism: role of the intracellular loop 3 of the D(2)R. *J Neurochem.* 2012;123:373–384.
93. Vallano A, Fernandez-Duenas V, Pedros C, Arnau JM, Ciruela F. An update on adenosine A2A receptors as drug target in Parkinson's disease. *CNS Neurol Disord Drug Targets.* 2011;10:659–669.
94. LeWitt PA, Guttman M, Tetrud JW, et al. Adenosine A2A receptor antagonist istradefylline (KW-6002) reduces "off" time in Parkinson's disease: a double-blind, randomized, multicenter clinical trial (6002-US-005). *Ann Neurol.* 2008;63:295–302.
95. Antonelli T, Fuxe K, Agnati L, et al. Experimental studies and theoretical aspects on A2A/D2 receptor interactions in a model of Parkinson's disease. Relevance for L-dopa induced dyskinesias. *J Neurol Sci.* 2006;248:16–22.
96. Chiodi V, Ferrante A, Ferraro L, et al. Striatal adenosine-cannabinoid receptor interactions in rats over-expressing adenosine A2A receptors. *J Neurochem.* 2016;136:907–917.

CHAPTER

12

Adenosine Control of Striatal Function—Implications for the Treatment of Apathy in Basal Ganglia Disorders

Sergi Ferré
National Institutes of Health, Baltimore, MD, United States

OUTLINE

Adenosine Receptors in Neurodegenerative Diseases.
DOI: http://dx.doi.org/10.1016/B978-0-12-803724-9.00012-0

2017 Published by Elsevier Inc.

INTRODUCTION

The central adenosine system and adenosine receptors play a fundamental role in the modulation of striatal function. As reviewed in this chapter, this is mostly achieved by the strategic localization of different adenosine receptors heteromers in different neuronal striatal elements, mainly in the two populations of striatal efferent neurons that give rise to the "Go" and "NoGo" striatal efferent pathways and in the corticostriatal glutamate terminals. The striatum includes different compartments that differentially process information related to the implementation and learning of reward- and punishment-guided behaviors. The main role of the ventral striatum can be synthesized as determining "whether to respond" while that of the dorsal striatum is "how to respond" to reward- and punishment-associated stimuli. The better understanding of the functional role of different striatal compartments allows a more syndromic subclassification of basal ganglia disorders with potentially significant new therapeutic approaches. As a relevant clinical example, "apathy" is a prevalent symptom in Parkinson's disease, schizophrenia, and substance use disorders (SUD) and depends mostly on dysfunction of the ventral striatal compartment. It is proposed that striatal adenosine receptor heteromers can provide important therapeutic targets for the treatment of apathy.

DOPAMINE CONTROL OF STRIATAL FUNCTION

Classically, the basal ganglia have been considered as subcortical structures involved in motor function, but it is now generally stated that they form part of largely segregated parallel brain circuits that process not just sensorimotor but also "cognitive" and "limbic" information.[1,2] With these ill-defined terms it is at least recognized that different striatal compartments have different functions. The striatum, the main input structure of the basal ganglia, is the main target of the ascending mesencephalic dopamine system, with the densest dopamine innervation and the highest density of dopamine receptors in the brain. Ascending dopamine systems originate in the substantia nigra pars compacta (SNc) and ventral tegmental area (VTA), and their mostly apparent differential functional roles depend on their innervation of functionally different striatal compartments. In fact, SNc and VTA constitute a conventional arbitrary separation of a continuum of mesencephalic dopamine cells.[2] Nevertheless, to understand the functional role of different striatal compartments we need to understand the basis of striatal dopamine function. As here briefly reviewed, dopamine in the striatum *increases the responsiveness to reward-associated stimuli* and *elicits reinforcement.*

Psychostimulants have been fundamental in our understanding of dopamine function, because of their ability to produce an increase in central dopamine neurotransmission. Thus, classical psychostimulants like cocaine and amphetamine elicit psychomotor-activating and reinforcing effects.[3] The terms "psychostimulant" or "psychomotor stimulant" are to be distinguished from "general central nervous system stimulant" such as strychnine and pentylenetetrazol or picrotoxin, antagonists of the inhibitory neurotransmitter receptors for glycine and GABA, respectively.[4,5] Psychomotor activation is a major pharmacological effect of psychostimulants, while general central nervous system stimulants fail to increase psychomotor activity at doses below those that produce convulsions. We should also distinguish between "psychomotor" and simply "motor" responses, implying a behavioral response to reward-associated stimuli (see below) and activation of a fixed motor program, respectively.[3] The two most commonly identified psychostimulant-induced psychomotor responses in the experimental animal are "locomotion" and "stereotypy" and it has been generally but faultily accepted that forward locomotion and stereotypy induced by psychostimulants are motor responses that result from increased dopamine transmission in the ventral striatum and the dorsal striatum, respectively. Both types of responses have a characteristic dependence on external (environmental or somatic) stimuli, and what psychostimulants do is to produce an increased responsiveness to those stimuli.[6] The opposite is seen with genetic or pharmacological blockade of dopamine[3,7] or with selective lesions of the ascending dopamine pathways or targeted dopamine denervation of striatal areas,[8–10] where the animals develop sensory neglect, inattention to relevant stimuli, rather than a simple motoric or sensory impairment.

With the pioneering work of Wolfgang Schultz it was demonstrated that mesencephalic dopamine cells are mostly involved with the processing of a particular kind of salient stimuli: Reward-associated stimuli.[11] By selectively increasing responsiveness to reward-associated stimuli, dopamine promotes orienting and approaching responses to those stimuli, with psychomotor activation and reward-oriented behavior.[12] Concomitantly, dopamine is directly involved in reinforcement, in the learning ("stamping-in") of stimulus-reward and reward-response associations that follows the receipt of reward.[12] The reinforcement of stimulus-reward associations establishes signals that guide and orient to rewards (discriminative stimulus) or which become rewards themselves (conditioned rewarding stimulus). The stamping-in of reward-response associations promotes the learning of the optimal sequential response, the action skill that leads to the reward.

Two main different temporal operating modes of dopamine neuronal function have been described: a fast, subsecond-scale, phasic response, and a prolonged, minute-scale, tonic dopamine modulation.[11] The fast operating mode codes for a reward prediction error, which provides a rapid response to reward-related signals and can significantly contribute to the role of dopamine in reinforcement. A large number of studies have shown that these dopamine signals represent reward predictions in a manner that closely matches behavioral preferences, including the preference for large, probable, or immediate rewards over small, improbable, or delayed rewards.[13–19] If a reward is larger than predicted, dopamine neurons are strongly excited (positive prediction error); if a reward is smaller than predicted or fails to occur at its appointed time, dopamine neurons are phasically inhibited (negative prediction error). The second operating mode is determined by signals of proximity, probability, and effort value of rewards[20,21] and therefore can significantly contribute to the role of dopamine in the degree of responsiveness to reward-oriented behaviors. The two operating modes seem to be functional attributes of all dopamine cells, in spite of the generally accepted dichotomy of the VTA and SNpc.[11,19] As mentioned before, there are no clear boundaries between these cells groups, which in fact have very similar main inputs: GABA inputs from the striatum and pallidal complex, glutamate and acetyl-choline inputs from the pedunculopontine nucleus and a serotonergic innervation from the dorsal raphe nucleus.[22,23] On the other hand, quite different functional striatal compartments can be delineated in terms of their substantially different glutamate inputs.[2,24]

In humans and nonhuman primates, we can distinguish three main functional striatal compartments, the ventral, the rostral dorsal, and the caudal dorsal striata. The ventral striatum concept has expanded from its initial inclusion of VTA-innervated areas, mostly the nucleus accumbens (NAc; with its two compartments core and shell) and the olfactory tubercle, to the striatal areas receiving glutamate inputs from the ventromedial prefrontal cortex (vmPFC), orbitofrontal cortex (OFC), and anterior cingulate cortex (aCC). In fact, SNpc provides partial and predominant dopamine inputs to both OFC- and aCC-innervated striatal areas.[2] Furthermore, the ventral striatum receives afferent projections from the insular cortex, amygdala, and hippocampus. Taken together, the ventral striatum occupies over 20% of the striatum in nonhuman primates and about the same in humans. That includes the NAc and the most medial and anterior part of the head of the caudate nucleus and putamen.[2] The rostral dorsal striatum comprises the head and body of the caudate nucleus and anterior putamen and receives afferent projections from frontal and parietal association cortices. The caudal dorsal striatum comprises the tail of the caudate nucleus and posterior part of the putamen and receives afferent projections from sensorimotor cortices.[25]

The ventral striatum forms part of decision-making brain circuits involved in reward valuation tasks, which determine and store reward values (often named as "subjective values of rewards") and constantly chose the reward to be obtained by a process of maximizing utilities associated with different options, the highest benefit/cost ratio.[26] "Delay discounting" (DD), "effort discounting" (ED), and "probability discounting" (PD) refer to the empirical finding that both humans and animals value immediate, low-effort, and high probability rewards more than delayed, high-effort, and low probability rewards. Standard theories of economic decision-making do not distinguish between decisions related to different types of costs, such as delay or effort costs.[27] Thus, a choice is made after a valuation stage, regardless of the nature of the cost. Large number of behavioral and clinical studies indicate that DD, ED, and PD are independent variables possibly involving corticostriatal circuits with different ventral striatal compartments, differentially connected to different prefrontal cortical areas.[28,29]

The main role of the ventral striatum, classically labeled as an interface between motivation and action,[30] can be synthesized as determining "whether to respond" while that of the dorsal striatum is "how to respond" to reward-associated stimuli. The rostral dorsal and caudal dorsal striata receive afferent projections from frontal and parietal association cortices and sensorimotor cortices, respectively.[1,25] The insightful model of basal ganglia function developed by Kim and Hikosaka,[31] based on their elegant experiments on gaze orienting and learning of sequential motor responses in nonhuman primates, highlights the simultaneous and yet differential processing of reward-oriented behaviors and reinforcement by the rostral and caudal dorsal striata. The model also implies that all dopamine-dependent functions, reward-oriented behavior and learning of stimulus-reward and reward-response associations, are simultaneously processed by rostral dorsal and caudal dorsal striatal areas. In relation to reinforcement, rostral dorsal areas are predominantly involved in an initial, more controlled, "volitional" (contingent on the outcome), accurate, and more labile learning, while caudal dorsal areas are involved in a slower and more "automatic" (noncontingent on the outcome) and long-lasting learning.[31] The same functional dichotomy has been demonstrated in the rodent striatum but with a medial-lateral distribution; with dorsomedial and dorsolateral striatum being preferentially involved with outcome-dependent and independent learning, respectively.[32–34]

An increasing number of experimental data also indicates that mesencephalic dopamine cells also process aversive-related stimuli (for recent review, see Holly and Miczek[35]). Electrophysiological experiments have found evidence for the existence of two separated populations of dopamine neurons that respond differently to aversive stimuli. Most dopamine cells respond by decreasing their activity, and these are cells that also code for positive reward prediction errors, increasing their firing upon presentation of reward-related stimuli or with the termination of an aversive-related stimulus and decreasing their firing upon omission of reward-related stimuli.[36,37] In fact, experimental evidence demonstrates that this majority of dopamine cells codes both positive and negative reward prediction errors.[38,39] The other subpopulation increases its activity upon presentation of an aversive stimulus, and it seems to be specifically localized in the most medial and posterior part of the ventral tegmental area (pmVTA).[36,37,40] This area projects to a specific striatal area, also the posteriomedial part of the shell of the nucleus accumbens (pmNAc shell),[41,42] specifically involved in threat-related behaviors.[43] The infralimbic cortex innervates both pmVTA and pmNAc shell as well as the amygdala, and this circuit plays a key role in "fear" extinction, in the suppression of threat conditioning.[44,45] Importantly, according to Moscarello and LeDoux,[46] active avoidance learning, with the elicitation of a behavioral response that avoids the interaction with the aversive stimulus, requires the suppression of threat conditioning. In summary, *dopamine promotes approach by directly increasing the responsiveness to reward-related stimuli, but also indirectly by decreasing the withdrawal reaction from previously conditioned aversive stimuli.*

DIFFERENTIAL DOPAMINE CONTROL OF THE DIRECT AND INDIRECT STRIATAL EFFERENT PATHWAYS, THE "GO" AND "NOGO" PATHWAYS

A mechanism by which dopamine is also directly involved with negative reinforcement is emerging from the study of the functional role of the two main striatal neuronal populations. Mesencephalic dopamine and cortical glutamate inputs converge in the GABA medium spiny neuron, which constitutes more than 95% of the neuronal population.[47] There are two subtypes of medium spiny neurons, which define the two main striatal efferent pathways, which connect the striatum with the output structures of the basal ganglia, substantia nigra pars reticulata, and internal segment of the globus pallidus. The striatonigral neuron constitutes the direct efferent pathway and expresses dopamine D_1 receptors (D_1R) and the striatopallidal neuron gives rise to the indirect pathway and expresses D_2 receptors (D_2R)[47] (Fig. 12.1). Striatopallidal neurons from the dorsal or ventral striatum connect with the pallidal complex, more specifically with the external segment of the globus pallidus and the ventral pallidum, respectively[47] (Fig. 12.1). The

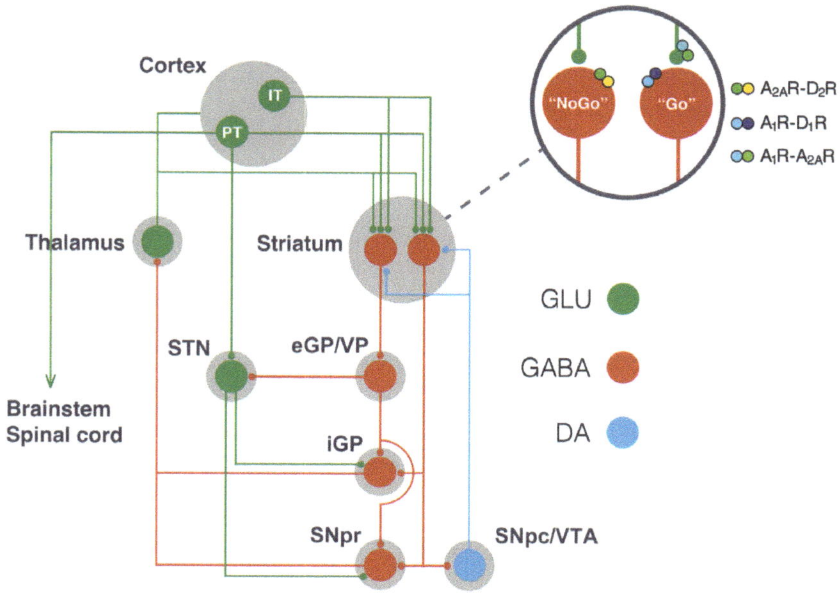

FIGURE 12.1 Scheme of the cortical–striatal–thalamic–cortical circuits showing the specific localization of $A_{2A}R–D_2R$ and $A_1R–D_1R$ heteromers in the striatopallidal and striatonigral neurons, which give rise to the "NoGo" and "Go" pathways, respectively. $A_1R–A_{2A}R$ are preferentially localized in corticostriatal glutamate terminals that make synaptic contact with the striatonigral neurons. *eGP/VP*, external segment of the globus pallidus/ventral pallidum; *iGP*, internal segment of the globus pallidus; *SNpr*, substantia nigra pars reticulata; *SNpc*, substantia nigra pars compacta; *VTA*, ventral tegmental area; *IT and PT*, intratelencephalic and pyramidal tract neurons; *GLU, GABA, and DA*, glutamate, GABA, and dopamine neurons.

output structures of the basal ganglia connect with the thalamus and the thalamus back to the cortex, constituting the well-known parallel cortical–striatal–thalamic–cortical circuits.[47] But apart from this well-known, largely segregated, parallel processing, a nonappreciated substantial convergence of cortical inputs takes place in the striatum, in the striatonigral and striatopallidal neurons. This convergence includes: Inputs from the two types of cortical pyramidal neurons, the intratelencephalic and pyramidal tract neurons, coming from the same cortical area[48] (Fig. 12.1); inputs from different although close and related cortical areas, such as from homologous primary motor and somatosensory cortical areas[49]; and even from different and separated sensorial areas, providing the frame for multisensory integration.[50] Convergent corticostriatal glutamate inputs provide the convergent input necessary for establishing the dopamine-mediated stamping-in of stimulus-reward and reward-response associations that follows the receipt of reward.

There are no obvious qualitative or quantitative differences in the glutamate inputs to the striatonigral and striatopallidal neurons, but specially from recently developed optogenetic techniques it has been possible to dissect a differential role of the direct and indirect pathways in mediating "Go" or "NoGo" responses upon exposure to reward- and aversive-related stimuli, which depend on their different connecting output structures and different phenotypes. Differential connectivity implies that activation of striatonigral neurons leads to facilitation, while activation of striatopallidal neurons leads to suppression of body movements. On the other hand, the differential affinities of D_1R and D_2R for endogenous dopamine and their respective predominant expression in striatonigral and striatopallidal neurons provide a fine-tuning device by which bursts and pauses of dopamine neurons can differentially influence their activity.[51,52] Dopamine has significantly higher affinity for D_2R than for D_1R. Therefore, D_2R are more activated than D_1R by basal dopamine levels and are more sensitive to the effects of dopamine pauses, while D_1R are more sensitive to dopamine bursts, to conditions of higher dopamine release. Bursts of dopamine neurons produce conditions of high dopamine release, which activate stimulatory D_1R and cause the direct pathway to promote high-value reward-associated movements, whereas dopamine pauses produce conditions of low dopamine release, which remove activation of inhibitory D_2R and cause the indirect pathway to suppress low-value reward-associated or high-value punishment-associated movements.[53–58] In agreement, strong dopamine receptor activation promotes potentiation of corticostriatal synapses onto the direct pathway and learning from positive outcomes (positive reinforcement) and striatal D_1R receptor blockade selectively impairs movements to rewarded targets.[59–62] On the other hand, weak dopamine receptor activation promotes potentiation of corticostriatal synapses onto the indirect pathway and learning from negative outcomes (negative reinforcement) and striatal D_2R blockade selectively suppresses movements to nonrewarded targets.[59–62]

In summary, opposite dopamine signaling in the striatum is directly involved in approach to or withdrawal from reward-associated and punishment-associated stimuli, respectively, and to positive and negative reinforcement, by its ability to differentially modulate the activity of the "Go" and "NoGo" pathways. More specifically, a predominant activity of the "Go" or the "NoGo" pathways in the ventral striatum should determine "whether to respond" to or "whether to withdraw" from reward- and punishment-associated stimuli; a predominant activity of the "Go" or the "NoGo" pathways in the dorsal striata (rostral and caudal) should determine "how to respond" to or "how to withdraw" from those stimuli. We should then expect that the role of the tonic operating mode of dopamine neurons extents from an increased responsiveness to reward-oriented behaviors to punishment-oriented responses, which should be determined by the ratio of activation of the "Go" and "NoGo" pathways.

DIFFERENTIAL ADENOSINE CONTROL OF THE "GO" AND "NOGO" PATHWAYS. THE ADENOSINE–DOPAMINE RECEPTOR HETEROMERS

Since their discovery, receptors have mostly been considered as single functional units. However, in recent years, a fast-growing list of G protein–coupled receptors (GPCR) forming receptor oligomers has emerged.[63–66] Receptor oligomers are defined as macromolecular complexes composed of at least two (functional) receptor units (protomers) with biochemical properties that are demonstrably different from those of its individual components.[65] A first important concept that arises from the new field of GPCR oligomerization is that the pentameric structure constituted by one GPCR homodimer and one heterotrimeric G protein provides a main functional unit, and oligomeric entities can be viewed as multiples of dimers.[66] More specifically, GPCR heteromers are being considered as heterotetramers, formed of two different homodimers, each able to signal with their preferred G protein.[66–70]

Within GPCR heteromers, each GPCR protomer can convey allosteric modulations of ligands binding to the other different GPCR protomer (ligand-independent allosteric modulation). Furthermore, GPCR heteromers can convey allosteric interactions between orthosteric ligands, altering their affinity or intrinsic efficacy. A ligand binding to one GPCR unit in the complex can lead to changes in the properties of a ligand binding to a different GPCR unit.[66] Well-characterized examples of these allosteric phenomena occur within the adenosine A_{2A} receptor ($A_{2A}R$)–D_2R heteromers. In the brain both $A_{2A}R$ and D_2R are particularly expressed in the same neuron, the striatopallidal neuron (Fig. 12.1). We have now a large amount of experimental evidence for the existence of a predominant striatal population of both $A_{2A}R$ and D_2R forming functionally and pharmacologically significant heteromers that modulate the function of the striatopallidal neurons, of the "NoGo" striatal efferent pathway.[70–73] These $A_{2A}R$–D_2R heteromers provide a frame for the apparent preferential role of $A_{2A}R$ and D_2R receptors in the psychostimulant effects of the nonselective adenosine A_1R and $A_{2A}R$ antagonist caffeine.[74] Large number of experiments have shown that $A_{2A}R$ agonists and antagonists induce the same qualitative psychomotor depressant and activating effects as D_2R antagonists and agonists, respectively,[74–80] that $A_{2A}R$ agonists and antagonists selectively counteract and potentiate the psychomotor-activating effects of D_2R agonists, respectively,[81–84] but also that selective $A_{2A}R$ but not A_1R antagonists counteract the psychomotor depressant effects of reserpine and D_2R antagonists.[78,80,85–87]

Initial studies about the psychomotor-activating effects of caffeine or selective $A_{2A}R$ antagonists dealt with general locomotor activity and at their clinical translational level were mostly related with Parkinson's disease. In fact, the allosteric mechanisms within the $A_{2A}R$–D_2R heteromer provided the rationale for the use of $A_{2A}R$ antagonists in Parkinson's disease.[88–90] Those initial studies implicitly considered $A_{2A}R$–D_2R heteromers in the dorsal striatum, but a large number of studies indicated that not only dorsal but also ventral striatopallidal neurons (which project to the external segment of the globus pallidus and the ventral pallidum, respectively) express A_2R and $A_{2A}R$–D_2R heteromers.[73,91–98] Then, more recent studies also dealt with measures of specific reward-oriented behavior, with caffeine and $A_{2A}R$ antagonists increasing the responsiveness to specific reward-associated stimuli, such as those associated with regular food,[79,80] sucrose solutions,[99] and those eliciting maternal behavior[100] and intracranial self-stimulation.[101] The work by Salamone's group has specifically addressed the role of adenosine and $A_{2A}R$ in effort-related choice behavior. Direct administration in the NAc of $A_{2A}R$ agonists altered effort-related choice behavior in a manner closely resembling the effects of interference with ventral striatal dopamine neurotransmission, decreasing the degree of responsiveness ("effort") to reward-associated stimuli. Furthermore, $A_{2A}R$ antagonists reversed the ED effects of D_2R antagonists.[80,102]

Recent studies suggest that $A_{2A}R$–D_2R heteromers constitute a heterotetrameric structure with $A_{2A}R$ and D_2R homodimers coupled to their cognate $G_{s/olf}$ and $G_{i/o}$ protein, respectively.[70,72,103] The heterotetrameric structure allows multiple simultaneous and reciprocal interactions between adenosine and dopamine and exogenous $A_{2A}R$ and D_2R ligands.[70,72,103] Two most salient interactions are the ability of adenosine or exogenous $A_{2A}R$ ligands to decrease the affinity and intrinsic efficacy of dopamine or exogenous D_2R ligands (allosteric $A_{2A}R$–D_2R interaction) and the ability of D_2R agonist-mediated and $G_{i/o}$ protein-dependent counteraction of $A_{2A}R$ agonist-mediated and $G_{s/olf}$-dependent activation of adenylyl-cyclase (adenylyl-cyclase $A_{2A}R$–D_2R interaction). In fact, it can be postulated that the canonical G_s–G_i interaction at the adenylyl-cyclase level depends on heteromerization of G_s- and G_i-coupled homodimers.[67–69] One more recently discovered interaction is a negative allosteric modulation between an orthosteric $A_{2A}R$ agonist and an orthosteric $A_{2A}R$ antagonist binding simultaneously to the $A_{2A}R$ homodimer within the $A_{2A}R$–D_2R heterotetramer.[70] Then, negative allosterism between caffeine and endogenous adenosine, more than competitive antagonism, seems to be mostly involved with the psychomotor effects of caffeine.[70] In summary, the $A_{2A}R-D_2R$ heteromer acts as an integrative device that allows very elaborated interactions between adenosine and dopamine controlling the function of the striatopallidal neuron, of the "NoGo" pathway. Preferential $A_{2A}R$ versus D_2R activation leads to an increased activity of the "NoGo" pathway with a signal of not responding to reward-associated stimuli or to withdrawing from nonrewarded or aversive stimuli; preferential D_2R versus $A_{2A}R$ activation leads to a decreased activity of the "NoGo" pathway, which promotes responding to reward-associated stimuli and avoids withdrawal from nonrewarded or aversive stimuli.

But adenosine also modulates the "Go" pathway and again in an opposite way than dopamine, which stimulates striatonigral neurons by activation of D_1R. In this case the interaction takes place between A_1R and D_1R (Fig. 12.1). Previous studies demonstrated the ability of A_1R to complex with D_1R in transfected cells and striatal neurons in culture, which allows allosteric modulations between A_1R and D_1R ligands within the complex.[104–106] These modulations implied decrease in the binding of D_1R agonists and antagonists that were also demonstrated in striatal tissue[107,108] as well a G_s-G_i canonical interaction at the adenylyl-cyclase level, by which A_1R agonists counteract D_1R agonist-mediated cAMP accumulation and the consequent AMPA receptor phosphorylation and immediate-early gene expression.[104,108,109] Unpublished results from our research group demonstrate a similar heterotetrameric molecular structure of A_1R-D_1R heteromers to that of $A_{2A}-D_2R$ heteromers, as well as their localization in the rat striatum (in preparation). At a neuronal level, striatal A_1R activation led to a selective significant counteraction of D_1R-mediated activation of striatonigral neurons.[106] At the behavioral level, A_1R agonists and antagonists selectively depress and potentiate psychomotor activation induced by D_1R agonists, respectively.[81,82,108,110]

In summary, although less studied than the $A_{2A}-D_2R$ heteromer, the A_1R-D_1R heteromer seems to provide a similar kind of integrative device of adenosine and dopamine signals that controls the function of the "Go" pathway. The segregation of adenosine and dopamine receptor subtypes in the striatonigral and striatopallidal neurons constitutes one of the most noticeable examples of the functional implications of GPCR subtypes in the brain, allowing dopamine to produce opposite neuronal effects in the two functionally different efferent striatal neurons and allowing adenosine to counteract those effects. The segregation is even present in the terminals of the striatonigral and striatopallidal neurons, where evidence for respective A_1R-D_1R and $A_{2A}R-D_2R$ interactions has been confirmed.[111–113] Obviously, A_1R-D_1R and $A_{2A}-D_2R$ heteromers constitute important therapeutic targets for the treatment of the many neuropsychiatric disorders which pathogenesis depend on striatal dysfunction (see below).

ADENOSINE CONTROL OF THE STRIATAL GLUTAMATE TERMINAL. THE $A_1R-A_{2A}R$ HETEROMER

$A_1R-A_{2A}R$ heteromer is another possible heterotetramer that functions very similarly to the $A_{2A}R-D_2R$. In this case A_1R is the one coupled to $G_{i/o}$ proteins. In the striatum $A_1R-A_{2A}R$ heteromers are selectively localized in corticostriatal glutamate terminals, where they exert a fine-tune control of glutamate release.[114–116] The $A_1R-A_{2A}R$ heteromer works as an adenosine concentration-dependent switch. Thus, adenosine has more affinity for A_1R than for $A_{2A}R$ and low concentrations of adenosine will activate A_1R, which inhibits glutamate release; higher concentrations also activate $A_{2A}R$ which allosterically shuts down A_1R signaling and promotes G_s signaling, activation of cAMP and glutamate release.[114] Electrophysiological studies in striatal slices and immunocytochemical (synaptosomes and glutamate nerve terminal preparations) and immunohistochemical (electron microscopy preparations) experiments identified a preferential localization of $A_1R-A_{2A}R$ heteromers in the coticostriatal terminals making synaptic contact with the D_1R-containing neurons[115] (Fig. 12.1). This would indicate a preferential presynaptic adenosinergic control of the activity of the "Go" versus the "NoGo" pathway.

A proof of concept for using GPCR heteromers to pharmacologically dissect distinct subpopulations of receptors came from experiments that compared the effects of several $A_{2A}R$ antagonists for their ability to produce psychomotor activation or to block glutamate release induced by cortical stimulation.[117] Thus, psychomotor activation depends on postsynaptic $A_{2A}R$ forming $A_{2A}R-D_2R$ heteromers in the striatopallidal neuron, while blockade of presynaptic corticostriatal neurotransmission depends on presynaptic $A_{2A}R$ forming heteromers with adenosine A_1Rs in terminals of cortical neurons.[114,115] Based on their potencies for blocking striatal glutamate release and for inducing psychomotor activation (measured as locomotor activity) in rats, two $A_{2A}R$ antagonists, SCH-442416 and KW-6002, were found to have preferential pre- or postsynaptic activities, respectively.[117] Parallel experiments in transfected cells demonstrated that the pre- and postsynaptic effects of these $A_{2A}R$ antagonists depend on their differential affinity for binding to $A_{2A}R$ heteromers. SCH-442416 bound with lower affinity to $A_{2A}R$ when coexpressed with D_2R than with A_1R and KW-6002 showed the best relative affinity for $A_{2A}R$ coexpressed with D_2R.[117] These results represented a specific ligand-independent allosteric modulation of the $A_{2A}R$ protomer on a specific ligand, SCH-442416, then demonstrating the possibility of pharmacologically dissecting, of selective targeting, receptor heteromers.[66] The expected differences in affinity of SCH-442416 for $A_{2A}R$ in the presence and absence of D_2R have been reproduced in striatal tissue from wild-type mice and conditional striatal D_2R knockout mice (in preparation).

The possibility of targeting $A_1R-A_{2A}R$ heteromers was also used to identify an important contributor to the reinforcing effects of cannabinoids.[118,119] We initially reported the apparent paradoxical result that the $A_{2A}R$ antagonist MSX-3 decreases THC and anandamide self-administration in squirrel monkeys at a relatively low dose, while a threefold higher dose produces the opposite effect.[118] Based on results obtained in rats,[117] it was hypothesized that the different dose-dependent effects of MSX-3 could be related to a slightly selective presynaptic effect at lower doses with an overriding postsynaptic effect at larger doses. This hypothesis was confirmed by testing the effects of SCH-442416 and KW-6002.[119] SCH-442416 produced a significant shift to the right of the THC self-administration dose-response curves, consistent with antagonism of the reinforcing effects of THC. On the other hand, KW-6002 produced a significant shift to the left, consistent with potentiation of the reinforcing effects of THC. These results show that selectively blocking presynaptic $A_{2A}R$ could provide a pharmacological approach to the treatment of marijuana dependence, and underscore corticostriatal glutamate neurotransmission as a possible main mechanism contributing to the rewarding effects of THC. At a more general level, these results also show that while the concept of using GPCR heteromers to target specific cell types is relatively new, it is a promising approach for targeting specific cell types to modulate specific symptoms of SUD.

DOPAMINE AND STRIATAL DYSFUNCTION. APATHY VERSUS AKINESIA

Dysfunction of the central dopamine system is involved in a variety of disorders, including Parkinson's disease, schizophrenia and SUD. The functional separation of striatal compartments in ventral, rostral dorsal and caudal dorsal striata allows a more syndromic subclassification of those disorders with potentially significant new therapeutic approaches. Parkinson's disease and nonhuman primate models of Parkinson's disease provide the clearest illustration. The cardinal motor symptoms of Parkinson's disease, bradykinesia, rigidity, and tremor,[120] have been classically attributed to dysfunction of the skeletomotor system, the brain circuits involved in the execution and coordination of body movements. Contemporary theories embracing parallel cortical–striatal–thalamic–cortical circuits in the pathogenesis of this disorder emphasize the particular involvement of the "motor circuit," which includes motor cortical areas.[121] In fact, in Parkinson's disease, dopamine cell degeneration tends to occur initially and predominantly in the lateral part of the SNpc, which projects mainly to the caudal dorsal striatum. Thus, there is a predominant deficit of the more "automatic" versus "volitional" action skills and most sequential psychomotor responses need to be performed with full attention.[31] Nevertheless with more advanced stages of Parkinson's disease the function of the more rostral striatum becomes also compromised, with deficits in "volitional" actions skills.[31] And with further (or preferential) ventral degeneration of the dopamine mesencephalic nuclei we move to the pathology of the ventral striatum, to apathy,[122] as also demonstrated experimentally in the nonhuman primate.[123,124]

Clinically, apathy has been defined as "a syndrome consisting on loss of motivation not attributable to disturbances in emotion, intellect, or consciousness."[125] However, the same as with the concept of "impulsivity," it is becoming obvious that apathy is a multifaceted concept that includes dissociable constructs that should correspond to dissociable neurobiological correlates.[126] One of the tenets of the present chapter is that some if not all those dissociable correlates correspond to corticostriatal circuits involving the different functional striatal compartments and their "Go" and "NoGo" pathways. In fact, attuned with the role of dopamine in reward-associated behavior in all striatal compartments, recent studies even allow conceptualizing Parkinson's disease bradikinesia in a motivational frame.[127,128] Nevertheless, as defined clinically, apathy is a common nonmotor symptom of Parkinson's disease[129] that correlates negatively with dopamine innervation in the ventral striatum.[122,130,131] In fact, a deficit in the dopamine modulation of the ventral striatum should translate, first, in a deficit in the responsiveness with a global inability to respond to reward- and punishment-associated stimuli (attuned with the "whether to respond" vs "how to respond" functions of ventral vs dorsal striatum). Second, it should lead to dysfunction of reward valuation, in alterations (increase) in DD, ED, and PD (attuned with the ventral striatum as forming part of corticostriatal circuits involved in reward valuation tasks). Indeed, unmedicated patients with Parkinson's disease have shown increases in DD and ED.[128,132] The possible association in Parkinson's disease of apathy with increase in DD, which is one major operational indicator of choice impulsivity,[133] demonstrates that the multifaceted impulsivity and apathy concepts are not just opposite poles of just a one-dimensional construct,[126] as suggested in the literature.[122,134]

Interestingly, apathy is also a major negative symptom of schizophrenia, classically considered as a disorder associated with central hyperdopamine tone. Several studies have found evidence for selective dysfunction of the ventral striatum in schizophrenia, specifically hypoactivation with reward-associated stimuli.[135–138] Ventral striatal activation during reward anticipation was in fact found to be selectively and inversely correlated with apathy but not with other negative symptoms.[135,138] Two additional findings give a clue for the mechanisms of apathy in schizophrenia, which seem to be dopamine-independent or at least not related to a decrease in the dopamine tone. First, there is a reduced functional connectivity between the OFC and the ventral striatum[136]; second, there is consistent evidence that schizophrenic patients suffer from selective deficits in learning from positive outcomes, with intact learning from negative outcomes.[139] Therefore, the apathetic schizophrenic patient seems to have a selective decreased activation of the "Go" pathway, a reduction in the ratio of activation of "Go" versus "NoGo" pathways secondary to impaired cortical–ventral striatal connectivity.[139]

A similar situation would also be present in the patient with SUD, a decreased "Go"/"NoGo" pathway activation, also with reduced ventral striatal activation to reward stimuli (which can basically only be activated by the addictive drugs).[140] Apathy is a well-known symptom in SUD, although it has been scarcely addressed experimentally.[141–143] The SUD patient is motivated to procure the drug but tends to be withdrawn and apathetic when exposed to nondrug-related activities.[141] In this case, however, the pathogenesis seems to follow from an initial reduction in D_2R density, leading to an increased activity of the "NoGo" pathway. The tonic decrease in feedback activation of the vmPFC, OFC, and aCC leads to additional dysfunction of the decision-making cortical–ventral striatal circuits.[140,144] These changes lead to a similar situation to the nonmotor symptoms in patients with Parkinson's disease, to apathy and choice impulsivity, as demonstrated by several studies indicating increase DD in patients with SUD.[133,144]

TARGETING ADENOSINE RECEPTOR HETEROMERS FOR THE TREATMENT OF APATHY

As mentioned above, the allosteric mechanisms within the $A_{2A}R$–D_2R heteromer provided the first rationale for the use of $A_{2A}R$ antagonists for the motor symptoms in Parkinson's disease,[88–90] where the hyperactivity of the striatopallidal neuron included in the cortical–striatal–thalamic–cortical "motor circuit" constitutes the main pathogenic mechanism responsible for the akinetic symptoms.[121] Clinical studies so far indicate an important potential value of $A_{2A}R$ antagonists as cotreatment with L-dopa or D_2R agonists.[145,146] Although not yet proved, reduction of the doses of dopamine agonists should reduce long-term secondary effects, particularly L-dopa-induced dyskinesia, which depends strongly on phenotypic changes in the striatonigral neuron.[147,148]

The role of $A_{2A}R$ antagonists in the treatment of bradikinesia in Parkinson's disease has been comprehensibly addressed by recent reviews.[145,146] This review wants to stress a nonexplored role of adenosine and adenosine receptor heteromers in the treatment of apathy in neuropsychiatric disorders, mainly Parkinson's disease, schizophrenia, and SUD. As mentioned above, $A_{2A}R-D_2R$ heteromers are localized in all striatal compartments and quite an extensive preclinical work indicates that $A_{2A}R$ antagonists could improve the apathetic symptoms associated with dopamine deficiency in the ventral striatum in Parkinson's disease.[102] The same as for the treatment of bradikinesia, $A_{2A}R$ antagonists could potentiate D_2R signaling upon concomitant L-dopa treatment, increasing "Go"/"NoGo" pathway activation by selectively decreasing the hyperactivated "NoGo" pathway in the ventral striatal circuitry.

A similar rational could be expected in SUD, where $A_{2A}R$ blockade could restore a downregulated D_2R-mediated apathy (see above). In this frame, SUD would also be associated to a decreased ventral striatal "Go"/"NoGo" pathway activation in the ventral striatum due to a reduction on D_2R signaling. The theoretical frame by Balleine's group also proposes the role of a decrease in "Go"/"NoGo" pathway activation in SUD, but related to selective decrease in the activity of the "Go" pathway in the "automatic-learning"-related dorsal striatum.[149] In this frame, $A_{2A}R$ antagonism successfully counteracted methamphetamine-induced the striatal functional deficit.[149] However, the possible application of $A_{2A}R$ antagonists in SUD posits a potential serious drawback, since they could only be applied under drug-free conditions. We already discussed about the potentiating effect of a postsynaptic striatal $A_{2A}R$ antagonist on THC-induced self-administration in squirrel monkeys.[119] In addition, there is an extensive literature that demonstrates the ability of caffeine and also selective $A_{2A}R$ antagonists to potentiate the psychomotor and reinforcing effects of prototypic psychostimulants, such as cocaine and amphetamine.[74] On the other hand, the same as for THC self-administration in squirrel monkeys, selective presynaptic $A_{2A}R$ blockade has been shown to counteract cocaine-seeking behavior in rodents.[150]

Finally, the same as for SUD, using $A_{2A}R$ antagonists to treat apathy in schizophrenia would probably be a wrong approach. If anything, selective activation of striatal presynaptic $A_{2A}R$ could counteract a hypoactive "Go" pathway-mediated apathy (see above). Although not yet available, a presynaptic $A_{2A}R$ agonist could be discovered by using the same experimental models that led to the discovery of the selective striatal presynaptic properties of SCH-442416.[117] But there are other targets that should be considered in order to obtained a selective potentiation of hypoactive "Go" pathway in the ventral striatum, by using ligands binding to receptors forming heteromers with D_1R, such A_1R, but also D_3R or histamine H_3R.[67,105,106,108,151–154] In summary, striatal adenosine receptor heteromers can provide important therapeutic targets for apathy, a very prevalent nonmotor symptom in Parkinson's disease, a main negative symptom of schizophrenia, and a not well-recognized symptom of SUD.

Acknowledgments

Work supported by the intramural funds of the National Institute on Drug Abuse.

References

1. Alexander GE, DeLong MR, Strick PL. Parallel organization of functionally segregated circuits linking basal ganglia and cortex. *Annu Rev Neurosci*. 1986;9:357–381.
2. Haber SN, Behrens TE. The neural network underlying incentive-based learning: implications for interpreting circuit disruptions in psychiatric disorders. *Neuron*. 2014;83:1019–1039.
3. Wise RA, Bozarth MA. A psychomotor stimulant theory of addiction. *Psychol Rev*. 1987;94:469–492.
4. Huang RQ, Bell-Horner CL, Dibas MI, Covey DF, Drewe JA, Dillon GH. Pentylenetetrazole-induced inhibition of recombinant gamma-aminobutyric acid type A (GABA(A)) receptors: mechanism and site of action. *J Pharmacol Exp Ther*. 2001;298:986–995.
5. Dutertre S, Becker CM, Betz H. Inhibitory glycine receptors: an update. *J Biol Chem*. 2012;287:40216–40223.
6. Miller R, Beninger RJ. On the interpretation of asymmetries of posture and locomotion produced with dopamine agonists in animals with unilateral depletion of striatal dopamine. *Prog Neurobiol*. 1991;36:229–256.
7. Szczypka MS, Kwok K, Brot MD, Marck BT, Matsumoto AM, Donahue BA, et al. Dopamine production in the caudate putamen restores feeding in dopamine-deficient mice. *Neuron*. 2001;30:819–828.
8. Ljungberg T, Ungerstedt U. Sensory inattention produced by 6-hydroxydopamine-induced degeneration of ascending dopamine neurons in the brain. *Exp Neurol*. 1976;53:585–600.
9. Marshall JF, Berrios N, Sawyer S. Neostriatal dopamine and sensory inattention. *J Comp Physiol Psychol*. 1980;94:833–846.
10. Miyashita N, Hikosaka O, Kato M. Visual hemineglect induced by unilateral striatal dopamine deficiency in monkeys. *Neuroreport*. 1995;6:1257–1260.
11. Schultz W. Getting formal with dopamine and reward. *Neuron*. 2002;36:241–263.
12. Wise RA. Dopamine, learning and motivation. *Nat Rev Neurosci*. 2004;5:483–494.
13. Floresco SB, West AR, Ash B, Moore H, Grace AA. Afferent modulation of dopamine neuron firing differentially regulates tonic and phasic dopamine transmission. *Nat Neurosci*. 2003;6:968–973.
14. Floresco SB, St Onge JR, Ghods-Sharifi S, Winstanley CA. Cortico-limbic-striatal circuits subserving different forms of cost-benefit decision making. *Cogn Affect Behav Neurosci*. 2008;8:375–389.

15. Satoh T, Nakai S, Sato T, Kimura M. Correlated coding of motivation and outcome of decision by dopamine neurons. *J Neurosci.* 2003;23:9913–9923.
16. Morris G, Arkadir D, Nevet A, Vaadia E, Bergman H. Coincident but distinct messages of midbrain dopamine and striatal tonically active neurons. *Neuron.* 2004;43:133–143.
17. Tobler PN, Fiorillo CD, Schultz W. Adaptive coding of reward value by dopamine neurons. *Science.* 2005;307:1642–1645.
18. Roesch MR, Calu DJ, Schoenbaum G. Dopamine neurons encode the better option in rats deciding between differently delayed or sized rewards. *Nat Neurosci.* 2007;10:1615–1624.
19. Kobayashi S, Schultz W. Influence of reward delays on responses of dopamine neurons. *J Neurosci.* 2008;28:7837–7846.
20. Howe MW, Tierney PL, Sandberg SG, Phillips PE, Graybiel AM. Prolonged dopamine signalling in striatum signals proximity and value of distant rewards. *Nature.* 2013;500:575–579.
21. Hamid AA, Pettibone JR, Mabrouk OS, Hetrick VL, Schmidt R, Vander Weele CM, et al. Mesolimbic dopamine signals the value of work. *Nat Neurosci.* 2016;19:117–126.
22. Lavoie B, Parent A. Pedunculopontine nucleus in the squirrel monkey: projections to the basal ganglia as revealed by anterograde tract-tracing methods. *J Comp Neurol.* 1994;344:210–231.
23. Mori S, Matsuura T, Takino T, Sano Y. Light and electron microscopic immunohistochemical studies of serotonin nerve fibers in the substantia nigra of the rat, cat and monkey. *Anat Embryol.* 1987;176:13–18.
24. Haber SN. The place of dopamine in the cortico-basal ganglia circuit. *Neuroscience.* 2014;282C:248–257.
25. Haber SN. The primate basal ganglia: parallel and integrative networks. *J Chem Neuroanat.* 2003;6:317–330.
26. Kable JW, Glimcher PW. The neurobiology of decision: consensus and controversy. *Neuron.* 2009;63:733–745.
27. Tversky A, Kahneman D. The framing of decisions and the psychology of choice. *Science.* 1981;211:453–458.
28. Prévost C, Pessiglione M, Météreau E, Cléry-Melin ML, Dreher JC. Separate valuation subsystems for delay and effort decision costs. *J Neurosci.* 2010;30:14080–14090.
29. Stopper CM, Floresco SB. Contributions of the nucleus accumbens and its subregions to different aspects of risk-based decision making. *Cogn Affect Behav Neurosci.* 2011;11:97–112.
30. Mogenson GJ, Jones DL, Yim CY. From motivation to action: functional interface between the limbic system and the motor system. *Prog Neurobiol.* 1980;14:69–97.
31. Kim HF, Hikosaka O. Parallel basal ganglia circuits for voluntary and automatic behaviour to reach rewards. *Brain.* 2015;138:1776–1800.
32. Voorn P, Vanderschuren LJ, Groenewegen HJ, Robbins TW, Pennartz CM. Putting a spin on the dorsal-ventral divide of the striatum. *Trends Neurosci.* 2004;27:468–474.
33. Yin HH, Knowlton BJ. The role of the basal ganglia in habit formation. *Nat Rev Neurosci.* 2006;7:464–476.
34. Balleine BW, O'Doherty JP. Human and rodent homologies in action control: corticostriatal determinants of goal-directed and habitual action. *Neuropsychopharmacology.* 2010;35:48–69.
35. Holly EN, Miczek KA. Ventral tegmental area dopamine revisited: effects of acute and repeated stress. *Psychopharmacology.* 2016;233:163–186.
36. Brooks AM, Berns GS. Aversive stimuli and loss in the mesocorticolimbic dopamine system. *Trends Cogn Sci.* 2013;17:281–286.
37. Abraham AD, Neve KA, Lattal KM. Dopamine and extinction: a convergence of theory with fear and reward circuitry. *Neurobiol Learn Mem.* 2014;108:65–77.

38. Steinberg EE, Keiflin R, Boivin JR, Witten IB, Deisseroth K, Janak PH. A causal link between prediction errors, dopamine neurons and learning. *Nat Neurosci*. 2013;16:966–973.
39. Chang CY, Esber GR, Marrero-Garcia Y, Yau HJ, Bonci A, Schoenbaum G. Brief optogenetic inhibition of dopamine neurons mimics endogenous negative reward prediction errors. *Nat Neurosci*. 2016;19:111–116.
40. Lammel S, Lim BK, Malenka RC. Reward and aversion in a heterogeneous midbrain dopamine system. *Neuropharmacology*. 2014;76:351–359 Pt B.
41. Vertes RP. Differential projections of the infralimbic and prelimbic cortex in the rat. *Synapse*. 2004;51:32–58.
42. Quiroz C, Orrú M, Rea W, Ciudad-Roberts A, Yepes G, Britt JP, et al. Local control of extracellular dopamine levels in the medial nucleus accumbens by a glutamatergic projection from the infralimbic cortex. *J Neurosci*. 2016;36:851–859.
43. Richard JM, Castro DC, Difeliceantonio AG, Robinson MJ, Berridge KC. Mapping brain circuits of reward and motivation: in the footsteps of Ann Kelley. *Neurosci Biobehav Rev*. 2014;:351–359:76 Pt B.
44. Vidal-Gonzalez I, Vidal-Gonzalez B, Rauch SL, Quirk GJ. Microstimulation reveals opposing influences of prelimbic and infralimbic cortex on the expression of conditioned fear. *Learn Mem*. 2006;13:728–733.
45. Knapska E, Macias M, Mikosz M, Nowak A, Owczarek D, Wawrzyniak M, et al. Functional anatomy of neural circuits regulating fear and extinction. *Proc Natl Acad Sci U S A*. 2012;109:17093–17098.
46. Moscarello JM, LeDoux JE. Active avoidance learning requires prefrontal suppression of amygdala-mediated defensive reactions. *J Neurosci*. 2013;33:3815–3823.
47. Gerfen CR. Basal ganglia. In: Paxinos G, ed. *The Rat Nervous System*. Amsterdam: Elsevier Academic Press; 2004;445–508.
48. Kress GJ, Yamawaki N, Wokosin DL, Wickersham IR, Shepherd GM, Surmeier DJ. Convergent cortical innervation of striatal projection neurons. *Nat Neurosci*. 2013;16:665–667.
49. Hoffer ZS, Alloway KD. Organization of corticostriatal projections from the vibrissal representations in the primary motor and somatosensory cortical areas of rodents. *J Comp Neurol*. 2001;439:87–103.
50. Reig R, Silberberg G. Multisensory integration in the mouse striatum. *Neuron*. 2014;83:1200–1212.
51. Bromberg-Martin ES, Matsumoto M, Hikosaka O. Dopamine in motivational control: rewarding, aversive, and alerting. *Neuron*. 2010;68:815–834.
52. Macpherson T, Morita M, Hikida T. Striatal direct and indirect pathways control decision-making behavior. *Front Psychol*. 2014;5:1301.
53. Frank MJ. Dynamic dopamine modulation in the basal ganglia: a neurocomputational account of cognitive deficits in medicated and nonmedicated Parkinsonism. *J Cogn Neurosci*. 2005;17:51–72.
54. Hikosaka O. Basal ganglia mechanisms of reward-oriented eye movement. *Ann N Y Acad Sci*. 2007;1104:229–249.
55. Hikida T, Kimura K, Wada N, Funabiki K, Nakanishi S. Distinct roles of synaptic transmission in direct and indirect striatal pathways to reward and aversive behavior. *Neuron*. 2010;66:896–907.
56. Hikida T, Yawata S, Yamaguchi T, Danjo T, Sasaoka T, Wang Y, et al. Pathway-specific modulation of nucleus accumbens in reward and aversive behavior via selective transmitter receptors. *Proc Natl Acad Sci U S A*. 2013;110:342–357.
57. Kravitz AV, Tye LD, Kreitzer AC. Distinct roles for direct and indirect pathway striatal neurons in reinforcement. *Nat Neurosci*. 2012;15:816–818.
58. Danjo T, Yoshimi K, Funabiki K, Yawata S, Nakanishi S. Aversive behavior induced by optogenetic inactivation of ventral tegmental area dopamine neurons is mediated

by dopamine D2 receptors in the nucleus accumbens. *Proc Natl Acad Sci U S A.* 2014;111:6455–6460.
59. Frank MJ, Seeberger LC, O'Reilly RC. By carrot or by stick: cognitive reinforcement learning in parkinsonism. *Science.* 2004;(306):1940–1943.
60. Voon V, Pessiglione M, Brezing C, Gallea C, Fernandez HH, Dolan RJ, et al. Mechanisms underlying dopamine-mediated reward bias in compulsive behaviors. *Neuron.* 2010;65:135–142.
61. Nakamura K, Hikosaka O. Role of dopamine in the primate caudate nucleus in reward modulation of saccades. *J Neurosci.* 2006;26:5360–5369.
62. Shen W, Flajolet M, Greengard P, Surmeier DJ. Dichotomous dopaminergic control of striatal synaptic plasticity. *Science.* 2008;321:848–851.
63. Pin JP, Neubig R, Bouvier M, Devi L, Filizola M, Javitch JA, et al. International Union of Basic and Clinical Pharmacology. LXVII. Recommendations for the recognition and nomenclature of G protein-coupled receptor heteromultimers. *Pharmacol Rev.* 2007;59:5–13.
64. Milligan G. The prevalence, maintenance, and relevance of G protein-coupled receptor oligomerization. *Mol Pharmacol.* 2013;84:158–169.
65. Ferré S, Baler R, Bouvier M, Caron MG, Devi LA, Durroux T, et al. Building a new conceptual framework for receptor heteromers. *Nat Chem Biol.* 2009;5(3):131–134.
66. Ferré S, Casadó V, Devi LA, Filizola M, Jockers R, Lohse MJ, et al. G protein-coupled receptor oligomerization revisited: functional and pharmacological perspectives. *Pharmacol Rev.* 2014;66:413–434.
67. Guitart X, Navarro G, Moreno E, Yano H, Cai NS, Sánchez-Soto M, et al. Functional selectivity of allosteric interactions within G protein-coupled receptor oligomers: the dopamine D1-D3 receptor heterotetramer. *Mol Pharmacol.* 2014;86:417–429.
68. Navarro G, Quiroz C, Moreno-Delgado D, Sierakowiak A, McDowell K, Moreno E, et al. Orexin-corticotropin-releasing factor receptor heteromers in the ventral tegmental area as targets for cocaine. *J Neurosci.* 2015;35:6639–6653.
69. Ferré S. The GPCR heterotetramer: challenging classical pharmacology. *Trends Pharmacol Sci.* 2015;36:145–152.
70. Bonaventura J, Navarro G, Casadó-Anguera V, Azdad K, Rea W, Moreno E, et al. Allosteric interactions between agonists and antagonists within the adenosine A2A receptor-dopamine D2 receptor heterotetramer. *Proc Natl Acad Sci U S A.* 2015;112: E3609–E3618.
71. Ferré S, O'Connor WT, Fuxe K, Ungerstedt U. The striopallidal neuron: a main locus for adenosine-dopamine interactions in the brain. *J Neurosci.* 1993;13:5402–5406.
72. Ferré S, Bonaventura J, Tomasi D, Navarro G, Moreno E, Cortés A, et al. Allosteric mechanisms within the adenosine A(2A)-dopamine D(2) receptor heterotetramer. *Neuropharmacology.* 2016;104:154–160.
73. Azdad K, Gall D, Woods AS, Ledent C, Ferré S, Schiffmann SN. Dopamine D2 and adenosine A2A receptors regulate NMDA-mediated excitation in accumbens neurons through A2A-D2 receptor heteromerization. *Neuropsychopharmacology.* 2009;34:972–986.
74. Ferré S. Mechanisms of the psychostimulant effects of caffeine: implications for substance use disorders. *Psychopharmacology.* 2016;233:1963–1979. Available from: http://dx.doi.org/10.1007/s00213-016-4212-2.
75. Ferré S, Rubio A, Fuxe K. Stimulation of adenosine A2 receptors induces catalepsy. *Neurosci Lett.* 1991;130:162–164.
76. Kanda T, Shiozaki S, Shimada J, Suzuki F, Nakamura J. KF17837: a novel selective adenosine A2A receptor antagonist with anticataleptic activity. *Eur J Pharmacol.* 1994;256:263–268.
77. Rimondini R, Ferré S, Ogren SO, Fuxe K. Adenosine A2A agonists: a potential new type of atypical antipsychotic. *Neuropsychopharmacology.* 1997;117:82–91.

78. Shiozaki S, Ichikawa S, Nakamura J, Kitamura S, Yamada K, Kuwana Y. Actions of adenosine A2A receptor antagonist KW-6002 on drug-induced catalepsy and hypokinesia caused by reserpine or MPTP. *Psychopharmacology*. 1999;147:90–95.
79. Randall PA, Nunes EJ, Janniere SL, Stopper CM, Farrar AM, Sager TN, et al. Stimulant effects of adenosine antagonists on operant behavior: differential actions of selective A2A and A1 antagonists. *Psychopharmacology*. 2011;216:173–186.
80. Nunes EJ, Randall PA, Podurgiel S, Correa M, Salamone JD. Nucleus accumbens neurotransmission and effort-related choice behavior in food motivation: effects of drugs acting on dopamine, adenosine, and muscarinic acetylcholine receptors. *Neurosci Biobehav Rev*. 2016;37:2015–2025.
81. Popoli P, Giménez-Llort L, Pezzola A, Reggio R, Martínez E, Fuxe K, et al. Adenosine A1 receptor blockade selectively potentiates the motor effects induced by dopamine D1 receptor stimulation in rodents. *Neurosci Lett*. 1996;218:209–213.
82. Rimondini R, Ferré S, Giménez-Llort L, Ogren SO, Fuxe K. Differential Effects of Selective Adenosine A1 and A2A Receptor Agonists on Dopamine Receptor Agonist-Induced Behavioral Responses in Rats. 1998;347:153–158.
83. Strömberg I, Popoli P, Müller CE, Ferré S, Fuxe K. Electrophysiological and behavioural evidence for an antagonistic modulatory role of adenosine A2A receptors in dopamine D2 receptor regulation in the rat dopamine-denervated striatum. *Eur J Neurosci*. 2000;12:4033–4037.
84. Ferré S, Popoli P, Giménez-Llort L, Rimondini R, Müller CE, Strömberg I, et al. Adenosine/dopamine interaction: implications for the treatment of Parkinson's disease. *Parkinsonism Relat Disord*. 2001;7:235–241.
85. Kanda T, Shiozaki S, Shimada J, Suzuki F, Nakamura J. KF17837: a novel selective adenosine A2A receptor antagonist with anticataleptic activity. *Eur J Pharmacol*. 1994;256:263–268.
86. Pardo M, Lopez-Cruz L, Valverde O, Ledent C, Baqi Y, Müller CE, et al. Adenosine A2A receptor antagonism and genetic deletion attenuate the effects of dopamine D2 antagonism on effort-based decision making in mice. *Neuropharmacology*. 2012;62:2068–2077.
87. Minor TR, Hanff TC. Adenosine signaling in reserpine-induced depression in rats. *Behav Brain Res*. 2015;286:184–191.
88. Ferré S, Fuxe K, von Euler G, Johansson B, Fredholm BB. Adenosine-dopamine interactions in the brain. *Neuroscience*. 1992;51:501–512.
89. Müller CE, Ferré S. Blocking striatal adenosine A2A receptors: a new strategy for basal ganglia disorders. *Recent Pat CNS Drug Discov*. 2007;2:1–21.
90. Armentero MT, Pinna A, Ferré S, Lanciego JL, Müller CE, Franco R. Past, present and future of A(2A) adenosine receptor antagonists in the therapy of Parkinson's disease. *Pharmacol Ther*. 2011;132:280–299.
91. Ferré S, O'Connor WT, Snaprud P, Ungerstedt U, Fuxe K. Antagonistic interaction between adenosine A2A receptors and dopamine D2 receptors in the ventral striopallidal system. Implications for the treatment of schizophrenia. *Neuroscience*. 1994;63:765–773.
92. Ferré S. Adenosine-dopamine interactions in the ventral striatum. Implications for the treatment of schizophrenia. *Psychopharmacology*. 1997;133:107–120.
93. Hauber W, Münkle M. Motor depressant effects mediated by dopamine D2 and adenosine A2A receptors in the nucleus accumbens and the caudate-putamen. *Eur J Pharmacol*. 1997;323:127–131.
94. Pinna A, Wardas J, Cristalli G, Morelli M. Adenosine A2A receptor agonists increase Fos-like immunoreactivity in mesolimbic areas. *Brain Res*. 1997;759:41–49.
95. Svenningsson P, Le Moine C, Kull B, Sunahara R, Bloch B, Fredholm BB. Cellular expression of adenosine A2A receptor messenger RNA in the rat central nervous system with special reference to dopamine innervated areas. *Neuroscience*. 1997;80:1171–1185.

96. Svenningsson P, Le Moine C, Aubert I, Burbaud P, Fredholm BB, Bloch B. Cellular distribution of adenosine A2A receptor mRNA in the primate striatum. *J Comp Neurol*. 1998;399:229–240.
97. Svenningsson P, Hall H, Sedvall G, Fredholm BB. Distribution of adenosine receptors in the postmortem human brain: an extended autoradiographic study. *Synapse*. 1997;27(4):322–335.
98. Ishiwari K, Madson LJ, Farrar AM, Mingote SM, Valenta JP, DiGianvittorio MD, et al. Injections of the selective adenosine A2A antagonist MSX-3 into the nucleus accumbens core attenuate the locomotor suppression induced by haloperidol in rats. *Behav Brain Res*. 2007;178:190–199.
99. Brianna Sheppard A, Gross SC, Pavelka SA, Hall MJ, Palmatier MI. Caffeine increases the motivation to obtain non-drug reinforcers in rats. *Drug Alcohol Depend*. 2012;124:216–222.
100. Pereira M, Farrar AM, Hockemeyer J, Müller CE, Salamone JD, Morrell JI. Effect of the adenosine A2A receptor antagonist MSX-3 on motivational disruptions of maternal behavior induced by dopamine antagonism in the early postpartum rat. *Psychopharmacology*. 2011;213:69–79.
101. Lazenka MF, Moeller FG, Negus SS. Effects of caffeine and its metabolite paraxanthine on intracranial self-stimulation in male rats. *Exp Clin Psychopharmacol*. 2015;23:71–80.
102. Salamone JD, Correa M, Nunes EJ, Randall PA, Pardo M. The behavioral pharmacology of effort-related choice behavior: dopamine, adenosine and beyond. *J Exp Anal Behav*. 2012;97:125–146.
103. Navarro G, Aguinaga D, Moreno E, Hradsky J, Reddy PP, Cortés A, et al. Intracellular calcium levels determine differential modulation of allosteric interactions within G protein-coupled receptor heteromers. *Chem Biol*. 2014;21:1546–1556.
104. Ferré S, Torvinen M, Antoniou K, Irenius E, Civelli O, Arenas E, et al. Adenosine A1 receptor-mediated modulation of dopamine D1 receptors in stably cotransfected fibroblast cells. *J Biol Chem*. 1998;273:4718–4724.
105. Ginés S, Hillion J, Torvinen M, Le Crom S, Casadó V, Canela EI, et al. Dopamine D1 and adenosine A1 receptors form functionally interacting heteromeric complexes. *Proc Natl Acad Sci U S A*. 2000;97:8606–8611.
106. Ferre S, O'Connor WT, Svenningsson P, Bjorklund L, Lindberg J, Tinner B, et al. Dopamine D1 receptor-mediated facilitation of GABAergic neurotransmission in the rat strioentopenduncular pathway and its modulation by adenosine A1 receptor-mediated mechanisms. *Eur J Neurosci*. 1996;8:1545–1553.
107. Sakiyama Y, Hatano K, Kato T, Tajima T, Kawasumi Y, Ito K. Stimulation of adenosine A1 receptors decreases in vivo dopamine D1 receptor binding of [11C]SCH23390 in the cat striatum revealed by positron emission tomography. *Ann Nucl Med*. 2007;21:447–453.
108. Ferré S, Rimondini R, Popoli P, Reggio R, Pèzzola A, Hansson AC, et al. Stimulation of adenosine A1 receptors attenuates dopamine D1 receptor-mediated increase of NGFI-A, c-fos and jun-B mRNA levels in the dopamine-denervated striatum and dopamine D1 receptor-mediated turning behaviour. *Eur J Neurosci*. 1999; 11:3884–3892.
109. Hobson BD, O'Neill CE, Levis SC, Monteggia LM, Neve RL, Self DW, et al. Adenosine A1 and dopamine d1 receptor regulation of AMPA receptor phosphorylation and cocaine-seeking behavior. *Neuropsychopharmacology*. 2013;38:1974–1983.
110. Schwienbacher I, Fendt M, Hauber W, Koch M. Dopamine D1 receptors and adenosine A1 receptors in the rat nucleus accumbens regulate motor activity but not prepulse inhibition. *Eur J Pharmacol*. 2002;444:161–169.

111. Dayne Mayfield R, Larson G, Orona RA, Zahniser NR. Opposing actions of adenosine A2a and dopamine D2 receptor activation on GABA release in the basal ganglia: evidence for an A2a/D2 receptor interaction in globus pallidus. *Synapse.* 1996;22:132–138.
112. Florán B, Barajas C, Florán L, Erlij D, Aceves J. Adenosine A1 receptors control dopamine D1-dependent [(3)H]GABA release in slices of substantia nigra pars reticulata and motor behavior in the rat. *Neuroscience.* 2002;115:743–751.
113. Mango D, Bonito-Oliva A, Ledonne A, Cappellacci L, Petrelli R, Nisticò R, et al. Adenosine A1 receptor stimulation reduces D1 receptor-mediated GABAergic transmission from striato-nigral terminals and attenuates l-DOPA-induced dyskinesia in dopamine-denervated mice. *Exp Neurol.* 2014;261:733–743.
114. Ciruela F, Casadó V, Rodrigues RJ, Luján R, Burgueño J, Canals M, et al. Presynaptic control of striatal glutamatergic neurotransmission by adenosine A1-A2A receptor heteromers. *J Neurosci.* 2006;26:2080–2087.
115. Quiroz C, Luján R, Uchigashima M, Simoes AP, Lerner TN, Borycz J, et al. Key modulatory role of presynaptic adenosine A2A receptors in cortical neurotransmission to the striatal direct pathway. *Scientific World J.* 2009;9:1321–1344.
116. Quiroz C, Orrú M, Rea W, Ciudad-Roberts A, Yepes G, Britt JP, et al. Local Control of extracellular dopamine levels in the medial nucleus accumbens by a glutamatergic projection from the infralimbic cortex. *J Neurosci.* 2016;36:851–859.
117. Orrú M, Bakešová J, Brugarolas M, Quiroz C, Beaumont V, Goldberg SR, et al. Striatal pre- and postsynaptic profile of adenosine A(2A) receptor antagonists. *PLoS One.* 2011;6:e16088.
118. Justinová Z, Ferré S, Redhi GH, Mascia P, Stroik J, Quarta D, et al. Reinforcing and neurochemical effects of cannabinoid CB1 receptor agonists, but not cocaine, are altered by an adenosine A2A receptor antagonist. *Addict Biol.* 2011;16:405–415.
119. Justinová Z, Redhi GH, Goldberg SR, Ferré S. Differential effects of presynaptic versus postsynaptic adenosine A2A receptor blockade on Δ9-tetrahydrocannabinol (THC) self-administration in squirrel monkeys. *J Neurosci.* 2014;34:6480–6484.
120. Jankovic J. Parkinson's disease: clinical features and diagnosis. *J Neurol Neurosurg Psychiatry.* 2008;79:368–376.
121. DeLong MR, Wichmann T. Basal Ganglia Circuits as Targets for Neuromodulation in Parkinson Disease. *JAMA Neurol.* 2015;72:1354–1360.
122. Tremblay L, Worbe Y, Thobois S, Sgambato-Faure V, Féger J. Selective dysfunction of basal ganglia subterritories: from movement to behavioral disorders. *Mov Disord.* 2015;30:1155–1170.
123. Brown CA, Campbell MC, Karimi M, Tabbal SD, Loftin SK, Tian LL, et al. Dopamine pathway loss in nucleus accumbens and ventral tegmental area predicts apathetic behavior in MPTP-lesioned monkeys. *Exp Neurol.* 2012;236:190–197.
124. Tian L, Xia Y, Flores HP, Campbell MC, Moerlein SM, Perlmutter JS. Neuroimaging analysis of the dopamine basis for apathetic behaviors in an MPTP-lesioned primate model. *PLoS One.* 2015;10:e0132064.
125. Marin RS. Apathy: a neuropsychiatric syndrome. *J Neuropsychiatry Clin Neurosci.* 1991;3:243–254.
126. Sinha N, Manohar S, Husain M. Impulsivity and apathy in Parkinson's disease. *J Neuropsychol.* 2013;7:255–283.
127. Mazzoni P, Hristova A, Krakauer JW. Why don't we move faster? Parkinson's disease, movement vigor, and implicit motivation. *J Neurosci.* 2007;27:7105–7116.
128. Chong TT, Bonnelle V, Manohar S, Veromann KR, Muhammed K, Tofaris GK, et al. Dopamine enhances willingness to exert effort for reward in Parkinson's disease. *Cortex.* 2015;69:40–46.

129. den Brok MG, van Dalen JW, van Gool WA, Moll van Charante EP, de Bie RM, Richard E. Apathy in Parkinson's disease: a systematic review and meta-analysis. *Mov Disord*. 2015;30:759–769.
130. Remy P, Doder M, Lees A, Turjanski N, Brooks D. Depression in Parkinson's disease: loss of dopamine and noradrenaline innervation in the limbic system. *Brain*. 2005;128:1314–1322.
131. Chaudhuri KR, Martinez-Martin P, Schapira AH, Stocchi F, Sethi K, Odin P, et al. International multicenter pilot study of the first comprehensive self-completed non-motor symptoms questionnaire for Parkinson's disease: the NMSQuest study. *Mov Disord*. 2006;21:916–923.
132. Al-Khaled M, Heldmann M, Bolstorff I, Hagenah J, Münte TF. Intertemporal choice in Parkinson's disease and restless legs syndrome. *Parkinsonism Relat Disord*. 2015;21:1330–1335.
133. Hamilton KR, Mitchell MR, Wing VC, Balodis IM, Bickel WK, Fillmore M, et al. Choice impulsivity: definitions, measurement issues, and clinical implications. *Personal Disord*. 2015;6:182–198.
134. Sierra M, Carnicella S, Strafella AP, Bichon A, Lhommée E, Castrioto A, et al. Apathy and impulse control disorders: yin & yang of dopamine dependent behaviors. *J Parkinsons Dis*. 2015;5:625–636.
135. Simon JJ, Biller A, Walther S, Roesch-Ely D, Stippich C, Weisbrod M, et al. Neural correlates of reward processing in schizophrenia-relationship to apathy and depression. *Schizophr Res*. 2010;118:154–161.
136. Simon JJ, Cordeiro SA, Weber MA, Friederich HC, Wolf RC, Weisbrod M, et al. Reward system dysfunction as a neural substrate of symptom expression across the general population and patients with schizophrenia. *Schizophr Bull*. 2015;41:1370–1378.
137. Strauss GP, Morra LF, Sullivan SK, Gold JM. The role of low cognitive effort and negative symptoms in neuropsychological impairment in schizophrenia. *Neuropsychology*. 2015;29:282–291.
138. Kirschner M, Hager OM, Bischof M, Hartmann MN, Kluge A, Seifritz E, et al. Ventral striatal hypoactivation is associated with apathy but not diminished expression in patients with schizophrenia. *J Psychiatry Neurosci*. 2015;40:140383.
139. Strauss GP, Waltz JA, Gold JM. A review of reward processing and motivational impairment in schizophrenia. *Schizophr Bull*. 2014;40(suppl):S107–S116.
140. Volkow ND, Wang GJ, Fowler JS, Tomasi D, Telang F. Addiction: beyond dopamine reward circuitry. *Proc Natl Acad Sci U S A*. 2011;108:15037–15042.
141. Verdejo-García A, Bechara A, Recknor EC, Pérez-García M. Executive dysfunctionin substance dependent individuals during drug use and abstinence: an examination of the behavioral, cognitive and emotional correlates of addiction. *J Int Neuropsychol Soc*. 2006;12:405–415.
142. Verdejo-García A, Pérez-García M. Substance abusers' self-awareness of the neurobehavioral consequences of addiction. *Psychiatry Res*. 2008;158:172–180.
143. Gjini K, Qazi A, Greenwald MK, Sandhu R, Gooding DC, Boutros NN. Relationships of behavioral measures of frontal lobe dysfunction with underlying electrophysiology in cocaine-dependent patients. *Am J Addict*. 2014;23:265–271.
144. Belcher AM, Volkow ND, Moeller FG, Ferré S. Personality traits and vulnerability or resilience to substance use disorders. *Trends Cogn Sci*. 2014;18:211–217.
145. Jenner P. An overview of adenosine A2A receptor antagonists in Parkinson's disease. *Int Rev Neurobiol*. 2014;119:71–86.
146. Müller T. The safety of istradefylline for the treatment of Parkinson's disease. *Expert Opin Drug Saf*. 2015;14:769–775.

147. Berthet A, Bezard E. Dopamine receptors and L-dopa-induced dyskinesia. *Parkinsonism Relat Disord*. 2009;15(suppl 4):S8–S12.
148. Ferré S, Lluis C, Lanciego JL, Franco R. Prime time for G-protein-coupled receptor heteromers as therapeutic targets for CNS disorders: the dopamine D_1-D_3 receptor heteromer. *CNS Neurol Disord Drug Targets*. 2010;9:596–600.
149. Furlong TM, Supit AS, Corbit LH, Killcross S, Balleine BW. Pulling habits out of rats: adenosine 2A receptor antagonism in dorsomedial striatum rescues meth-amphetamine-induced deficits in goal-directed action. *Addict Biol*. 2015. Available from: http://dx.doi.org/10.1111/adb.12316.
150. O'Neill CE, LeTendre ML, Bachtell RK. Adenosine A2A receptors in the nucleus accumbens bi-directionally alter cocaine seeking in rats. *Neuropsychopharmacology*. 2012;37:1245–1256.
151. Fiorentini C, Busi C, Gorruso E, Gotti C, Spano P, Missale C. Reciprocal regulation of dopamine D1 and D3 receptor function and trafficking by heterodimerization. *Mol Pharmacol*. 2008;74(1):59–69.
152. Marcellino D, Ferré S, Casadó V, Cortés A, Le Foll B, Mazzola C, et al. Identification of dopamine D1-D3 receptor heteromers. Indications for a role of synergistic D1-D3 receptor interactions in the striatum. *J Biol Chem*. 2008;283:26016–26025.
153. Moreno E, Hoffmann H, Gonzalez-Sepúlveda M, Navarro G, Casadó V, Cortés A, et al. Dopamine D1-histamine H3 receptor heteromers provide a selective link to MAPK signaling in GABAergic neurons of the direct striatal pathway. *J Biol Chem*. 2011;286:5846–5854.
154. Moreno E, Moreno-Delgado D, Navarro G, Hoffmann HM, Fuentes S, Rosell-Vilar S, et al. Cocaine disrupts histamine H3 receptor modulation of dopamine D1 receptor signaling: σ1-D1-H3 receptor complexes as key targets for reducing cocaine's effects. *J Neurosci*. 2014;34:3545–3558.

Index

Note: Page numbers followed by "*f*" and "*t*" refer to figures and tables, respectively.

B

C

FSC
www.fsc.org
MIX
Papier | Fördert gute Waldnutzung
FSC® C083411

Kolibri 360 GmbH
Ferdinand-Jühlke-Straße 7
99095 Erfurt, Deutschland
produktsicherheit@kolibri360.de